T0305265

Construction Microeconomics

Construction Microeconomics

Christian Brockmann

University of Utah
Salt Lake City, USA

WILEY Blackwell

Registered Offices
John Wiley & Sons, Inc., 111 River Street, Hoboken, NJ 07030, USA
John Wiley & Sons Ltd, The Atrium, Southern Gate, Chichester, West Sussex, PO19 8SQ, UK

For details of our global editorial offices, customer services, and more information about Wiley products, visit us at www.wiley.com.

Wiley also publishes its books in a variety of electronic formats and by print-on-demand. Some content that appears in standard print versions of this book may not be available in other formats.

Library of Congress Cataloging-in-Publication Data
Names: Brockmann, Christian, 1954- author.
Title: Construction microeconomics / Christian Brockmann.
Description: Hoboken, NJ : Wiley-Blackwell, 2023. | Includes index.
Identifiers: LCCN 2022050327 (print) | LCCN 2022050328 (ebook) | ISBN
    9781119828785 (cloth) | ISBN 9781119831914 (adobe pdf) | ISBN
    9781119831921 (epub)
Subjects: LCSH: Contractors' operations. | Construction industry—Economic
    aspects.
Classification: LCC TA210 .B76 2023 (print) | LCC TA210 (ebook) | DDC
    692/.8—dc23/eng/20221026
LC record available at https://lccn.loc.gov/2022050327
LC ebook record available at https://lccn.loc.gov/2022050328

Cover Design: Wiley
Cover Image: © joe daniel price/Getty Images
Printed and bound by CPI Group (UK) Ltd, Croydon, CR0 4YY

C9781119828785_040123

Brief Contents

Contents

Foreword by Gerard de Valence

Construction economics (CE) is the application of economic theory and techniques to the construction industry. However, because of the distinctive characteristics of the organization of construction processes, the structure of construction markets, and the management of construction firms, this is not a straightforward process.

The physical nature of the product, the variability of demand, the method of price determination by auction, and the contractual relationships between owners, contractors, and suppliers, requires adaptation of economic theory to adequately reflect the industry.

Therefore, CE combines industry-specific knowledge of the construction industry and its characteristics with economic theory to analyze and understand industry structure and dynamics. Christian Brockmann's knowledge of construction is based on his experience in designing and managing projects over a period of 40 years, including several megaprojects, and this experience deeply informs his book. *Construction Microeconomics* is an important milestone in the theoretical analysis of the construction process, the construction industry, and construction firms.

This is the first book to focus on microeconomic aspects of construction, and that focus allows an extended discussion of topics beyond those found in previous CE books. Microeconomics studies the interaction of producers and consumers of goods and services in specific industries and markets, and the tools and techniques used are well-known. In a simple exchange market, where the transaction is complete, the analysis of demand and supply is relatively straightforward, but construction markets are not like that. The book introduces the idea of contract goods that are delivered over time and priced by bids from contractors, and the analysis of how those characteristic features of construction affect the behavior of owners and contractors in *Construction Microeconomics* is both original and insightful.

Microeconomic theory can be applied to a range of issues found in the procurement and contracting of projects. Topics such as competitive and monopolistic or monopsonistic markets, auction theory, game theory, and buyer and supplier power are relevant, as is the economics of contracts. Other related topics are production theory and the theory of the firm, transaction cost economics and its application to property rights, agency theory, moral hazard, and incentives.

Among construction economists there are different views on construction markets, the behavior of firms, and the nature of construction products and processes. In particular, bidding for work in auctions and contracting under uncertainty for owners who are risk averse raises complex issues around marginal costs and prices, incentives and behavior, and information asymmetry and bargaining power that this book discusses in detail, drawing on developments in industry economics to support the points made and the approach taken.

The library of CE books has been growing recently, with the publication of several edited volumes of research that cover a wide range of topics. A book that focuses on the operation and organization of construction from a microeconomic perspective is a welcome addition, building on previous work and bringing a new perspective to issues and topics that are fundamental to CE. Christian Brockmann has written a readable, accessible book that will be of interest not just to academic researchers but to industry and policy makers as well, and his *Construction Microeconomics* is a significant contribution to the development of CE.

Gerard de Valence
Kurmond, July 2022

Preface

You are stepping out onto the observation deck of the Empire State Building in Manhattan, New York, and feel overwhelmed by the dazzling array of buildings all around you. Many high-rises are scratching the sky next to you. Others are of medium size somewhere below, and as you strain your eyes, you see a few buildings that seem to be just one story high (Figure 0.1).

To the south is the Flatiron Building, which was one of the tallest buildings at its opening in 1902. Not far beyond and further south is One World Trade Center, built at the site of the terrorist attacks of 9/11. You know that Wall Street with the New York Stock Exchange is close by, but you cannot see it. A little bit to the west and far away, Lady Liberty holds high the flame of freedom. Quite a way out, you can distinguish the Verrazzano-Narrows Bridge.

Turning east, there is the sea of buildings constituting Brooklyn and Queens; in between lie the Brooklyn bridge as a historical landmark and the complex of the UN headquarters. The planes on the horizon are starting from or landing at JFK International Airport.

Up north, Central Park is settled among other groups of skyscrapers, some slender as a pencil. Recognizable to you is 432 Park Avenue and One57. Your eye is by now used to all the skyscrapers, but these two seem to be rising from a plot of land not larger than the size of a towel. Central Park is clearly delineated by concrete, stone, steel, and glass. Millions of people live in the confined space of New York City and every single person expects electricity when switching on a light, fresh water when turning the faucet, a connection when dialing a telephone number, and gas when preparing a meal; the internet provides music and information all the time. You know that all these utility lines are below the streets, underground together with the metro, invisible from the Empire State Building.

What might Manhattan have looked like some 250 years ago at the time of the Declaration of Independence? How did it look in 1609, when the Dutch started the first European settlement, or in 1524, when Giovanni da Verrazzano sailed into the natural harbor to meet the native Indians, the Lenape?

What engineering and what economic forces were at work in the transition from pastures to the built environment of today – what knowledge and what innovation? Surely, the expansion of our understanding in civil engineering design and construction was necessary, but without the motivation of owners, designers, and contractors, nobody would have raised a finger. Without financing from banks on Wall Street, without investment from the government, not much would have been possible. Lady Liberty invited the huddled masses to come to New York and provide the labor force required for construction. She also

Figure 0.1 Looking south from the observation deck of the Empire State Building. *Source:* Arthur/ Adobe Stock.

symbolizes democracy and capitalism as institutions framing all economic activities over the 400 years of building growth in Manhattan. In the beginning, land was freely available, while its scarcity today exacts the construction of pencil towers. Over this period, the border between the natural and the built environment shifted decisively toward the latter. The water and air quality deteriorated at the same time, and an increase in CO_2 emissions into the atmosphere affected the climate.

If these questions and problems are relevant to you, then this book is for you. It addresses economic problems that you will face as an interested individual or as a contributor to the built environment. It focuses on owners who initiate construction projects and on contractors who constitute the construction sector that transforms ideas and plans into infrastructure and buildings. Owners must at least provide an idea, land, and financial resources. Next, they can ask designers, project managers, quantity surveyors, lawyers, insurers, and bankers to develop the first idea into plans and a contract. All this requires institutions and functioning markets and a contractor for implementation. The contractor combines inputs such as land, labor, materials, and equipment to produce the building or infrastructure.

Economics consists of two major parts, micro- and macroeconomics. With the help of microeconomic theory, we try to answer questions about decision-making by single entities (individuals, firms, or parts of the government). Some entities try to maximize their benefits as consumers, others try to optimize their profits as producers. Together, they constitute the demand and supply side of a market. What happens on construction markets depends most of all on the decisions by owners and contractors. Microeconomics focuses on quantity and price as a result of decision-making by consumers and producers.

Macroeconomics relies on aggregated data from national or supranational accounts. Instead of the price for a specific one-family house, the average price of comparable one-family houses becomes of interest. The main goals of macroeconomics are economic growth, full employment, and limited inflation. To achieve these goals, societies decide on economic institutions between the extremes of pure market or pure planned economies. Governments use fiscal (tax) and monetary policies as tools.

A monograph with the title Construction Microeconomics unsurprisingly concentrates on microeconomics and neglects many macroeconomic aspects. This sharpens the focus and allows presenting the theories and applications with greater depths. Most of all, it provides the space to develop microeconomic theory for the construction sector.

There are many excellent textbooks and monographs on microeconomics, some written by Nobel Prize–winning authors. The wish to add to this body of knowledge something meaningful would be presumptuous, and I refrain from it. However, paradigmatic for this literature are manufacturing industries, and the question arises whether we can use the same concepts in the construction sector. My personal experience is negative. While working as construction project manager some 35 years ago, I felt a keen lack of basic economic knowledge in my training as civil engineer. I started to study economics besides work, and the idea was to learn in the evening and to apply the concepts in the morning. My enthusiasm for economics in the evening was as great as my frustration in the morning: The concepts from manufacturing did not fit into the framework of construction. Among the main reasons are singular construction goods versus standardized manufacturing goods; movable factories instead of stationary ones; labor intensity in place of automation; and much higher uncertainty due to project organization in construction compared with process organization in manufacturing. The days of the Ford Model T (Tin Lizzie) are long passed in manufacturing as it moves into the direction of customization, while construction moves toward standardization. However, there remains a sizable and insurmountable gap.

A gap calls for a bridge, and this is what I aim to achieve with Construction Microeconomics. I will use general economics as a quarry to extract building material for the bridge, and I will use my skills as a civil engineer and project manager, as well as my knowledge as an economist, to construct it. As a result, construction will meet microeconomics.

While this is a textbook, it also contains some new thoughts based on research. Otherwise, I could not possibly hope to build the bridge. Colleagues have provided books on construction economics (Chapter 1), but a book specializing in construction microeconomics has not been available.

Economic reasoning, as presented in textbooks or as impersonated by winners of the Nobel Memorial Prize in Economics, is deeply embedded in Western thinking with its historical roots in Europe (Pribram 1983). While I hope that all readers will be critical, I invite those from other cultural backgrounds to be even more so. Microeconomic concepts are adopted around the world, but there will be differences in what is optimal. Here is a gap that I cannot even dream to bridge.

I will treat economics as a separate subject in the coming chapters, although I am fully aware that this cuts important ties to other domains which exert a strong influence, and vice versa. We are living on this planet, and we live off its natural resources. The environment provides our food, our clothes, our built environment, the many goods we use daily, as well as the air we breathe and the water we drink. When we produce goods by transforming inputs into outputs of higher value, we use the natural environment as a source and as a sink. The greater the population of the world, the more resources we take from the environment and the more waste we send back to it.

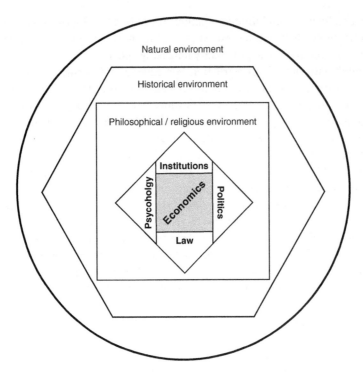

Figure 0.2 Economic embeddedness.

Economic theory is also embedded in ongoing philosophical discussions. Especially pertinent are the topics of ontology, epistemology, philosophy of the mind, and ethics. The questions and answers arising from these discussions form the way we perceive the world and how we construct economic models. Religion informs our thinking at the same time. Religion, ethics, and perception are not globally uniform.

Another domain of influence is the history of economic thought itself. What I will present to you was developed over many centuries by many great thinkers. Some of the conflicts during these centuries of evolutions and revolutions still lead to different opinions today. Thought and action developed around the world in different ways throughout history forming different cultures, beliefs, institutions, and policies.

Politics, law, and economics strongly influence each other. What we today call economics was addressed by our forefathers as political economy, stressing the relationship between the two subjects. Fundamental to all economics are decisions about property rights and their definition is the task of law.

Psychology plays an ever-larger role in developing economic models because the actors in economic models are humans. Actual human behavior determines economic outcomes. Thaler (2015) thus distinguishes between humans and econs, with econs being part of many economic models and humans living in the real world and populating the models of behavioral economics. Unfortunately, most models that economists have developed over the centuries are based on econs. For lack of something better, I will often have to stick with them.

The idea that Figure 0.2 shall convey is that we are all living in this one world; we have developed as a species over time until today. In the process, we have transformed the

Figure 0.3 Economics cannot be modeled as a machine. *Source:* Kaihsu Tai/Wikimedia Commons/ Public domain".

natural environment and we have developed ourselves as well. As humans, we try to make sense by advancing ideas and theories to understand our world. Driven by cognitive development, we create and change our institutions, politics, and laws. Institutions contain culture, regulation, and norms. Psychology describes the minds that we use as a tool for understanding and action. Economics is embedded in all these domains, is impacted by them and impacts them in turn. This is how I see economics.

Other scholars describe economics as a separate entity with causality at its core. An extreme view is that of a mechanistic model. One example is the partial model of the monetary flow in an economy by Phillips from 1949. He built a machine called Moniac – monetary national income analogue computer. Figure 0.3 shows a model of the machine: Everything in this model is determined. This is how I do not see economics.

For the treatment of microeconomics in this text, I will neglect to a large degree the natural, cultural, and cognitive environment, as well as more specific institutions, policies, and laws. Cutting all these links is almost inexcusable. However, it has the advantage of concentrating on microeconomics, and it is up to us to use that advanced knowledge by reconnecting it with all the other aspects. Concentration, not importance, forces me to omission; the embeddedness of economics in the overall environment is very important!

As an excuse, I can offer that I will link microeconomics and construction. On the one hand, this is an enlargement of focus; on the other hand, it is a further specialization. Construction microeconomics remains just one element of our world, and this is the building block that I offer to you. I feel it has its own beauty and purpose; yet, it remains just a building block. I am sorry, dear reader, but I must put on your shoulders the burden to work through this book and to connect any new economic understanding with your own life. Maybe you share my appreciation of the complexity and diversity of our lives as human beings, as economists, engineers, and managers. Diversity allows for different receptions of my text as each reader will place the information and arguments in her or his own context.

References

Pribram, K. (1983). *A History of Economic Reasoning*. Baltimore: Johns Hopkins University Press.

Thaler, R. (2015). *Misbehaving: The Making of Behavioural Economics*. New York: Norton.

1

Introduction

In the preface, I identified a gap between mainstream and construction economics. To estimate the size of the gap requires knowledge in both economics and construction. As this is primarily a textbook, I do not take it for granted that all readers can gauge the size of the gap. If the statement sounds a bit plausible, please content yourself with it for the moment. Once you have read the book to the end you will be in a better position to cast judgment. For me as author, the gap seems frightening. Luckily, there is always someone to cheer us on (Figure 1.1).

Let us assume you want to go on a trip from Yangon in Myanmar to Surabaya in Indonesia. This is not the typical one, like London to Paris, where you hop on the Eurostar train at St. Pancras in London and you step off at the Gare du Nord in Paris. There is a wider gap to cross in Southeast Asia. You might want to get some travel books or check some websites

Figure 1.1 The gap between construction and microeconomics. *Source:* Gabriel Utasi.

Construction Microeconomics, First Edition. Christian Brockmann.
© 2023 John Wiley & Sons Ltd. Published 2023 by John Wiley & Sons Ltd.

to prepare for the trip. You know your starting point and your goal, but you must devise a way of getting from here to there. You can look for some general idea how to behave when travelling in that region. Thinking ahead, you find that your plans and hopes for the trip are a bit different than might be typical, so you look for specific advice.

Well, instead of a real trip, I invite you on a mental trip from microeconomics to construction. In this Chapter 1, I will try to give you all the information that you might want to prepare for the actual trip, which starts in Chapter 2.

A number of textbooks on construction economics are on the market. These books take a general and introductory approach. It also means that some colleagues have already narrowed the gap. However, a gap remains between microeconomics and construction. Chapter 1.1 will review the existing literature and its contribution to reaching the goal. It describes the promontory where the cyclist is standing, the status quo. On the other side is the goal which promises sound microeconomic understanding in the context of construction. Practical outcomes of the understanding could be a better house for your money, a higher profit for the contractor, or maybe a win-win situation for owner and contractor.

Chapter 1.2 introduces some tools and presentations that I will use throughout the book. Chapter 1.3 summarizes the methodological approach, i.e. the means that I will employ to help us get to the finish line. Always looking at the goal, I introduce some helpful theories in Chapter 1.4 with their basic concepts. In Chapter 1.5., I will provide a framework for what you can expect based on your interest in construction. The you in the above sentences describes a heterogenous audience, so Chapter 1.6 addresses six different groups of readers and how each one might benefit from the book. Finally, Chapter 1.7 will offer a guideline for the book by outlining the contents of the main chapters and their importance for different audiences.

1.1 Navigating the Maze of Economic Literature

There exists a plethora of textbooks on economics, and this can cause confusion. Which of the books are helpful, and what do they offer? A first distinction is between mainstream economics and sectorial economics. The former primarily considers manufactured goods that are typically rather homogenous. The latter focuses on a sector and its corresponding goods. This could be, among other services, retail, banking, or construction. Construction goods are far less homogenous than manufactured goods. Besides the differences between mainstream and sectoral economics, we can distinguish different levels of detailing and different tools (Figure 1.2). From introductory to advanced texts, mathematics as a tool becomes more and more important.

1.1.1 Economics

Economic theorising goes back many centuries. Much early thought concerned agriculture. With the first Industrial Revolution, the focus started to shift toward manufacturing. A seminal work was Adam Smith's *An Inquiry into the Nature and Causes of the Wealth of Nations* (Smith 1776). The subject is the individual (person, firm or institution). Much later (1936), John Maynard Keynes published his magnum opus, *The General Theory of Employment, Interest and Money*. He shifted the emphasis from the individual to the national level; hence,

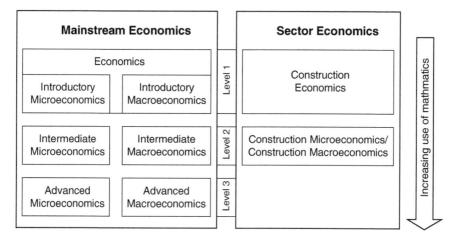

Figure 1.2 Categorizing textbooks on economics.

we distinguish between microeconomics with a focus on the individual and macroeconomics using aggregated data from national accounts. Models of economic behavior describe individual entities (microeconomics) but they also form the basis for macroeconomics when aggregating for example the total output of all firms.

Textbooks on economics cover both micro- and macroeconomics on an introductory level. At Stanford University, the corresponding course carries the abbreviation ECON 1 with Taylor as instructor using his own textbook (Taylor and Weerapana 2011). Other similar textbooks are available from (chronological order):

- Samuelson and Nordhaus (2009, first published 1948)
- Begg et al. (2020, first published in 1984)
- Stiglitz and Walsh (2002, first published 1993)
- Krugman and Wells (2017, first published in 2005)
- Mankiw (2017a, first published 2006)
- Sowell (2015)
- Acemoglu et al. (2017, first published in 2015)

These books cover a wide range of topics at the introductory level; most are around 1000 pages. Samuelson (1970), Stiglitz (2001) and Walsh and Krugman (2008) have won the Nobel Memorial Prize in Economic Sciences. The year indicating the first edition signals the way the book is organized. The later this year, the more contemporary is the approach. As research advances our understanding of economics over time, the newer textbooks tend to include more recent topics and findings.

1.1.2 Microeconomics

Recommendable textbooks on microeconomics grouped in chronological order and level are:

- Acemoglu et al. (2016) – basic
- Nicholson and Snyder (2017) – basic

- Krugman and Wells (2018) – basic
- Nicholson and Snyder (2014) – intermediate
- Varian (2014) – intermediate
- Tirole (2000) – advanced, industrial organization with focus on imperfect markets
- Hirshleifer et al. (2005) – advanced, focus on pricing
- Beattie et al. (2009) – advanced, focus on production
- Jehle and Reny (2011) – advanced focus on strategic behavior
- Muñoz-Garcia (2017) – advanced, focus on intuitive assumptions
- Elsner et al. (2015) – advanced, focus on complexity

Basic microeconomic textbooks do not carry the arguments much further than books on economics. They have the advantage of fewer pages. Intermediate textbooks are more mathematical in their approach, basically working with calculus and set theory. As both are a prerequisite for engineers, the level of mathematics is comprehensible. In addition, advanced books on microeconomics usually have a specific focus. I have used mainly the books by Krugman and Wells, Varian, Tirole, and Jehle and Reny.

1.1.3 Macroeconomics

Macroeconomics are not part of this book on construction microeconomics but maybe there are some avid readers who like to venture a bit outside the boundaries. There will also be some references to macroeconomics in this text.

- Dornbusch et al. (2011) – basic
- Mankiw (2017b) – basic
- Krugman and Wells (2018) – basic
- Blanchard (2020) – intermediate
- Romer (2019) – advanced

Much of what applies to textbooks on microeconomics also applies to those in macroeconomics.

1.1.4 Construction Economics

There are a number of textbooks on construction economics and the first one by Hillebrandt is a classic. The two edited books by Ruddock and de Valence contain much more specialized and advanced topics. However, their downside is a lack of coherence.

- Hillebrandt (1974, last edition 2000)
- Cooke (1996)
- Gruneberg (1997)
- Gruneberg and Ive (1997)
- Runeson (2000)
- Ruddock (ed. Ruddock 2009)
- De Valence (ed. 2011a)
- Myers (2017)
- Gruneberg and Francis (2019)

As I stated in the introduction to this chapter, all of these authors have narrowed the gap between mainstream and construction economics, but the remaining one is still considerable. Except for the edited books, all others are on the level of basic economics, and many particularities of construction are not addressed in detail. All of these books constitute the status quo. I have profited very much from the ideas laid out in these books and feel greatly indebted to the authors. To the serious scholar of construction economics, I recommend all of them. As construction economics is still an evolving field of studies, however, different ideas allow to stretch the horizon and to test hypotheses.

1.2 Tools and Presentations

Throughout the book, I will use some common features as tools. These include definitions, people, assumptions, business cases, observations as well as summaries. The presentations include figures and tables. I have also added some comics to lighten the mood at times. You should not read more into the comics than the context allows.

1.2.1 Definitions

Sometimes, when I read academic books, I find that definitions are missing. Therefore, I will introduce definitions formally and highlight them in the text as shown below.

 Introducing definitions = (def.) Definitions are indicated in this way. Typically, I will create definitions that most of all apply to this book. My definitions will be nominal, i.e. name-giving definitions. Names and nominal definitions are context-sensitive, and they can be useful or not.

We learn our everyday language by trial and error during the different stages of our socialization. In schools and universities, we learn a more professional use of specific terms. However, these terms often lack codification or general acceptance. This might lead to misunderstandings, especially when different professional languages mix. In construction, this could be the professional jargon of economists, managers, or engineers. Accordingly, we need definitions to formulate our thoughts more precisely.

Philosophers distinguish between nominal and real definitions. The latter describe the essential attributes of a term – "*What is a home?*" for example. This is quite a challenging task and, in my opinion, it is best left to philosophers. Real definitions are either true or false, and this makes them so difficult to formulate. Nominal definitions are much easier to state – they are either useful or not, but they must be useful to the user – "*What is a house?*" They are also practical, as they replace a long explanation by a single term or by a name. To avoid confusion, the defined term should be similar to everyday language. As nominal definitions need to be useful to the person developing them, this person has some freedom when formulating a definition.

Nominal definition = (def.) In modern linguistics, only nominalist definitions occur, that is to say, words are given meaning as shorthand symbols or labels that cut a long story short.

Popper (1945 /2003, p. 16)

Thus, we find a long story as the right-hand part of the definition and a shorthand label (term) to the left. I will formally emphasize definitions in the following way: Term = (def.) long explanation.

1.2.2 Economic Scholars

Often, I will rely on research in economics as advanced by specific authors. To me, some of these scholars are giants, and I climb onto their shoulders to look in my direction of construction economics. I will refer to those authors by introducing them with a picture, some basic data, and often a succinct quote. I hope to stimulate your interests to dig deeper into the works of these important economists: There is a whole world to discover!

 Economic scholar: Picture and short information about an author, sometimes with a quote.

Here: Adam Smith (1723–1790) Founding father of capitalist economics

Source: Cadell and Davies/Wikimedia Commons/Public domain

1.2.3 Assumptions

Many of the theoretical models in microeconomics are derived from rigid assumptions and apply only when these assumptions are observed. If we do not understand the assumptions, we will often fail in using the models. Thus, assumptions are of paramount importance for microeconomic models. I will single them out in the following way for perfectly competitive markets. As you can see, assumptions can sometimes be numerous.

 Assumptions for perfectly competitive markets:

1) *There are many consumers and producers.*
2) *The goods in the market are homogenous.*
3) *Market transparency is complete.*
4) *Market transaction are without costs (no transaction costs).*
5) *Market barriers for entry and exit do not exist.*
6) *The market price can adjust freely and instantaneously.*

1.2.4 Case Studies

Part A of this book explains microeconomic theory without adjustment to construction. While I will provide some applications, readers might at times wonder about the importance of the concepts for construction engineers and managers. To avoid too much wondering, I introduce each chapter with a case study to highlight the importance of the chapter's topic for construction.

Case study

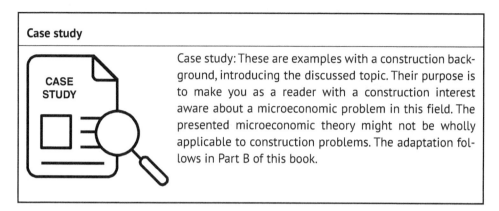

Case study: These are examples with a construction background, introducing the discussed topic. Their purpose is to make you as a reader with a construction interest aware about a microeconomic problem in this field. The presented microeconomic theory might not be wholly applicable to construction problems. The adaptation follows in Part B of this book.

1.2.5 Observations

Practitioners rely on stories and observations by using analog communication. From an academic point of view, stories provide only anecdotal evidence, i.e. evidence that we cannot trust academically. Geertz (1973) has introduced the term thick description to ethnographic methodology. People in construction form a very specific tribe, an ethnographic unit. Following the academic Geertz, a description of construction microeconomics requires use of stories or observations to highlight some theoretical aspects and to create a context. I will introduce them formally as follows:

 Observations = (obs.) Observations are stories that further explain a topic. Like all stories, they are subjective. However, they might and should be exemplary.

By using stories, I want to make sure that readers with a construction background can see the relevance of the microeconomic concepts discussed and how they relate to their everyday life.

1.2.6 Summaries

Sometimes, it is nice to sum up a sequence of arguments, thoughts, or a chapter. This is indicated as follows:

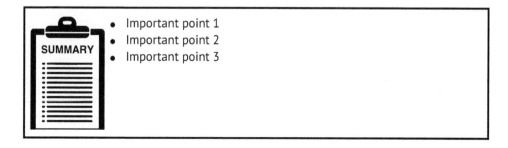

SUMMARY
- Important point 1
- Important point 2
- Important point 3

1.3 Methodological Approach

I have already described my goal and the starting point (status quo) for construction micro-economics. As this plots a journey from point A to point B, I owe you an explanation of how I plan to undertake it by describing my methodological approach.

1.3.1 Laws and Regularities

Civil engineers, project managers, and economists have quite different views because of their professional backgrounds. Civil engineering students endeavor to understand the world of objects. One of the most important fields they encounter is Newtonian mechanics. Here, Newton's laws of motion reign supreme assigning absolute values to mass, distance, and time. These laws of motion serve civil engineers well to design complex structures even 350 years after their publication in 1687 (Figure 1.3).

Newton's thoughts became paradigmatic in defining the term natural law. However, this does not allow us today to think of natural laws as universally applicable. Neither relativity theory nor quantum mechanics are compatible with Newton's laws. Each theory has its domain; civil engineers work with large bodies at rest and for such problems Newton provides the tools.

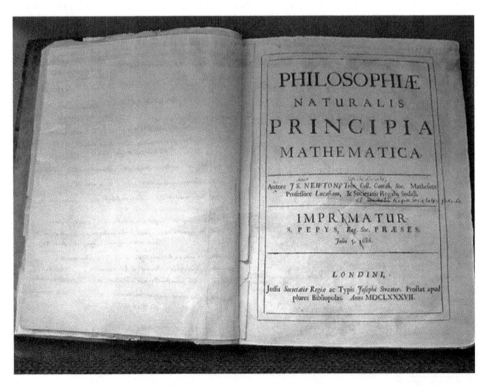

Figure 1.3 Newton's *Philosophiæ Naturalis Principia Mathematica* (first edition, 1687). *Source:* Andrew Dunn/Wikimedia Commons/CC BY-SA 2.0"

Newton's laws and physics are more generally of fundamental importance in philosophy of science. Here, we find two ways of discovering knowledge, by deduction and induction. There exists an asymmetry between the two. An inductive approach is characterized by observation of data and a subsequent summation into a more general statement. These statements take the form of a universal proposition such as the famous *"all swans are white"*. A single observation of a black swan gives us irrefutable proof that the universal proposition is wrong. This leads to the principle of falsification (Popper 2002, first published in 1935). As it is not possible to give proof by induction, we must work with hypotheses and their falsification (deductive approach). If hypotheses such as Newton's laws withstand all serious attempts of falsification, then the hypotheses can be understood as corroborated. After more than 350 years, Newton's laws are well corroborated within its domain, but we have also learned that they are not universally valid. By falsification, we can never gain absolute knowledge, we are rather edging closer by increasing the verisimilitude (closeness to knowledge). The basic understanding is then that all knowledge is conjectural and hypothetical.

> Verisimilitude = (obs.) When I was project director for the design and construction of the 55 km long BangNa expressway in Bangkok, we decided on a thoroughly new design. This would allow us to make better use of materials. To test civil engineering hypotheses with the goal to determine the ultimate bearing capacity, we analysed a full-scale test span (44,40 m × 27,20 m). After loading the bridge span to the ultimate level, the span should have collapsed, but it did not. There were cracks visible, yet knowledge of the true ultimate bearing capacity remained elusive. However, we got closer to knowledge and were able to provide an example for increasing verisimilitude.

In sum, we do not know a single universally applicable law; assuming their existence is a leap of faith. This shall not discredit a metaphysical discussion of the problem, where we face the validity of arguments. Following this path would lead us deep into ontology, philosophy of the mind, and epistemology. I will not follow that path here.

I would like to describe instead the scientific status of economics. If we accept that laws are based on regularities, then the question of their character arises. Two basic possibilities are imaginable: (i) law of causality and (ii) principle of causality. I can leave open the question of how to understand causality because it does not impact the argument. Typical answers are that nature is organized in the way of causality, that it is a human habit of interpreting the world (Hume), or that it is a way of human thinking (Kant). The idea of verisimilitude opposes the belief that we can understand the causal organization of nature, even if it should exist. Whether it is habitual or cognitive matters little if we take an instrumental view of causality.

Laws of causality, on the one hand, posit that the same causes always have the same effects, (i.e. this view supposes the existence of universal laws). The principle of causality, on the other hand, requests only that for each event there must be a cause and for the same event there might be different causes. The difference between the two is contingency.

Civil engineers tend to understand their domain as deterministic (based on Newton's laws), but few are the historians that understand history as deterministic. Neither Hegel's Weltgeist nor Marx's path of historical materialism as deterministic approaches have a large group of defenders. Historians interpret the course of history today as contingent. We can explain the course but we cannot predetermine it because the possible outcomes are too variable. Explanation is possible, but prediction is not. Returning to civil engineering, we are luckier, as we can explain and predict to a degree that serves our purposes.

If we accept the principle of causality for economics, there are three possible forms of regularities:

- Deterministic regularities
- Probabilistic regularities
- Hedged regularities

Civil engineers are well aware of deterministic regularities such as Newton's laws and probabilistic ones such as the strength of concrete. The typical regularities that we encounter in economics are hedged regularities or, as economists would phrase it: They apply under a ceteris paribus clause – meaning that one varies one parameter and all other things remain unchanged. A point in case is the law of supply and demand (i.e. that markets determine an efficient equilibrium price and quantity). The ceteris paribus clause (the hedging) refers mostly to rational behavior of consumers and producers, complete information, and perfectly competitive markets with homogenous goods. Only if these assumptions hold will an increase in demand lead to a greater quantity produced and a higher price.

Roberts (2004) makes the point that there cannot be any laws in social sciences since the number of hedged conditions is infinite given the complexity of social systems. While I agree with the observation in its strict sense, I do not care about it. Whether a regulation is called a law or not is not important. What is important to understand is what effect the regularity has. Not all causal influences are taken into consideration when formulating a ceteris paribus clause – there are too many. Thus, we must interpret the results with caution. Theories, hypotheses, and models in economics can explain and predict but not with the same accuracy as natural laws.

Few, if any, will ever be in a position to contradict $F = mg$ (i.e. if we predict a certain force that a body with mass m will exert at the surface of the world, we will be exact). The same does not hold true when we try to predict the effects of a recession on unemployment. Of course, we would expect a decrease of demand followed by a decrease in production and a corresponding decrease of the factors of production. As labor (employment) is one of these factors, we would assume a decrease as well. The question is: How many people will lose their jobs? Figure 1.4 (Mankiw 2017b) shows six predictions for the US economy given ex-ante at different times in 1981, 1982, and 1983 plus the actual development (ex-post). The prediction in 1981 assumed a maximum unemployment rate of 7.5%; it turned out to be almost 11% at the end of 1982, resulting in misery for millions of people.

We can learn from this data set that predictions are rather accurate in the short run (one or two quarters) and not good at all in the long run. However, this applies especially to problems of limited duration, like the shown increase in unemployment during the first years of the Reagan administration. Overall, the Reagan administration tried to bring down inflation. Economics would predict a rise in unemployment – and that is what happened.

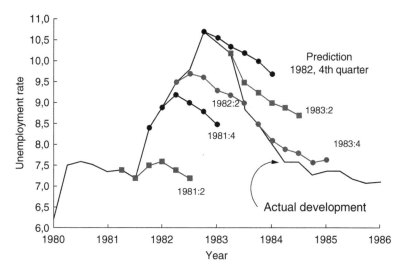

Figure 1.4 Prediction of the effect of a recession on unemployment (Mankiw 2017b). *Source: Makroökonomik*, 4th edition, 2000, Stuttgart: Schaeffer-Poeschel, page 428.

Economic models allow us to predict general relationships between action and result, but the detailed prediction of the specific outcome is doubtful. Now, if we compare the complexity of the environments, we have on the one side a brick placed on top of a wall exerting a force of $F = mg$ and on the other hand a workforce of more than 80 million with the most diverse ideas, wishes, and hopes plus maybe 10 million employers and manifold additional influences. The difference of settings has led Lave and March (1993, p. 2) to complain: "*God has chosen to give the easy problems to the physicists.*"

In sum, economics offers hedged regularities with explanatory content and limited predictive capabilities. This makes it so important to clearly spell out the ceteris paribus assumptions for each model. In the example about the law of supply and demand, one assumption was that goods in a market must be homogenous. As this does not apply to the construction sector with its singular products, we already see an argument for the need of construction microeconomics. Hedged regularities allow us to understand the functioning of economics.

1.3.2 Focus and Goals

Mainstream and construction economics concentrate on the efficiency of production. Here, the allocation of goods is important. There is nothing wrong with this concentration as long as we keep in mind that life also has other aspects. An aspect that finds less interest in microeconomics is equity, the distribution of income (Piketty 2013). Sen (1999) stresses personal freedom as a framework to evaluate economic outcomes and not only growth of the gross domestic product, a materialistic stance. He uses five types of freedom for his analysis: (i) political freedom, (ii) economic advantages, (iii) social chances, (iv) transparency, and (v) security. This provides a different focus on equity. If you are interested, you are encouraged to discover the arguments, but this cannot be goal of construction microeconomics.

Economics and especially microeconomics study market efficiency and thus provide specific answers to such questions. Civil engineering also has a focus by specialising in safe and user-friendly design and construction for the built environment. However, I would assert that the reach of economics is more encompassing than that of civil engineering. We need only to think of women and their rights 150 years ago. Their economic freedom was very limited – they were not free to sign contracts. Changes since then have affected not only the economic but also the political and social standing of women. These spheres are interconnected and progress in one entails development in the others. The social acceptance of women engaging in work outside home (Rosie the Riveter) required the right to enter into contracts and the income provided for economic independence. Work requires education, and this is, in turn, a very strong determinant in reducing population growth (Sen 1999). Overpopulation leads to contamination (e.g. CO_2 emissions) of our environment beyond its capacity: Economics are vital. I would follow the thoughts of Stigler (portrait below) that economics touches us all more directly than construction. You may also think of Bill Clinton's successful campaign slogan in 1992, "*It's the economy, stupid!*"

 George J. Stigler (1911–1991), Nobel laureate 1982

"Whether one is a conservative or a radical, a protectionist or a free trader, a cosmopolitan or a nationalist, a churchman or a heathen, it is useful to know the causes of economic phenomena".

Stigler (1982, p. 61)

Source: KEYSTONE Pictures USA/Keystone Press/Alamy Stock Photo

---

Microeconomics uses mostly neoclassical concepts, while Keynesian economics are influential in macroeconomics. For the travel from microeconomics to construction microeconomics, I will rely on the broad shoulders of many classical and neoclassical economists. You will encounter some of them in the portraits interspersed throughout the book. In recent times, others have widened the neoclassical perspective, namely:

- Coase, North, Williamson (institutional economics)
- Simon, Kahneman, Shiller, Thaler (behavioral economics)
- Tirole, Roth, Shapley (game theory)
- Nash, Milgrom, Wilson (auction theory)
- Akerlof, Spence, Stiglitz (economics of information)

1.3.3 Descriptive and Normative Economics

The last sentence in the discussion about regularities stated that the study of (micro-) economics allows us to understand the functioning of the economy. This is the result of descriptive or positive economics as approach. The findings are typically expressed as conclusions of models. These models are based on assumptions, and the conclusions follow by strict logic from the assumptions. Mathematics serve as a tool to guarantee stringency. Thus, descriptive economics are based on the rationality of science, and the goal is to find answers that are right or wrong. The model of supply and demand gives us correct answers if and only if the assumptions hold. Descriptive economics work with hypotheses, data collection, falsification, theories, and models.

In normative economics, values play a prominent role. This approach tries to find solutions to everyday problems. When I began writing this book, the COVID-19 pandemic was still taking its toll around the world. Schools or businesses struggled with how to open in the usual way, and that depended on many values – how much weight we give to scientific knowledge, how much we care about the lives of others, how much we fear becoming infected, how important material goods are, how important social contacts are, and more. It should not come as a surprise that there were disagreements on those values. Normative economics gives advise often based on descriptive models by including values.

For example, an answer to the question of how rent controls will affect supply on the housing market and the quality of housing can be answered by descriptive models. In the future we can check whether the predictions were right or wrong. The question of whether rent controls are a good approach in cities where rents continuously increase, so that people with smaller incomes are driven out of desirable neighborhoods, involves ethical values as well as how rent controls affect the supply and cost of housing.

1.4 Theoretical Background

The theoretical models of microeconomics draw on other fields of studies with their own models. Important to construction microeconomics are: (i) industrial organization, (ii) new institutional economics, (iii) game theory, (iv) auction theory, (v) economics of information, (vi) law and economics, and (vii) behavioral economics.

1.4.1 Industrial Organization

Industrial organization focuses on the interaction between firms, industries, and markets. Bain (1968) contributed strongly to what is known as the Harvard tradition with the structure/conduct/performance paradigm (SCP). According to this paradigm, the specific market structure of an industry – with its level of competition, product differentiation, cost, and degree of vertical integration – determines the conduct of firms (pricing, innovation, investment, advertising), and from this follows market performance (efficiency, profit, innovation rate). Director and Stigler, on the other hand, started the Chicago tradition stressing the functioning of markets with enough real or potential competition. This leads to a strong mistrust of government intervention. Success, according to the Chicago tradition, is not a question of SCP with low competition but of excellence of some outstanding firms (Tirole 2000).

Important aspects of SCP are monopoly power of a single firm in a market or oligopolistic structures with few competitors. Such structures lead to strategic behavior of the firms when setting prices. They are deviations from the ideal model of a perfectly competitive market, the only one guaranteeing an efficient market outcome. It should be clear that the proponents of the Harvard tradition worry much more about the competition in actual markets than the proponents of the Chicago tradition.

1.4.2 New Institutional Economics

New institutional economics (NIE) stress institutions over individual actors (Brockmann 2021). Institutions shape our lives in many ways. The simplest definition is that institutions determine the rules of the game (North 1991). Schotter (1981) introduced a formal definition:

"Institution = (def.) A regularity R in the behavior of members of a population P when they are agents in a recurrent situation Γ is an institution if and only if it is true that it is common knowledge in P that

(1) everyone conforms to R;

(2) everyone expects everyone else to conform to R; and

(3) either everyone prefers to conform to R on the condition that the others do, if Γ is a coordination problem, in which case uniform conformity to R is a coordination equilibrium; or

(4) if anyone ever deviates from R, it is known that some or all of the others will also deviate and the payoffs associated with the recurrent play of Γ using these deviating strategies are worse for all agents than the payoff associated with R".

Schotter (1981, p. 11)

This definition clarifies that game theory can be a preferred means for the analysis of institutions. There are five different types of rules (or regularities):

1) There are conventions observed by self-control.
2) We have ethical rules with which we comply by imperative obligation; Kant's categorical imperative is an example with regard to generally applicable laws.
3) Others impose the observation of customs.
4) Control by others enforces following private formal rules; organizational rules belong to this group.
5) Law enforcement agencies demand observance of public laws.

The degree of observance of rules or institutions can explain differences in welfare between countries (Acemoglu and Robinson 2012).

Neoclassical economics work with a model of man called *homo economicus*. This actor knows the future and because of this knowledge can choose the best alternative without spending time or resources. Few economists see this as a realistic description but models like supply and demand provide reasonable results based on the behavior of a homo economicus.

NIE assumes different characteristics for economic actors. They do not know the future and accordingly must rely on incomplete information. In addition, limited rationality describes their behavior and decision-making. They try to act rationally but sometimes fail. Maximising benefits becomes impossible; instead, the actors of NIE employ satisficing, the choice of an acceptable solution (Simon 1955). Furthermore, market exchanges are no longer without price; there are transaction costs (Williamson 1985). Anyone who has been an economic actor will feel much more comfortable with the assumptions of NIE. Behavioral economics nudge the model of man in an even more realistic direction by describing in which ways we fail to behave rationally (Thaler 2015).

The modeling of actors in NIE led to three different theoretical approaches. The inclusion of external effects such as the free use of water from a river for cooling of a plant that played no role in neoclassic thinking led to the (i) property rights theory; the negligence of market costs to the (ii) transaction cost theory and incomplete information to the (iii) principal-agent theory.

The theory of property rights assumes profit or utility maximising behavior and transaction costs for writing and enforcing contracts. Connected to property rights is the theory of incomplete contracts (Grossman and Hart 1986). Picot et al. (2015) cite construction contracts as prime examples of incomplete contracts. The main conclusion is that incomplete consideration of property rights leads to external effects and therefore to inefficient factor allocation. Risk allocation in construction contracts is a constant concern, and the property rights theory demands that the allocation of risks is complete. As risks are events in the future and nobody can foresee the future, there will always be some risks that are not part of the contract. According to the theory, this will lead to inefficiencies.

The theory of transaction costs adds transaction costs to production costs. Wallis and North (1986) researched the amount of transaction costs of private enterprises as a percentage of GDP. They found an increase from 20% in 1870 to 40% in 1970: Transaction costs are substantial and increase with economic development. Efficiency must consider both, production and transaction costs as the latter depend on different institutional arrangements. Actors are limitedly rational, opportunistic, and risk averse. In consequence, efficient institutions minimize transaction costs, and at the same time they are safeguarding against opportunistic behavior. It is difficult to determine transaction costs exactly. If transaction costs include all items connected to writing and enforcing the contract (the owner's responsibilities), then production costs, on the other hand, encompass design and construction.

The principal–agent theory focusses on contracts that regulate exchanges. These can be exchanges between organizations (owner and contractor) or within an organization (owner and manager or any other hierarchical relationship). The exchange requires coordination and allows for motivational problems, which we can solve through incentives, controls and information. The actors are utility maximising, limitedly rational with asymmetrical information. They have different risk appetites, and they tend to display opportunistic behavior. Who would not think of the antagonisms in construction projects between owner and contractor, designer, or project manager or those between contractors and suppliers or subcontractors?

1.4.3 Game Theory

Game theory studies multi-personal decision-making (Gibbons 1992). It provides rigorous models that describe situations of conflict and cooperation based on rational decision-making (Tadelis 2013).

Static games describe situations where players move simultaneously while sequential actions by the players characterize dynamic games. Important is also the degree of information that can be either complete or incomplete.

Game theory is especially helpful when analysing bargaining and auctions. We find auctions in construction bidding; as such, they are at the center of economic action determining actual prices. Private owners complement auctions often with contract negotiations to arrive at the contract prize ex-ante (before construction). During construction, there are often further changes to the contract that require adjustments of the price ex-post.

1.4.4 Auction Theory

Most of us are familiar with exchange markets where we buy a good in exchange of money. The goods are on display, and we just have to pick. To the seller, it does not matter who the buyer is, and the individual seller again is of no great importance to the buyer. This describes interactions in perfectly competitive markets. An extreme is an internet market where seller and buyer never meet. An invisible hand organizes the exchange. This is a microeconomic view; marketing takes a different approach by stressing customer satisfaction.

However, economics also knows matching markets where the market participants matter and the markets are designed for specific purposes. Auction theory knows four main market designs: (i) first price, sealed-bid, (ii) second price, sealed-bid, (iii) Dutch auctions starting with a high price and ending with the highest offer, and (iv) English auctions starting at a low price and ending with the highest offer.

Jehle and Reny (2011, p. 427) give the following reasons for the use of auctions:

> *"In most real-world markets, sellers do not have perfect knowledge of market demand. Instead, sellers typically have only statistical information about market demand. Only the buyers themselves know precisely how much of the good they are willing to buy at a particular price."*

Such a situation lends itself to auctions and it is typical for construction markets.

1.4.5 Behavioral Economics

Microeconomic models make at times use of quite stringent assumptions. An example is the metaphor of the homo economicus. Few of us act strictly rational when making economic decisions. Marketing, for example, knows impulse purchases. Kahneman and Tversky published a paper in 1979 on prospect theory. This theory describes actual behavior in decision-making and contradicts the model of utility maximization that is central to the microeconomic model of supply and demand. Of special interest is the behavior of humans (instead of econs) with regard to search heuristics, cognitive biases, and fallacies.

Behavioral economics casts strong criticism on neoclassical microeconomic models. As it is a rather young discipline, it has not yet achieved theoretical coherence regarding modeling. As I stated before, this is unfortunate, and we must rely for the moment on the criticized microeconomic models. However, behavioral economics provide much food for thought.

1.4.6 Economics of Information

It does not need much proof for the statement that information is crucial in decision-making. This understanding goes back to Bacon's *ipsa scientia potestas est* – knowledge by itself is power (1597), and power influences market outcomes.

We have already seen that game theory distinguishes between complete and incomplete information. As game theory is used as tool in auction theory, complete or incomplete information play a prominent role there as well. It is also decisive in bargaining situations. A special form of information advantage is asymmetric information where one side has more information and can use it against the other side. It might lead to market breakdowns as Akerlof (1970) showed in his famous article on "lemons" – used cars of bad quality.

Whoever has worked in the construction sector be it as owner or contractor has experienced the pervasive lack of information. Ambiguity persists in many decision-making situations, and often the information is asymmetric. This typically leads to inefficiencies or market failure. An example of such a market failure is called adverse selection and identifies a suboptimal choice. Signalling to the market, especially by contractors, and screening of the market by the owner are tools to overcome the market failure.

1.4.7 Law and Economics

Construction goods are contract goods and not exchange goods as chocolate bar. It requires a contract to describe the responsibilities of owner and contractor during the construction period. Typically, the main responsibilities consist of delivering a structure without defects (contractor) and payment according to contract (owner). It requires little imagination that the economic outcome of a construction project depends on the quality of the contract.

Unfortunately, most construction contracts are incomplete because of incomplete knowledge and task complexity. Thus, incomplete contracts are of the highest interest in construction.

1.5 What You Can and Cannot Expect

Construction economics is not a well-defined field (Ofori 1994). De Valence (2011b) agrees with this view and identifies five development paths for construction economics:

1) Construction project economics
2) Construction industry economics
3) Environmental economics applied to building and construction
4) Industrial organization applied to building and construction
5) Macroeconomics applied to building and construction

In using a nominal definition, we have some freedom when defining *construction economics*. It is not my intention to provide a final definition but one that is useful for the purpose of this book. I do not want to provide arguments which of the five options is the most appropriate, I simply choose to analyse construction industry economics. Figure 1.5 provides a classification of economic disciplines to show the interrelatedness

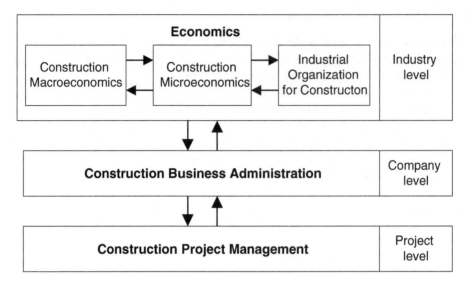

Figure 1.5 Classification of economic disciplines within construction.

of construction microeconomics. A full classification system is provided by the American Economic Association (2022) through its JEL Classification System.

There are four different levels for the economic analysis of the construction sector: project, company, industry, and national / international. Most practitioners in construction will gather their first experiences on the project level, be it on site or in the firm's office.

The tasks are typically described in construction project management books, and they include make-or-buy decisions, scheduling, estimating, logistics, health/safety/environment (HSE), resource planning, construction technology, site installation as well as the management, technical, and legal organization (Brockmann 2021). All tasks are interrelated.

On the firm level, these tasks are coordinated and augmented by further activities. Among them are strategy development, organization, personnel, marketing, investment and financing, accounting, communication, and innovation. These activities are typically discussed within the discipline of construction business administration.

Microeconomics addresses the decision-making of consumers and suppliers as they interact through markets in general. The focus is on the individual decision. Construction microeconomics reduces the breadth of analysis by limiting it to the construction sector. All macroeconomics has a microeconomic foundation, but it concentrates on the aggregate level. Topics like economic growth, employment, and inflation on a national or supranational level take preference. Construction macroeconomics instead tries to answer the contribution of the construction sector to growth, employment, and inflation. Industrial organization has an orientation similar to microeconomics but it focuses more narrowly on market structures and organizational behavior. In this text, I will switch between microeconomic and industrial organization analysis. Economics combines both micro- and macroeconomic perspectives while taking a more general approach looking at the interfaces between disciplines.

The essence of the discussion is that from the top of Figure 1.5 to the bottom, more and more details are added and more and more connections to other topics are ruptured. It is a move from the general to the specific.

For any practitioner, one of the most observable phenomena is the scarcity of resources. There is seldom enough money in the budget for a project manager to make the work crew, the boss, and the owner simultaneously happy. Most often, the work crew expects higher bonuses, the boss higher profits, and the owner a laissez-faire attitude toward claims. Scarcity is at the core of all construction work as well as all other economics.

1.6 Audience

When I studied mainstream economics, I had the feeling that much of the contents had no bearing on the built environment. When I read books on construction economics, this feeling was replaced by a wish for more adaptation of the general economic principles to construction and more depth. I hope to help those along the way who share my perceptions. In the end, I see microeconomics as a tool to advance the welfare of society by creating a better built environment and furthering the goals of the actors.

1.6.1 Students

You are the main audience. Students in civil engineering, project management, quantity surveying, real estate, and construction economics need advanced theoretical knowledge to understand the development of the built environment. Existing monographies do not differentiate between micro- and macroeconomics, thus causing, for example, confusion about the structure of construction markets. Concentration on microeconomics avoids confusion and allows for more detailed information. Adjusting microeconomic theory to heterogenous contract goods prepares students to apply microeconomics to construction problems.

1.6.2 Lecturers

Lecturers have in the past had to adapt microeconomics to construction by themselves. This book steps into that gap to provide introductory stories and observations that cater to the interests of the students, which will make teaching easier and more fun. It provides lecturers access to a range of new concepts, ideas, and models.

1.6.3 Academics

The new concepts, ideas, and models contain unpublished information contributing to an extension of existing knowledge. As always, new thoughts provide starting points for additional research. An example is a game theoretical approach using auction theory to determine the market price. This is based on the research of the Nobel Prize winners in economics for 2020, Milgrom and Wilson.

1.6.4 Contractors

Some experts in construction companies looking for a deeper understanding of construction markets will find new and specific perspectives. This will allow us to rethink and restructure market strategies. Contractors' associations can develop new arguments for public discussions.

1.6.5 Owners

Owners find an economic framework providing a fresh perspective on their decision-making. In addition, they obtain information about the behavior and problems of contractors that permits them to tailor their market approach, predict market outcomes, and understand the economic impact of contractual arrangements.

1.6.6 Policymakers

Policymakers make decisions on the macroeconomic level. However, all macroeconomics is based on microeconomic models. The provided concepts and ideas can assist policymakers in making sound decisions.

1.7 Structure of the Text

The whole text is separated into two parts, following this introduction. Part A addresses microeconomic theory without much reference to the construction sector. This is pure microeconomics and it includes Chapters 2–9 (principles, consumers, producers, perfect markets, imperfect markets, factor markets, information, and game theory/auctions). Part B contains the adaptation of microeconomics to construction (construction sector, owners, contractors, construction goods, construction markets, contracting, imperfections, government, and public goods).

1.7.1 Basic Economic Principles

Chapter 2 (basic economic principles) starts Part A of this text, covering mainstream microeconomic theory. In this part, construction is not the main interest. It is more important to lay the foundations that we later can apply to construction in Part B.

It presents the basic ideas that most economists share. These include agreement on the importance of: scarcity and choice, opportunity costs, incentives, marginal decisions, markets, trade, and government. Of, course there might be disagreement in degrees but seldom on principles.

When reading the economic section of a newspaper, one might conclude that economists always disagree. This perception is wrong. What is discussed in the news is often economic policy (normative economics), and here disagreement abounds because not only different values but also different interests collide. In the worst case, a self-proclaimed expert serves the interests of a lobby group. We can use concrete and steel to construct the greatest variety of buildings. The principles (material science, structural analysis, and dimensioning) remain untouched by the goals of the owners. In a similar way, economic principles are widely accepted.

Presenting shared basic ideas allows for the introduction of key terms and concepts which are later necessary for a deeper understanding of construction microeconomics.

1.7.2 Consumers in Perfectly Competitive Markets

Microeconomics simplifies markets by just looking at two actors: consumers and producers or, in other terms, buyers and sellers. In perfectly competitive markets we find many consumers and many producers. The consumers try to maximize their benefits, which is called utility maximising in econ speech. They prefer more goods and more of one good but with decreasing utility and all this is encapsulated in a preference order. Consumers follow their preferences in a strictly rational way. Of course, they have a limited amount of money to spend, which sets up a budget constraint.

Budget constraints and individual preferences allow us to derive the individual market demand curves, and if we add up all the individual demand curves, we get the market demand curve. This is exactly the goal of consumer theory as we need the demand curve at a later stage for the supply and demand model.

1.7.3 Producers in Perfectly Competitive Markets

Producers provide the goods that consumers buy. Each side relies on the other. Profits serve as incentives for producers and models their behavior: They try to maximize profits. This is, of course, an assumption for modeling and not a statement about the behavior of every single producer. Some might have other motives.

Producers need a technology to transform inputs into outputs of a higher value, and this technology finds its expression in a production function telling us how much output a mix of inputs will generate. If we know the costs of all inputs, we can calculate the costs of the output, and this leads to a general cost theory.

In the end, behavior, technology, production function, and cost theory allow to model the supply curve of the producers giving us the missing part for the supply and demand model.

1.7.4 Interaction in Perfectly Competitive Markets

In Chapters 3 and 4, I introduce the actors in markets, i.e. consumers and producers. Now, in Chapter 6, they can start to interact! As the market is perfectly competitive, there are many actors on both sides, busily trading. The interplay produces a result: Market interaction determines price and quantity of trading. The big question remains whether this result is stable over time, and the answer is yes (under certain circumstances). Therefore, we say price and quantity are in equilibrium. This can be upset by external changes. A famous example was the reduction of oil production in the early 1970s with a consequent fourfold price hike.

Another important question is whether market outcomes are efficient, and the answer is again yes (under certain circumstances). In perfectly competitive markets, the result is efficient, i.e. there is no waste in the economy, all trading options are used. Nobody can be better off without someone else being worse off. Efficiency says nothing about equity, and if you care about equity, the results of perfectly competitive markets can seem a bit disappointing.

If we include a time dimension in the model, then we will find differences, especially between short-run and long-run supply curves.

We arrive at the market outcomes by using strictly logical reasoning within the microeconomic models and by observing initial assumptions. However, the assumptions might seem troublesome at times, so that we can ask whether the models are realistic. Whatever the answer, the supply and demand model can serve perfectly well as a benchmark.

1.7.5   Imperfect Markets

Here we go: Unfortunately, most markets are not perfectly competitive; there exist imperfections. The reason for such imperfections is found in different forms of market power, a dimension that was missing in the perfectly competitive market model. Often, companies have more power than consumers. This can take the form of a monopoly, where only one company produces the demanded good, or of an oligopoly, where few firms determine market supply.

The monopolist can determine its price and quantity by optimising profits looking only at the demand curve; it does not have to consider actions of other producers. A company in an oligopoly faces a more difficult situation – it also has to consider the actions of the few competitors. Definitely, a monopoly has more market power.

At least theoretically, it is also possible that some buyers have more market power than producers. The arguments in Part B will show that market power of owners plays an important role in construction. As discussed in Chapter 6, we will find that imperfect markets are less efficient than perfectly competitive markets.

1.7.6   Factor Markets

Producers use land, labor, and capital in the production process. These are called factors of production and they are traded on factor markets. The goods traded are used for productive purposes and not for consumptive ones. Thus, we face a change of perspective.

The most striking example is the labor market. Households offer labor and they consume goods. Up to now, we met them as consumers (demand); now we face them as labor (supply). The producers are demanding labor on factor markets to produce goods which they consequently offer on consumer markets.

Capital includes fixed assets such as buildings and equipment but also financial assets.

1.7.7   Information, Risk, and Uncertainty

Chapter 8 starts with a static model and then introduces time to differentiate between short and long run in perfectly competitive markets. A chapter later on imperfect markets, I look at the influence of market power. Now, we will consider the influence of the future. Risk and uncertainty are troubling ingredients of the future and this makes information most important. Nobody can foretell the future and accordingly nobody has perfect information about it.

However, some people have more information than others which is helpful to evaluate future outcomes. We call this an information asymmetry. This generates power differences.

1.7.8 Game Theory and Auctions

Game theory has become a widely used tool in economics. It helps us to analyse strategic decision-making in a formal way. Game theory is especially an important tool for designing and conducting auctions. As most contracting happens in the form of sealed-bid auctions, understanding the basics of game theory becomes important for those interested in construction microeconomics.

The owner decides on the market design, and contractors are typically asked to submit their bids in a sealed envelope at submission time. After opening the envelopes, the bids are ranked from low to high; usually, the contract is awarded to the lowest bidder. Some owners add a bargaining phase to the auction for further discussions and lowering of prices by using the information advantage gained from the various bids.

1.7.9 Construction Sector

Part B on applied microeconomics, i.e. construction microeconomics in a stricter sense, starts with Chapter 10 on the construction sector. If there is a need for construction microeconomics, then we must find some justification in the sector and its products.

Important are the size of the sector and its contribution to overall welfare. We will find that construction is a basic sector that provides shelter and enables communication, i.e. social life. It is also important because of its size as it employs many people and provides goods to many consumers (one-family homes) and even more to investors or producers. Quite a number of ideas in this chapter have a macroeconomic background.

As the sector provides the setting, it is beneficial to discover the actors within this framework which shape their behavior to some degree. The topic will be enlarged in Chapters 11 and 12.

1.7.10 Theory of the Owner

Up to this point, we have met consumers or buyers but not owners. The smaller amount of the output of the construction sector serves consumptive purposes, mostly these are one-family homes. Some people buy a new house or building from a developer; they are buyers. To repeat myself, they comprise a small number of actors in new-build markets. Most actors in construction fulfil investment needs and require the help of the construction sector. They build for their own purpose, employing their own financial resources on a plot of land in their possession. They will own the structure at the end, but we will also see that they determine the design and the construction process at least partially. Consequently, the largest demand in the construction sector comes from owners. We must distinguish between the construction and real estate sector to avoid confusion. Only in real estate, we have buyers.

Owning the design, the process, and the end product gives owners market power that consumers and buyers in other markets can only dream about and this has a profound impact on the construction sector.

1.7.11 Theory of the Contractor

An ongoing discussion in construction economics centers on the type of supply by contractors. Some authors claim that they provide a homogenous management service, others contend that they produce heterogenous structures. The discussion has a strong impact on the analysis of the construction market. Homogeneity allows economists to define the market as perfectly competitive; heterogeneity excludes this market type.

I will emphatically reject the idea of homogeneity. The definition of construction goods as transitional contract goods in Chapter 13 does not allow to reduce construction to management: There are simply too many stones, rebars, steel members, and concrete masses on a site to neglect the production aspect.

Heterogeneity imposes further problems to contractors. They cannot apply marginal cost pricing which is a core idea of mainstream production theory. Instead, they must use mark-up pricing.

1.7.12 Construction Goods

The construction sector owes much of its structure to the goods it provides. I will introduce an argument that these are quite different from other goods which serves as the main justification for a separate analysis of construction microeconomics. Mainstream economics, be it macro- or microeconomics, does not pay much attention to the very sizable construction sector. I do not think it is wrong to state that the sector is almost nonexistent in mainstream economic discussions. The foremost reason is the different type of goods produced.

Economics focuses on exchange goods, goods that we can take off a shelf and pay when leaving a store. Construction goods are contract goods, individually defined by a contract before production. The buyer is heavily involved in the production process, and the goods are heterogenous.

There is some discussion among construction economists as to whether the goods have more of a service or a product character. I will advance the argument that they are both, at the beginning a pure service and at the end a pure product, thus, they are transitional contract goods. This definition has far-reaching consequences.

1.7.13 Construction Markets

When owners play a special role as buyers of heterogenous goods, we also need to analyse the impacts on the markets. Construction markets have their own characteristics. The supply side of the market is perfectly competitive. Market entry and exit are free with negligible barriers. However, the goods are not homogenous, and the owners have a certain degree of power. This leads to the question whether the structure can be called a *monopsony*. The answer is a conditional no. Unless we look at a strong recession, the contractors have too many options to sign other contracts. Nevertheless, they are rather dependent on the owners. Looking at one contract, we certainly face a monopsony, looking at a period such as a year, few contractors are bound to one owner as employer.

Contractors very seldom have the chance to influence the price; they are price takers. Owners in most instances can choose between different prices and can also influence the price further by using information asymmetry in bargaining.

1.7.14 Contracting

Chapter 15 looks at contracting in construction markets, a transaction that is very different from a simple exchange. Owners determine the market quantity by the design, and they typically choose to procure construction contract goods through sealed-bid, first-price auctions. Contractors go into the auctions with uncertainty about the contract price; they do not know what other contractors are bidding, nor do they know for sure what difficulties they might encounter if they receive the bid. They need to bid low enough to have a better chance of winning the contract, but not so low that there's no profit in the job. This often enough leads to the winner's curse, when winning the auction means losing money. Until the signature is dry, the owner holds market power.

However, contracting is a three-phase sequential game and construction characterizes the second phase. Now the power of asymmetric information shifts to the contractor. Since construction contracts are incomplete, there is in most cases a need to renegotiate some contractual terms. Finally, the owner takes over the building or infrastructure with a formal handover. Hidden defects are most likely known to the contractor and the owner is at pains to find them. Information advantage shifts from one party to the other during the different phases of construction.

1.7.15 Market Imperfections

Many markets suffer from imperfections in addition to monopolies or oligopolies. Some of these imperfections arise from incomplete knowledge. The model of perfectly competitive markets assumes complete knowledge and construction markets are far away from this ideal. The main reason is the heterogeneity of construction goods.

Externalities are a general problem to the functioning of markets. Externalities are influences of factors of production that are not considered in price and quantity determination. Many industries – and also construction – use air and water free of charge to the detriment of others. In construction, we must add noise, dust, and traffic interruptions.

Corruption and collusion are not only hideous phenomena – they also distort the efficiency of markets. Corruption is the use of entrusted power for private gain; this can happen in the private and public sector. Collusion is the illegal cooperation of contractors to improve their profits to the disadvantage of the owner.

1.7.16 Government

The government plays an active role on construction markets, as described in Chapter 17. It certainly is an important contributor to demand as owner. In some countries, it also acts on the supply side as contractor. It exerts a strong influence by regulation from zoning plans to product standards. The number of laws, norms, and standards is great.

Besides normal taxes such as sales taxes, there are also special taxes such as land transfer or property taxes. Taxes always have a negative influence on the market outcome, but they also provide the means to enable the government to build infrastructure as an owner.

Construction markets are not only influenced by taxes but also by subsidies. Most subsidies try to enable homeownership, some also the construction of fixed assets in other industries. They can take the form of low-interest loans, tax deductions, or cash payments.

1.7.17 Public Construction Goods

Public goods have two characteristics: Nonrivalry and nonexcludability. A typical example is street lightning. Regardless of how many people use the street at night, they all profit from it. Highways can be public or private goods (tollways). Not all public highways have at all times the benefit of nonrivalry: in traffic jams, drivers certainly are competing for space. This is an example of a quasi-public good.

Public construction goods are so numerous and important that they deserve special analysis, provided in Chapter 18.

1.7.18 Conclusion

The conclusion of contrasting mainstream and construction microeconomics is that construction microeconomics is a very promising field of study and research. Mainstream microeconomics is an indispensable building block for the construction sector, but it does not provide enough knowledge for decision-making: We need construction microeconomics.

1.7.19 Synopsis

I have identified different groups among my envisaged audience, and I try to give some guidance in Table 1.1 regarding each chapter and its importance to a special group.

Table 1.1   Importance of chapters to special audiences.

Chapter	Students	Lecturers	Academics	Contractors	Owners	Policymakers
1	+	+++	++	+	+	+++
2	+++	+	+	+	+	+
3	+++	++	+	+	+	+
4	+++	++	+	+	+	+
5	+++	++	+	+	+	+
6	++	++	+	+	+	+
7	+++	++	+	+	+	+
8	++	++	+	+	+	+

Table 1.1 (Continued)

Chapter	Students	Lecturers	Academics	Contractors	Owners	Policymakers
9	++	++	+	+	+	+
10	+++	+++	+++	++	++	+
11	+++	+++	+++	+++	+++	+++
12	+++	+++	+++	++	+++	+++
13	+++	+++	+++	+++	++	+++
14	+++	+++	+++	+++	+++	+++
15	+++	+++	+++	+++	+++	+++
16	++	++	+++	+++	+++	+++
17	++	++	++	+++	+++	+++
18	+	++	+	+++	+++	+++
19	++	++	++	++	++	++

References

Acemoglu, D. and Robinson, J. (2012). *Why Nations Fail: The Origins of Power, Prosperity, and Poverty*. New York: Crown Business.

Acemoglu, D., Laibson, D., and List, J. (2016). *Microeconomics*. Harlow: Pearson.

Acemoglu, D., Laibson, D., and List, J. (2017). *Economics*. Harlow: Pearson.

Akerlof, G. (1970). The market for lemons: quality uncertainty and the market mechanism. *Quarterly Journal of Economics* 84 (3): 488–500.

American Economic Association (eds.) (2022). *JEL Classification Codes Guide*. https://www.aeaweb.org/jel/guide/jel.php (accessed December 2021).

Bacon, F. (1597). *Meditationes sacrae*. Londini: Excusum impensis Humfredi Hooper.

Bain, J. (1968). *Industrial Organization*. New York: Wiley.

Beattie, B., Taylor, R., and Watts, M. (2009). *The Economics of Production*. Malabar: Krieger.

Begg, D., Vernasca, G., Fischer, S., and Dornbusch, R. (2020). *Economics*. London: McGraw-Hill.

Blanchard, O. (2020). *Macroeconomics*. Harlow: Pearson.

Brockmann, C. (2021). *Advanced Construction Project Management: The Complexity of Megaprojects*. Hoboken: Wiley.

Cooke, A. (1996). *Economics and Construction*. Basingstoke: Macmillan.

De Valence, G. (ed.) (2011a). *Modern Construction Economics*. Abingdon: Spon Press.

De Valence, G. (2011b). Theory and construction economics. In: *Modern Construction Economics* (ed. G. de Valence), 1–13. Abingdon: Spon Press.

Dornbusch, R., Fischer, S., and Startz, R. (2011). *Macroeconomics*. New York: McGraw-Hill.

Elsner, W., Heinrich, T., and Schwardt, H. (2015). *The Microeconomics of Complex Economies*. Oxford: Elsevier.

Geertz, C. (1973). *The Interpretation of Cultures*. New York: Basic Books.

Gibbons, R. (1992). *Game Theory for Applied Economics*. Princeton: Princeton University Press.

Grossman, S. and Hart, O. (1986). The costs and benefits of ownership: a theory of vertical and lateral integration. *Journal of Political Economy* 94 (4): 691–719.

Gruneberg, S. (1997). *Construction Economics: An Introduction*. Basingstoke: Macmillan.

Gruneberg, S. and Francis, N. (2019). *The Economics of Construction*. Newcastle: Agenda Publishing.

Gruneberg, S. and Ive, G. (1997). *The Economics of the Modern Construction Firm*. Basingstoke: Macmillan.

Hillebrandt, P. (2000). *Economic Theory and the Construction Industry*. Basingstoke: Macmillan.

Hirshleifer, J., Glazer, A., and Hirshleifer, D. (2005). *Price Theory and Applications: Decisions, Markets, and Information*. New York: Cambridge University Press.

Jehle, G. and Reny, P. (2011). *Advanced Microeconomic Theory*. Harlow: Pearson.

Kahneman, D. and Tversky, A. (1979). Prospect theory: an analysis of decision under risk. *Econometrica* 47 (2): 263–291.

Keynes, J. (1936). *The General Theory of Employment, Interest and Money*. London: Macmillan 1991.

Krugman, P. and Wells, R. (2017). *Economics*. New York: Worth.

Krugman, P. and Wells, R. (2018). *Microeconomics*. New York: Worth.

Lave, C. and March, J. (1993). *An Introduction to Models in the Social Sciences*. Lanham: University Press of America.

Mankiw, N. (2017a). *Principles of Economics*. Mason: Cengage Learning.

Mankiw, N. (2017b). *Principles of Macroeconomics*. Mason: Cengage Learning.

Muñoz-Garcia, F. (2017). *Advanced Microeconomic Theory: An Intuitive Approach with Examples*. Cambridge: MIT-Press.

Myers, D. (2017). *Construction Economics: A New Approach*. Abingdon: Routledge.

Newton, I. (1687). *Philosophiae Naturalis Principia Mathematica*. London: Juffu.

Nicholson, W. and Snyder, C. (2014). *Intermediate Microeconomics and its Application*. Ma-son: Cengage Learning.

Nicholson, W. and Snyder, C. (2017). *Microeconomic Theory: Basic Principles and Extensions*. Mason: Cengage Learning.

North, D. (1991). *Institutions, Institutional Change and Economic Performance*. Cambridge: Cambridge University Press.

Ofori, G. (1994). Establishing construction economics as an academic discipline. *Construction Management and Economics* 12 (4): 295–306.

Picot, A., Dietl, H., Franck, E. et al. (2015). *Organisation: Theorie und Praxis aus ökonomischer Sicht*. Stuttgart: Schäffer-Poeschel.

Piketty, T. (2013). *Le Capital au XXIe Siècle*. Paris: Éditions du Seuil.

Popper, K. (1935, 2002). *The Logic of Scientific Discovery*. Abingdon: Routledge.

Popper, K. (1945, 2003). *The Open Society and its Enemies - Volume Two: Hegel and Marx*. London: Routledge.

Roberts, J. (2004). There are no laws of the social sciences. In: *Contemporary Debates in Philosophy of Science* (ed. C. Hitchcock), 151–167. Oxford: Blackwell.

Romer, D. (2019). *Advanced Macroeconomics*. New York: McGraw-Hill.

Ruddock, L. (ed.) (2009). *Economics for the Modern Built Environment*. London: Taylor & Francis.

Runeson, G. (2000). *Building Economics*. Victoria: Deakin University Press.

Samuelson, P. and Nordhaus, W. (2009). *Economics*. New York: McGraw-Hill.

Schotter, A. (1981). *The Economic Theory of Social Institutions*. Cambridge: Cambridge University Press.

Sen, A. (1999). *Development as Freedom*. New York: Knopf.

Simon, H. (1955). A behavioral model of rational choice. *Quarterly Journal of Economics* 69 (1): 99–118.

Smith, A. (1776, 2012). *An Inquiry into the Nature and Causes of the Wealth of Nations*. Ware: Wordsworth Editions.

Sowell, T. (2015). *Basic Economics: A Common Sense Guide to the Economy*. New York: Basic Books.

Stigler, G. (1982). *The Economist as Preacher, and Other Essays*. Chicago: University of Chicago Press.

Stiglitz, J. and Walsh, C. (2002). *Economics*. New York: Norton.

Tadelis, S. (2013). *Game Theory: An Introduction*. Princeton: Princeton University Press.

Taylor, J. and Weerapana, A. (2011). *Principles of Economics*. Mason: Cengage Learning.

Thaler, R. (2015). *Misbehaving: The Making of Behavioral Economics*. New York: Norton.

Tirole, J. (2000). *The Theory of Industrial Organization*. Cambridge: MIT Press.

Varian, H. (2014). *Intermediate Microeconomics with Calculus: A Modern Approach*. New York: Norton.

Wallis, J. and North, D. (1986). Measuring the transaction sector in the American economy, 1870 – 1970. In: *Long-Term Factors in American Economic Growth* (ed. S. Engerman and R. Gallman), 95–161. Chicago: University of Chicago Press.

Williamson, O. (1985). *The Economic Institutions of Capitalism: Firms, Markets, Relational Contracting*. New York: Free Press.

Part I

Microeconomics

2

Basic Economic Principles

Case Study   Berlin (1994)

Four years after the reunification of Germany in 1994, Berlin was a city springing to life and bursting with construction activities. Hopes, dreams, and ambition exceeded action. For many years after the Second World War, a wall had divided the city into an eastern and a western part. In the eastern sector, a planned economy had caused long lines in front of stores and restaurants while the goods offered were just mediocre. In the western sector with its market economy, goods were plentiful and people enjoyed window shopping because they could not afford buying everything that was available. There was no unemployment in communist Germany while employment in capitalist Germany depended on the business cycle.

Air raids and bombardments during the Second World War had destroyed 75% of the buildings in Berlin. Reconstruction in the East and West also revealed differences in quantity and quality. In East Berlin, most buildings displayed 1990 grey facades and flaking plaster. Brown coal provided heating, and its pervasive smoke filled the streets in winter. At the same time, colorful, well-lit buildings brightened the streets of West Berlin.

The Berlin Wall divided the center of the city, and on both sides all buildings were removed within a certain area. The city center was empty, except for the wall. At long last, reconstruction began in 1994 after some planning. A subsidiary of Daimler-Benz decided to develop the real estate for a large area of Potsdamer Platz with the architect Renzo Piano as master planner. The first contracts were for the construction pits. The pits were large and deep, with a high groundwater level causing considerable engineering problems. After finishing the construction pits, tower cranes crowded the area, delighting the eyes (at least those of construction engineers and managers; (Figure 2.1)).

At the same time, I worked for Germany's fourth largest contractor with responsibility for all heavy civil engineering works in Berlin. This was my opportunity!

(continued)

Construction Microeconomics, First Edition. Christian Brockmann.

Figure 2.1   Dance of the tower cranes at Potsdamer Platz, Berlin. *Source:* Imago/Gueffroy/
Tip Berlin Media Group GmbH

The construction boom should provide for a lot of work, little competition, and considerable profits. As my own contract had a provision for a bonus of 3% from profits, I was looking forward to the challenge.

My enthusiasm fell when I saw all the incoming bids at submission date for the first construction pit, even though our price was competitive. It seemed like all the larger national and international contractors wanted the first bite from the cake. Later, the owner invited us for contract negotiations – not once but eight times. Each time, he would pressure me with information from other bids, asking me to lower the price.

The good thing is that there is no glory in construction pits, nobody wants credit for a hole in the ground, but owners need one to provide for basement floors. Because of this characteristic, construction pits offer the chance for innovation. They do not have to follow design strictly; they just have to follow performance criteria. Innovation, i.e. changes to the owner's design, allowed us to lower the price without compromising the profit. During the eight rounds of negotiations, the owner used his information advantage, and the competition caused much work, headaches, and innovation. One after another contractor dropped out of the negotiations, judging the demanded price being too low. Because of the complexity of the work, the consequences were not easy to ascertain. Uncertainty reigned, and when I signed the contract, I had my doubts about its profitability. Two years later, when the last accounts were closed, I knew we had done a fine job and gained a nice profit. We also had our private encounters with history: We demolished the last foundations of the Nazi's Supreme People's Court (a court of much injustice), and we had to stop the work temporarily when we found an unexploded bomb from the war in the clutches of our stone crusher.

> At this time, I often thought about the economic background behind my experiences. How do historical events influence the construction business cycle? In what way do different economic systems influence the well-being of people? Was the situation in Berlin a unique chance for economic observations with the development of two differing economic systems under the same starting conditions? How fast can the construction sector react to a raise in demand?
>
> What influences acted on the contract price? Why did the owner have so much more information than I did, and why did he always have negotiating leverage? How should I correctly determine a price for something that I have never produced before and that is highly complex? Did my incentive bonus lead me to reckless or cautious behavior? What caused and shaped innovation in this instance? Did competition play a major role?

The second to last paragraph of Case Study deals with questions that are macroeconomic in nature (regarding the whole economy) and the last paragraph with those that are microeconomic (concerning individuals or firms). Macro- and microeconomics are interacting in real-life problems, as the case of Berlin shows. However, the basis for the two is different. In macroeconomics we deal with aggregated data describing an average, and in microeconomics with specific data – e.g. the bonus, the contract price, the innovation for one project, etc.

Every person working in the built environment faces similar question, regardless of whether developer, owner, designer, or contractor. Politicians decide on the economic system and thus shape the possibilities of producers and consumers. Construction managers have developed their own approach, differing in many ways from mainstream economics. The question why construction takes a different approach will be the focus of Part B on applied construction microeconomics. Part A on microeconomic theory begins with this chapter and only once in a while refers to construction, but it lays the foundation of microeconomic thinking and shall serve as a benchmark to judge the efficiency of the construction sector.

Economics in general provides guidance to questions such as those raised in the case study, and this is why it is important for the built environment. Many economic theories are as relevant to construction as to all other sectors; differences show up at a more detailed level. The main consensual ideas (Section 2.1) cover the scarcity that we all face in economic life and the subsequent need to choose between different goods (Section 2.2), the question of rational decision-making (Section 2.3), the function of markets (Section 2.4), the advantages of trade (Section 2.5), and the role of government (Section 2.6).

2.1 Consensual Ideas

When following the news, one might get the impression that economists cannot agree on anything, but the opposite is quite true. Consensus reigns for a number of basic principles. Stiglitz and Walsh (2002), Mankiw (2017), and Krugman and Wells (2018) offer such basic principles. I have chosen those which reflect microeconomics and ordered them so that similar items line up; however, there is not always a match (Table 2.1).

Table 2.1 Comparison of consensual economic ideas.

Principle	Stiglitz/Walsh	Mankiw	Krugman/Wells
#1	Scarcity	People make choices	People make choices
#2	Supply and demand	Opportunity costs	Opportunity costs
#3	Innovation	Marginal decisions	Marginal decisions
#4	Incentives	Incentives	Incentives
#5	Comparative advantage	Comparative advantage	Comparative advantage
#6	Market efficiency	Market efficiency	Market efficiency
#7	Market failures	Market failures	Market failures
#8	Incomplete markets		Market equilibrium
#9	Financial markets		Spending = income
#10	Incomplete information		

Evidently, these basic ideas merit discussion, and they allow introducing key concepts. Thus, this chapter serves as preparation for the following chapters on microeconomic theory, a propaedeutic analysis.

2.2   Scarcity and Choice

Scarcity is a condition of life since Adam and Eve shared the apple from the tree of knowledge and were driven from paradise. While not everybody might believe in the myth of the fall of man, few are those who would not feel the want for something more in their daily lives. Given the prevalence of scarcity, humans organized in groups from the earliest days of human history to hunt or gather food as survival depended on it. Much later, economics became the science that tried to solve questions of production and allocation of goods. Already the father of capitalism, Smith (1776), stated that *"no complaint... is more common than that of a scarcity of money"*.

Only in paradise or similar utopias does scarcity not prevail. Consider the picture Land of Cockaigne from Pieter Bruegel (1567), where a soldier, a peasant, and a scholar representing three estates together with a servant enjoy the idleness of life. The lance, the flail, and the book are of no use; food and drink are plentiful (Figure 2.2).

There is absolutely no need for production, and waste is obvious but not harmful. To some, this world might be appealing, to others less so. Carnal joys are abundant, intellectual ones nonexistent.

However, in a nonutopian world like ours, scarcity governs production by determining prices and influencing demand, and economics is the discipline to explain how this happens. With the words of Sowell (2015, p. 2), this becomes even clearer: *"Without scarcity, there is no need to economize – and therefore no economics"*. Not due to its value to humankind is the price of diamonds much higher than the price of water but because of its

Figure 2.2 Land of Cockaigne, a utopia of waste. *Source:* The Yorck Project/Wikimedia Commons/ Public domain

relative scarcity; this is a scarcity caused by the supply side. A strong and urgent wish to buy a product can create scarcity from the demand side – the launch of a new iPhone with waiting lines in front of authorized retailers can illustrate the point.

Supply and demand determine not only prices but also the quantity of goods. For this, prices reflecting scarcity have four functions: (i) allocation of goods and factors of production; (ii) incentive to produce goods and services assuring that supply meets demands; (iii) information to make a production or purchase decision; and (iv) coordination of efficient consumption and production. All told, scarcity is the engine that drives supply and demand.

As we are not living in the Land of Cockayne, we must choose among the many possibilities against the background of our limited means. In this way, economics becomes the discipline interested in choices. There are two general types of decisions. Either/or decisions determine whether we buy something: Should you buy a house or keep renting your apartment? Facing a how much decision requires a determination whether you want some more: Should you move to a larger apartment, or stay in place? In both cases, microeconomics tries to provide help finding the best solution. By the way, doing nothing is also a choice; this becomes very clear when you look at the option of staying in your apartment in both situations above.

It was Alfred Marshall (1890) who modeled economic man with ever-increasing wants. And this is still the assumption used today: Scarcity exists because needs and wants are larger than available resources.

Alfred Marshall (1842–1924)

"*...economics is a study of mankind in the ordinary business of life; it examines that part of individual and social action which is most closely connected with attainment and with the use of the material requisites of wellbeing*". (1890, Chapter 1)

Marshall published the highly influential Principles of Economics in 1890. He popularized the graphic form of the supply and demand model by using marginal utility. He also introduced the idea of consumer surplus.

Source: Unknown author/Wikimedia Commons/Public domain

The idea of ever-increasing wants is not attractive; it smells of greed. This moral judgment is not at the forefront of economic analysis; rather, it is the question how the majority of people behave.

Wants = (obs.) I think one must distinguish between needs and wants. Needs are those commodities that ensure our bare survival. Fulfilment of wants makes our life easier and more enjoyable. It is an old question whether ease and joy come by gathering more and more goods or by restricting wants. Cynics proposed some 2500 years ago to lead a simple life without possessions. Remember that Alexander the Great granted Diogenes a wish and Diogenes in return asked Alexander to step out of the sun? Diogenes was a Cynic. Epicureans recommend a life of sustained pleasure. In the later form of hedonism, the goal of seeking pleasure became even more pronounced.

Today, we have to ask ourselves, how much more of everything we can afford in a world with many, many more people and much, much more productive capacity.

We have to distinguish between positive description and normative goal setting. Evidence is that most people strive for more material goods in our world; if this is true, than modeling man with unlimited wants makes sense. I wish it were different, but as a professor, I have an income that provides for a comfortable life, much more comfortable than that of many others, and I am far away from criticising those who have less and wish for more. We are definitely facing a dilemma here. We can either reduce scarcity by producing more or we can do it by wishing for less. This choice certainly depends on how much money we start with.

To sum up the discussion, I provide a box with the major points of this chapter.

SUMMARY

- Scarcity is a precondition of life, given unlimited needs and wants.
- Scarcity provides information.
- Scarcity determines prices.
- Consumers choose, reconciling their wants and scarce means.
- Producers choose among resources those promising the best results.

2.3 Decision-Making

Decision-making is a highly important subject in business administration and organizational behavior. It also deserves considerable attention in construction project management (Brockmann 2021). In economics, it is based on rational or intendedly rational behavior, maximising utility or profit (Chapters 3 and 4). Opportunity costs, incentives, and consideration of the last unit consumed or produced (which economists call marginal analysis) all influence decision-making.

2.3.1 Opportunity Costs

In economics we define a rational decision as one that delivers the subjectively best results under the prevailing circumstances. Since we must decide among different opportunities, the forgone opportunities in many cases contribute to the costs of a choice. Facing the choice between starting to work with a bachelor's degree or to continue studying, you must consider the cost of another two years at the university plus the money that you do not earn during those two years (i.e. the cost of not working). These are your opportunity costs in this case. If you would make 50 000 USD per year as civil engineer with a bachelor's degree and 65 000 USD with a master's degree, then you can calculate a break-even point under the assumption that you have 15 000 USD living expenses per year as a student plus 15 000 USD tuition. If you only offset income differences (15 000 USD/year) and university costs (30 000 USD/year), then you will break even after four years of work with a master's. However, you did not consider the full opportunity costs (Table 2.2).

There are high opportunity costs during your studies for a master's degree (80 000 USD/year or 160 000 USD total) and small advantages when working with a master's (15 000/year). After four years, the balance is still negative at 130 000 USD ($-2\,(80\,000) + 2\,(15\,000)$)

Without much trouble, we can calculate the break-even point n (years):

$$n = 160\,000 \text{ USD}/15\,000 \text{ USD} = 10.67 \text{ years}$$

This calculation holds only if a strong ceteris paribus clause applies, which means that all other things stay equal. This is quite an assumption, as most people get pay increases and advance to higher positions, and this is easier with a master's degree. The example allows defining opportunity costs.

Table 2.2 Opportunity costs bachelor versus master's degree.

	Year 1	Year 2	Year 3	Year 4
Bachelor	50 000 USD	50 000 USD	50 000 USD	50 000 USD
Master's direct costs/earnings	−30 000 USD	−30 000 USD	65 000 USD	65 000 USD
Master's indirect costs	−50 000 USD	−50 000 USD	−50 000 USD	−50 000 USD
Master's opportunity costs	−80 000 USD	−80 000 USD	15 000 USD	15 000 USD

Opportunity costs = (def.) Monetary value of the best alternative activity. By making a choice we forgo the realization of this alternative. In very simple terms, this is what we must give up in order to get what we want.

A rational actor will always choose between two alternatives, even if there are many. By comparing in pairs, we end up with the best and second-best and the costs of the second-best alternative constitute the opportunity costs. One might argue that suggesting making decisions about education by looking only at opportunity costs is a rather limited approach. I fully agree with such critique. However, other considerations such as the joy of learning are at the moment outside the theoretical domain of economics.

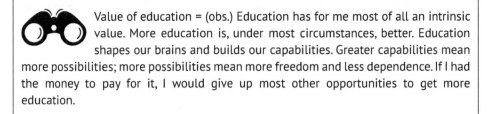 Value of education = (obs.) Education has for me most of all an intrinsic value. More education is, under most circumstances, better. Education shapes our brains and builds our capabilities. Greater capabilities mean more possibilities; more possibilities mean more freedom and less dependence. If I had the money to pay for it, I would give up most other opportunities to get more education.

2.3.2 Incentives

We all react to incentives, however, not as directly and unequivocally as Pavlov's dog. Incentives do not need to have a monetary value. Business economics also consider, among others, respect, fairness, social contacts, diversified activities, or the chance to make decisions. Given scarcity, however, profits are a very strong incentive.

When shaping decision-making situations, incentives play an important role. There are simple incentives such as a higher payer, more responsibility, or a better work environment. The efficiency of such incentives is discussed in publications on work and organizational psychology. Another example of more subtle shaping are opt-in or opt-out contract conditions. Consider a digital service with a monthly charge. Opt-in means that we must inform the service that we wish to extend the subscription. Opt-out means that we will keep getting charged unless we cancel the service. Here, we automatically extend the subscription unless we act. Carelessness often leads us not to opt out, even when we do not use the service any longer. Connecting the ending of a subscription to opt-out is good business. I am almost sure that you have experienced inertia in similar situations and paid for something longer than you cared for. My own last case was my mobile phone contract.

Inertia of intellect is similar to inertia of mass in such instances. Newton states in his first law that a body will stay in uniform motion unless an external force acts on it. Incentives are similar to such external forces. In the case of a subscription, the latest invoice might be the incentive to terminate the contract. There is a large difference in outcome between opt-in and opt-out solutions (Thaler and Sunstein 2008).

In one of the most well-known citations, Smith (1776, Chapter 2) describes the incentive of an entrepreneur: "*It is not from the benevolence of the butcher, the brewer, or the baker that we expect our dinner, but from their regard to their own self-interest. We address ourselves not to their humanity but to their self-love, and never talk to them of our own necessities, but of their advantages*".

2.3.3 Marginal Decisions

When making how-much decisions, we need to consider the benefits and costs of the last unit we want to add, and economists call this the marginal unit. In the example of the master's degree above with its associated opportunity costs, we added to our ongoing education one more unit. There were others preceding the master's degree: kindergarten, primary school, secondary school, and bachelor's degree. None of them played a role in decision-making, only the last marginal unit, the master's degree. In this case, we are the consumer and the university the producer of education, which is a somewhat awkward notation.

Producers face similar questions: Should a contractor sign another contract although he already employs most of his resources in other projects? He would need to hire equipment and subcontractors and buy material. In addition, he would have to provide site management and the head office would have to work overtime. This would only make sense if the benefits of the additional project would exceed costs, a decision at the margin.

SUMMARY
- We need to make economic decisions because of scarcity.
- Opportunity costs measure the full monetary value of a decision.
- Incentives influence our decision-making.
- Economic actors are free to choose incentives.
- If we decide on how much we want of a good, then we must look at the difference. between costs and benefits of the last unit.

2.4 Markets

Markets developed millennia ago to facilitate trade and thereby provide advantages to sellers and buyers. The seller offers something that she has in abundance and asks for something she is lacking. Since the invention of currency, the seller typically receives money. By this exchange both parties are better off. The market is a real or virtual place where certain rules prevail.

Economics are interested mostly in market exchanges; they are of central importance. When market rules do not apply to some goods, economists call these goods externalities and worry about them. It should be clear that some human feelings and actions are not considered as a good, such as friendship and love. Kant (1785, p. 42) gives us a guideline to distinguish the two areas: "*In the kingdom of ends everything has either a price or a dignity. What has a price can be replaced by something else as its equivalent; what on the other hand is above all price and therefore admits of no equivalent has a dignity*".

The following characteristics define markets:

1) Sellers of goods or services
2) Buyers with needs or wants
3) Goods or services
4) Market price
5) Market quantity
6) Market rules

We exchange in markets the property rights to a good. The seller transfers the property right of a good to the buyer and in return, the seller acquires the property rights to an amount of money. Therefore, property rights require a legal definition. Sellers, buyers, the quality of goods, and the market rules determine the market price and quantity. A transaction takes place only if seller and buyer agree on a market price.

 Property rights = (def.) The possession of a property right to an item allows this person to deal with the item in any way that she pleases as long as it happens within the boundaries of law. The idea of private property defines capitalism.

We would not need construction economics, if the characteristics of construction markets and goods traded in those markets were like those in mainstream markets. Its needs a careful analysis how seller, buyers, goods and services, prices and rules differ in construction markets, and I will analyse this in Part B beginning with Chapter 10.

Markets are so important because they are efficient in allocating scarce resources to meet the ends of a society. Criticism of capitalist markets seldom concerns efficiency (how much is produced?) but rather the distribution (who gets what and how much?) of the goods and services. In a historical competition (1917–1990), between market economies and planned economies, it became apparent that planned economies are less capable of providing goods. This became very manifest when East Germans crowded the stores and shops in West Germany after the raising of the Iron Curtain. Not so convincing are capitalistic market results about equity in a society. In 2020, the bottom 50% (165 million) of the US population owned 1,9% of the total net wealth and the top 1% (3,3 million) 30,5%. The Gini coefficient measures income inequality with a value of 0 signifying complete equality and a value of 1 complete inequality. The Organization for Economic Co-operation and Development (2021) provides global data on economic equality.

Markets are, unfortunately, not always providing efficient solutions. One such case is the use of air as a sink for pollution from production. In many cases, air is not considered a factor of production and it is common property (i.e. there are no individual property rights assigned). The result is that some producers pollute the air while many people suffer. Yet, a trade would be possible making both, the producers and the populace, better off. When markets are not functioning properly, governments can improve the results.

We call such markets perfectly competitive if there is no market power in the hands of any one party. In many instances, markets are not perfect but imperfect. Typical examples are monopolies and oligopolies (i.e. there are only one or very few sellers). Again, governments can also in such cases improve the general welfare by regulating the market. Antitrust regulation is one way of achieving better results.

Efficient markets need complete information on both sides. Often information is not complete. If one market side has more information than the other, we talk about asymmetric information, and this gives market power to the party with more information.

We can distinguish between different markets, such as markets for goods and services, factor markets (dealing in production inputs), or financial markets. If an industry produces at capacity, an increase in demand leads to higher prices. If production is below capacity, an increase in demand leads to an increase in production. Thus, the business cycle has a market impact.

A simplified circular flow diagram of an economy shows that someone's spending is someone else's income (Figure 2.3). The economy in this model consists of five actors: households, business, banks, government, and foreign countries. Households provide their labor and capital to business and receive in return wages or profits. Business produces goods and services for households and is recompensed by consumer spending. Banks accept savings from households and lend money for investments to business. Government collects taxes and spends it on goods and services. The economy imports from foreign

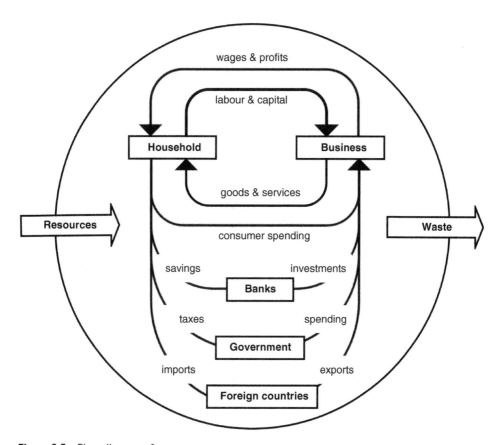

Figure 2.3   Flow diagram of an economy.

countries and must pay for these. Business receives income from exports. Materials are extracted from the earth and waste is deposited on the earth or in its orbit (Raworth 2017).

As I mentioned in the Foreword, there are external influences on this flow diagram not shown, e.g. politics, law, institutions, or psychology.

SUMMARY

- Seller and buyers meet in actual or virtual markets to trade voluntarily.
- Sellers and buyers determine the market price together.
- Markets are based on rules (institutions).
- Perfectly competitive markets are efficient, i.e. produce a maximum of goods.
- Perfectly competitive markets do not necessarily provide equitable outcomes.
- Governments can improve outcomes of imperfect markets.

2.5 Trade and Comparative Advantage

We trade in situations where both parties can be better off by exchange because each party has something that is more advantageous to the other than to themselves. There is a distinction between absolute and comparative advantage. When each one between two parties can produce something at lower costs than the other, both will definitely benefit from trading. In this case, each holds an absolute advantage over the other.

An example for such an absolute advantage is given when a general contractor (Firm-C) specialising in concrete works, signs a contract for a concrete office building with an adjoining factory set up with structural steel. Firm-C subcontracts the steel works to a specialized steel contractor (Firm-S) after winning the bid. Both are large firms and could do all the works by themselves, although at different costs. The bill of quantities shows a total of $10.000 \, \text{m}^2$ of concrete works (formwork, rebars, and concrete) and $1000 \, \text{t}$ of steel (members and welded joints). The relative costs for both firms are given in Table 2.3. The costs of concrete and steel works are:

$$\text{Firm C}: 10,000 \, \text{m}^3 \cdot 300 \, \text{USD/m}^3 + 1,000 \, \text{t} \cdot 4,000 \, \text{USD/t} = 7,000,000 \, \text{USD}$$

$$\text{Firm S}: 10,000 \, \text{m}^3 \cdot 450 \, \text{USD/m}^3 + 1,000 \, \text{t} \cdot 3,000 \, \text{USD/t} = 7,500,000 \, \text{USD}$$

Table 2.3 Cost for Firm-C and Firm-S.

	Costs of concrete works	Costs of steel works	Total costs
Firm-C	$300 \, \text{USD/m}^3$	4000 USD/t	7 000 000 USD
Firm-S	$450 \, \text{USD/m}^3$	3000 USD/t	7 500 000 USD
Firm-C (+S)	$300 \, \text{USD/m}^3$	3000 USD/t	6 000 000 USD

The owner awarded the contract to Firm-C in competitive bidding since the price is lower. Firm-C would face the following costs with Firm-S as subcontractor (Table 2.3):

$$\text{Firm-C}(+\text{Firm-S}): 10,000\,\text{m}^3 \cdot 300\,\text{USD/m}^3 + 1,000\,\text{t} \cdot 3,000\,\text{USD/t} = 6,000,000\,\text{USD}$$

Firm-C holds an absolute advantage over Firm-S regarding concrete works and Firm-S holds an absolute advantage concerning steel works.

Comparing the contract costs of 7 million USD with subcontracting (6 million USD) shows a difference of 1 million USD, ample space for negotiations between the two firms and mutual profits without making the owner worse off. It does not matter whether this is a national or international trade.

 Absolute advantage = (def.) A firm or a country holds an absolute advantage if it can produce a good with less input than others, e.g. it has a higher labor productivity.

Ricardo (1817) provided a well-known example of comparative trade advantages: It takes 100 workhours to produce one unit of cloth in England and 80 workhours to produce one unit of wine in Portugal. A trade would still be advantageous if Portugal could produce the same unit of cloth with 90 workhours assuming the wages were the same. In this case, Portugal would have had an absolute advantage by being able to produce wine and cloth cheaper than England. England would have no choice but trade, as it produced no wine in the nineteenth century. Portugal could save 10 workhours (90–80) by the trade and England would get wine. Both parties would profit. To be more detailed, we can assume that England produces two units of cloth with an input of 200 workhours. Next, it trades one unit of cloth against one unit of wine and becomes better off. Portugal could produce one unit of wine with 80 workhours and one unit of cloth with 90 workhours ($\Sigma = 170$ workhours). It could also produce two units of wine with an input of 160 workhours and trade one unit of wine for one unit of cloth with England. Portugal would enjoy the same amount of goods and save 10 workhours.

 Comparative advantage = (def.) A person, a firm or a country holds a comparative advantage if the opportunity costs of production are lower than those of its competitors.

How does this translate into our construction example? Consider a case where Firm-C has an absolute cost advantage and Firm-S a comparative cost advantage. A comparative advantage would exist if Firm-C could produce concrete and steel at lower costs compared to Firm-S and Firm-S is facing lower opportunity costs for steel works in terms of concrete works than Firm-C in the opposite case.

Let the productivity be for concrete works (formwork, rebar, concrete) 4.5 hours/m³ for Firm-C and 6.75 h/m³ for Firm-S and for steel works (members and joints) 10 h/t for Firm-C and 12 h/t for Firm-S. To find out whether a firm holds an absolute advantage, we need to look at the inputs. It is clear that Firm-C holds the absolute cost advantage since $4.5 < 6.75$ h/ m^3 and $10 < 12$ h/t if the input prices are the same for both companies. Higher productivity is the reason for the absolute advantage, as it produces the output with less input.

To determine whether one firm has a comparative advantage, we need to compare the opportunity costs of both firms. Opportunity costs are the costs of the second-best alternative. For this we must assume a limited capacity, say 100 000 manhours for each firm; the market demand consists of two tenders of equal volume, adding up to 20 000 m^3 and 2000 t.

Solution for no trades (Firm-C wins bid 1 and does not participate in bid because it has reached its capacity limit; therefore, Firm-S wins bid 2):

$$\text{Firm-C}: 10000\,m^3 \cdot 4,5\,h/m^3 + 1000t \cdot 10h/t = \quad 55000\,h$$
$$\text{Firm-S}: 10000\,m^3 \cdot 6,75\,h/m^3 + 1000t \cdot 12h/t = \quad 79500\,h$$

$$\text{Sum}: \qquad\qquad\qquad\qquad\qquad\qquad\qquad 134500\,h$$

If both firms insisted on no trading, they could only fulfil one contract because of capacity limitations. The opportunity costs are the forgone contract:

$$\text{Firm-C}: 10000\,m^3 + 6,75h/m^3 + 1000t\,^*12h/t = \quad 79500\,h$$
$$\text{Firm-S}: 10000\,m^{3\,*}\,4,5h/m^3 + 1000t\,^*10h/t = \quad 55000\,h$$

It is evident that the opportunity costs of Firm-S are lower. The firms can produce the output by Firm-C concentrating on concrete works and some steel works while Firm-S does the rest the steel works; the minimum input is:

$$20000\,m^3 \cdot 4.5h/m^3 + 1000\,t \cdot 10h/t + 1000\,t \cdot 12h/t = 112000\,h$$

In this case, Firm-C is working at full capacity as it holds an absolute advantage:

$$20000\,m^3 \cdot 4.5h/m^3 + 1000t \cdot 10h/t + 1000\,t = 100000\,h$$

Firm-S just uses a small part of its capacity, i.e. 12 000 hours. The savings of $134\,500 - 112\,000 = 22\,500$ hours can be used to make both firms better off.

In general, it is not possible that a firm with an absolute advantage in both activities has also a comparative advantage in both. The opportunity costs of one good are the reciprocal value of the other. Consequently, if a firm has low opportunity costs for one activity, it has high opportunity costs for the other; the inverse is true for the other firm unless by chance both face the same opportunity costs.

David Ricardo (1772–1823)

David Ricardo promoted the idea of comparative trade advantages by the example of cloth produced in England and wine produced in Portugal. He published them in 1817 in his book On the Principles of Political Economy and Taxation. He used a hypothetical approach to develop his ideas and has influenced several generations of economists.

Source: Juulijs/Adobe Stock

SUMMARY

- Trade helps to make people better off by voluntary exchange.
- Trade makes sense if two parties hold different absolute advantages.
- Trade also makes sense if one party has an absolute advantage and the other a comparative advantage.
- Trade makes no sense if a party can achieve its goals better alone.

2.6 Government

If markets were perfect, government would have no place in them. The better performance of market (small government influence) over planned economies (large government influence) is part of our historical records and data (Sowell 2015). Efficient harmony of balancing demands and supplies by myriads of decentralized decisions where pricing provides incentives and information is realistic only in perfectly competitive markets.

Unfortunately, these are rare. Construction approaches this ideal to a larger degree than manufacturing. In manufacturing, product differentiation and economies of scale are ways to limit competition. The most common type of markets are oligopolies, where some sellers face many buyers with the sellers wielding some market power (Petit 2013). In other markets, many buyers face a single seller – a monopolist – with an even stronger grip on the market. We are facing imperfect perfection in markets economies (and for those who like puns, perfect imperfection in planned economies). Under these circumstances, governments can improve market outcomes:

- In the case of monopolies by antitrust regulation, thus ensuring a larger quantity of a good at a lower price for buyers and overall a higher welfare for all participants in an economy.
- In the case of oligopolies by controlling illegal tacit or open collusion.
- In the case of perfect competition and the presence of so-called external factors, i.e. factors that are not part of markets. Examples are free use of air as sink for emissions or free use of water for cooling purposes. In such cases, the emitters and users have free access to inputs (air, water) and others pay the price. In the case of emissions, this is the general public, or even the healthcare system. In the case of water, it might be fishermen when fish die because of a higher water temperature.

Governments are not only responsible for the economy but also for social and legal life. They provide the rules by which we all have to abide and in democracies they have to follow the will of the sovereign – the people – in doing so. Rules or institutions pervade all aspects of life, not only the economy. According to Scott (2001), institutions rest on three pillars: regulations, norms, and cognition. The very foundation of a market economy is the allocation and enforcement of property rights and this we achieve by regulations. Culture provides norms for behavior. Cognition or the way how we see the world (sensemaking) pervades everything.

It is also true that governments mishandle situations, at times with very negative results. This affects the economy as all other aspects of life. We must search a balance in democracies between too much and too little action by the government. For this, politicians need good advice, and it is the task of economics to provide such advice in its own field.

Sometimes, economists complain that governments meddle in markets where they have nothing lost. Such complaints can be justified. On the other hand, economist should not meddle in politics outside their domain of competence (except as citizens). Reagan's statement that the government is not the solution but the problem is catchy, but in many cases, it is also wrong. As a side note, Reagan increased the state quota by 3.3% (i.e. he increased the cost of government spending to expand its role). He seemingly did not believe in his own statement.

The size of the world population and the corresponding economic production has led to a dangerous increase in CO_2 emissions. The phenomenon is called global warming, and in the scientific community it is considered a strongly corroborated theory that human action is responsible. This is a problem created by markets, and there is no other way than to solve it by global government cooperation. If we are not successful, then economic repercussions will be unavoidable.

If you are interested in different interpretations of government involvement by economists, you might compare two text where the authors seem to be on a mission. Sowell (2015) quite eagerly provides reasons for little government involvement, and Raworth (2017) argues as eagerly for more government involvement.

SUMMARY

- In perfect markets, government intervention is not helpful.
- Most markets are not perfect.
- In imperfect markets, governments can improve the outcome.
- Governments must always provide basic rules (e.g. property rights).

References

Brockmann, C. (2021). *Advanced Construction Project Management: The Complexity of Megaprojects*. Hoboken: Wiley-Blackwell.

Kant, I. (1785). *Groundwork for the Metaphysics of Morals*. Cambridge: Cambridge University Press (2012).

Krugman, P. and Wells, R. (2018). *Microeconomics*. New York: Worth.

Mankiw, N. (2017). *Principles of Economics*. Mason: Cengage Learning.

Marshall, A. (1890, 2013). *Principles of Economics*. London: Palgrave Macmillan.

Organization for Economic Co-operation and Development (2021). https://data.oecd.org/inequality/income-inequality.htm; .

Petit, N. (2013). The "oligopoly problem" in EU competition. In: *Handbook on European Competition Law* (ed. I. Lianos and D. Geradin), 259–349. Cheltenham: Edward Elgar.

Raworth, K. (2017). *Doughnut Economics: Seven Ways to Think like a 21st-Century Economist*. London: Random House.

Ricardo, D. (1817). *On the Principles of Political Economy and Taxation.* Mineola: Dover Publications (2004).

Scott, R. (2001). *Institutions and Organizations.* Thousand Oaks: Sage.

Smith, A. (1776). *An Inquiry into the Nature and Causes of the Wealth of Nations.* Ware: Wordsworth Editions (2012).

Sowell, T. (2015). *Basic Economics: A Common Sense Guide to the Economy.* New York: Basic Books.

Stiglitz, J. and Walsh, C. (2002). *Economics.* New York: Norton.

Thaler, R. and Sunstein, C. (2008). *Nudge: Improving Decisions about Health, Wealth, and Happiness.* New Haven: Yale University Press.

3

Consumers in Perfectly Competitive Markets

Case Study   Car or House?

CASE STUDY

You and your spouse have started work after graduating from a construction management program with master's degrees. After work, you sit together on the porch, watch the sun setting behind the hills with a glass of juice in your hand. You are both happy with the good paychecks that come at the end of the month.

Sooner than later, the discussion turns to spending the savings that you have put aside over the last couple of years. You know exactly how much you have saved as well as your current incomes, and you want to spend all your savings – this is the available budget. One of you wants to buy a used convertible, the other wants to use the savings as a down payment for a one-family home. Initially, the two of you do not agree on what you prefer, but after some further talking, you decide to spend the money for a house. In the end, you both find it more useful. While you look at the housing market on your tablet, you become more confused because of all the options. Well, no need to decide now, you can take your time.

At this point, your phone rings and it is your father, telling you that he just finished talking with an architect about the addition that he plans to his factory. After a while, you tell him about your decision to buy a house and he congratulates you.

After hanging up, you wonder why you decided to forgo the convertible. Too bad you cannot have both. In your economics class at the university, your learned about opportunity costs and you take out your class notes and run a quick check on your options. It seems that a house is the better option when considering the low interest rates, the rapid rise of house prices, and the rent that you pay now. Unfortunately, most of these variables can change in the future, so you become very much aware of the uncertainty in your calculations.

(continued)

Construction Microeconomics, First Edition. Christian Brockmann.
© 2023 John Wiley & Sons Ltd. Published 2023 by John Wiley & Sons Ltd.

> Well, time to go to bed! However, before you fall asleep, you wonder about the differences between your father's intention to enlarge the factory and your own to buy a house. One serves the purpose to produce more mechanical pumps and the other makes your life more enjoyable. A little drowsy already, you remember the difference between investment and consumer goods, but you have no idea what the consequences of the difference are. You promise yourself to look it up the next day.

The difference between a consumer and an investor depends on the use of a good. If you buy a house to live in by yourself, then you are consuming the service that the house provides. If you buy the house to rent it out, then you are making an investment. The pub owner who buys a barrel of beer is investing; when he serves you a glass of this beer, you are consuming. In Figure 2.3, there are households that consume and businesses that invest. Banks, governments, and foreign countries consume and invest at the same time.

Consumer

 Consumer = (def.) Person, household, or a group buying a good or service exclusively for private use. They are the end-users in the supply chain. They use the good or service up in the short run (beer) or in the long run (one-family home).

Economic actors decide in a certain context. Besides the private situation as discussed in the case study above, the market plays an important role. Analysing consumer theory, we typically start with perfectly competitive markets (Section 3.1). Consumer behavior describes the way how consumers react under these circumstances. Consumers can spend only within their budget, and this is called a budget constraint. They spend as they like – they are said to have preferences. We can transform preference orders into utility functions and into indifference curves, curves of constant utility (Section 3.2). If we assume that consumers try to maximize their utility, we can deduct from this their downward-sloping demand curve. By adding up all individual demand curves, we will find a market demand curve for a good, and this is what we are looking for: The first part of the supply and demand model (Section 3.3). If you want to know how all this works in detail, keep reading! Please pay special attention to all the assumptions that are part of the models. Some of them might strike you as unrealistic and consumer theory is indeed strongly discussed. You will find some further thoughts in Section 3.4. The demand curve as developed here is part of the neoclassical model and it will serve as a benchmark when we later look at the construction industry. In Chapter 11, you will find a theory of the owner which more realistically describes the demand side for the construction sector.

Up to now, I have used the unspecific terms buyer and seller to describe actors on markets. The term consumer is more limited as it applies only to goods for consumption purposes. Producers, the focus group in Chapter 4, can also be buyers but they cannot be consumers. They are buyers in factor markets, markets where production inputs are traded (Chapter 7). If a producer buys some ice cream to take home to the family, he switches his economic role from producer to that of a household. Mainstream microeconomics is not much interested in owners (Chapter 11) or contractors (Chapter 12), this is one more reason to develop a theory of construction microeconomics.

3.1 Perfectly Competitive Markets

The model of supply and demand describes two situations. The partial equilibrium defines what happens in a market for a specific good and the general equilibrium the interaction between prices and goods in all markets of an economy. The first one is a microeconomic model, the second one belongs to the interests of macroeconomics. In our context, the partial equilibrium model is important.

Léon Walras (1834–1910)

Walras established the theory of the general equilibrium using a mathematical approach from which Marshall later developed the theory of the partial equilibrium.

"*The market is like a lake stirred by the wind, in which the water continually seeks its equilibrium without ever achieving it*".

Walras (1874, p. 310).

Source: Unknown/Wikimedia Commons/Public domain

The model of supply and demand is an example of deductive economic reasoning. Starting with a few assumptions for the market, consumer behavior and producer behavior, we get to the well-known diagram of supply and demand for a perfectly competitive market (Figure 3.1). The model consists of the consumers' downward-sloping demand curve and the producers' upward-sloping supply curve. The price p^* and the quantity q^* designate the market equilibrium for a given good x_i. Whether the curves are straight lines or not is of little importance right now, but straight lines are simpler and for this reason preferable.

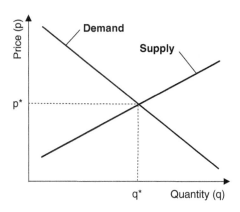

Figure 3.1 Equilibrium in a perfectly competitive market.

Figure 3.1 will not be a surprise; more interesting is the question, how does it come about? Results for other market configurations such as monopolies/monopsonies and oligopolies/oligopsonies are different. Such imperfect markets are discussed in Chapter 6. I think it is important to understand the implications of the model first and then maybe criticize the assumptions.

I will first highlight some hypotheses such as behavioral assumptions for consumers and producers, most often based on observation and experience. As human behavior varies greatly, such assumptions must generalize to a large degree. The point, however, is not the accuracy of the assumptions but the predictive accuracy of the model. Friedman (1953, p. 8) phrases such thoughts as follows:

> *"Viewed as a body of substantive hypotheses, theory is to be judged by its predictive power for the class of phenomena which it is intended to "explain". Only factual evidence can show whether it is "right" or "wrong" or better tentatively "accepted" as valid or "rejected". . ., the only valid test of the validity of a hypothesis is comparison of its predictions with experience."*

This view is highly influenced by Popper's critical rationalism (1935). While predictions based on the model of supply and demand are in general correct, they are not so in detail.

 Assumptions for perfectly competitive markets:

1) There are many consumers and producers.
2) The goods in the market are homogenous.
3) Market transparency is complete.
4) Market transaction are without costs (no transaction costs).
5) Market barriers for entry and exit do not exist.
6) The market price can adjust freely and instantaneously.

These assumptions have consequences. The large number of market participants guarantees that no individual can influence the price. The large number also makes it impossible to reach an agreement between consumers or between producers to fix a low or high price. All participants are price takers and quantity adjusters. All consumers will buy the quantity they can afford at a given price. All producers will supply the quantities they can produce economically at this price.

If goods are homogenous, then they are interchangeable. Consumers have no personal, temporal, or spatial preferences regarding a specific producer. They do not care whether they buy from producer X or Y.

Market transparency entails complete information about prices. Consequently, there is only one price. If a consumer offers a lower purchasing price than all others, no producer will sell to that person. If a producer demands a higher selling price, no consumer will buy from him. This is sometimes called the law of indiscriminate price.

Transaction costs are associated with contracts. Among them are costs for information gathering, negotiating, and concluding the transaction; they are neglected in the model.

If there were market entry barriers, the market could not possibly be completely competitive, as some competition would be kept out. Market exit barriers keep producers in the market even when they are not able to offer a competitive price. This leads to misallocation of scarce resources.

If the price could not fluctuate freely and instantaneously, then there would be periods of price discrimination separating the perfectly competitive market into submarkets with reduced competition.

3.2 Consumer Behavior

Consumers have certain characteristics in all markets, whether they are competitive or not. That is to say, market configurations do not affect consumer characteristics. These characteristics are again modeled as assumptions.

Assumptions for consumer behavior:

1) Goods are scarce.
2) Consumers face a budget constraint.
3) Consumers have preferences.
4) Consumers maximize their utility.
5) Utility has no saturation point; i.e. more goods are always better for each individual consumer.
6) Consumers decide rationally.

3.2.1 Budget Constraint

Consumers have incomes, be it from working, savings, or some form of investment. These incomes are limited while needs and wants are infinite. The discrepancy causes scarcity and consumers to choose between alternative uses of their income. We can influence scarcity by individual austerity or increased production. It remains up to the individual or society to reduce demand. We cannot observe self-constraint in consumption very often. Thus, economics pursues the goal of maximum production to satisfy the consumer needs and wants.

Figure 3.2 shows a two-dimensional model of a budget line. An individual (or a family) with a certain income (y), must decide whether to consume (c) all or save (s) some:

$$y = c + s \tag{3.1}$$

In this case, Susan decides to spend all her income in each period ($y = c$). Since the graph is two-dimensional, we can depict only two goods, and because of this technical reason, she will spend her income on two goods. Instead of looking at housing and bananas, I prefer referring to the second good as a bundle of all other goods; it is less precise but more meaningful. Assuming $y = c$ (i.e. no savings), the equation for the budget line (B) with prices (p) and quantities (q) becomes:

$$B = p_1 q_1 + p_2 q_2 \tag{3.2}$$

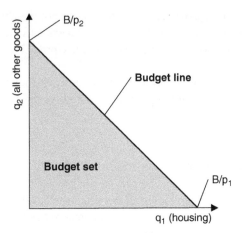

Figure 3.2 Budget line and budget set.

If we assume that she spends all money on housing, we get one endpoint of the budget line; we can calculate the other endpoint by assuming all money spent on the bundle of other goods. With these assumptions we get:

$$B = p_1 q_1 \rightarrow q_1 = B/p_1 \text{ or } B = p_2 q_2 \rightarrow q_2 = B/p_2 \tag{3.3}$$

The extremes are a bit theoretical as we need food and shelter simultaneously, but it allows drawing a complete budget line. The area between the budget line and the axes represents the budget set (i.e. all of Susan's spending possibilities). Choosing a point that is not on the line but part of the budget set would mean that she would save some amount. She cannot spend beyond the budget line because that is beyond her monetary abilities. This is true if we do not consider loans. For each choice on the budget line she will have no savings during the period; the line shows all possible combination of the two goods she can afford spending all her income. It is easy to imagine an n-dimensional space instead of the two-dimensional example to become more realistic, but it is hard to treat it graphically or even mathematically.

Two variables heavily influence the budget line: Income (y) and prices p. If Susan receives a pay increase of 10%, she can afford to spend 10% more on housing ($+0.1q_1$) or on the bundle of goods ($+0.1q_2$); the endpoints of the budget line shift outward on the axes (Figure 3.3, left side). Another possibility is an increase (or decrease) in prices. Let the price of housing (p_H) increase by 25%, then the affordable quantity will decrease to $q_1/1.25 = 0.8q_1$ (Figure 3.3, right side). For a pay increase, the budget set becomes larger, for a price increase it becomes smaller.

To find out what choice Susan will make we need to look at her preferences next.

3.2.2 Preferences and Utility Functions

Each individual has different preferences and can arrange them in an order, the preference order. The orders must have certain characteristics which take in our model of consumer behavior the form of assumptions.

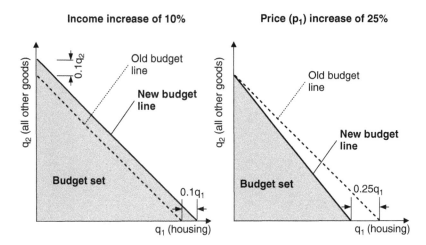

Figure 3.3 Shifts of the budget line due to income or price increases.

Assumptions for consumer preference orders:

1) They are complete.
2) They are reflexive.
3) They are transitive.
4) They are continuous.
5) They have no saturation point.
6) They are convex.

We can use mathematical symbols to describe preference orders when comparing two bundles of goods X^1 and X^2:

$$X^1 \succsim X^2 \; X^1 \text{ is at least equal to } X^2 \left(\text{weak preference} \right) \tag{3.4}$$

$$X^1 \succ X^2 \; X^1 \text{ is preferred over } X^2 \left(\text{strong preference} \right) \tag{3.5}$$

$$X^1 \sim X^2 \; X^1 \text{ is as good as } X^2 \left(\text{indifference} \right) \tag{3.6}$$

Completeness describes the fact that any two bundles from a set of bundles can be compared to each other. If we look at Tom, then completeness describes the fact that he can always come to a decision. Either $X^1 \succsim X^2$ holds or $X^2 \succsim X^1$ or both (indifference).

Reflexivity indicates that any bundle is at least as good as itself (weak preference). For all X we find $X \succsim X$.

Transitivity guarantees consistent decision making. If $X^1 \succsim X^2$ and $X^2 \succsim X^3$, then it must follow $X^1 \succsim X^3$. This is an application of logic to decision-making.

Continuity ensures that the preference order has no gap. The decisions of Tom are not erratic but continuous. For preferences to be continuous, wants must be divisible into infinitesimal parts. This looks like a rather innocent assumption, but it is indeed an audacious one.

Completeness, reflexivity, transitivity, and continuity make it possible that a utility function can represent the preference order. This allows for the transition from preferences to utility:

$$U(X^1) \geq U(X^2) \text{ if } X^1 \succsim X^2 \tag{3.7}$$

$$U(X^1) > U(X^2) \text{ if } X^1 \succ X^2 \tag{3.8}$$

$$U(X^1) = U(X^2) \text{ if } X^1 \sim X^2 \tag{3.9}$$

We can measure the preference order as well as the utility function only on an ordinal scale. Values describe an order but no distances between any two bundles of goods which would allow us to use a cardinal scale.

The lack of a saturation point means that more is better. If $X^1 \sim X^2$ and if we add one good to the bundle X^1, then follows $X^1 \succ X^2$.

All bundles which Tom is indifferent about are located on a continuous line, the indifference curve. The convex shape is due to rational choice in a given situation. If point A in Figure 3.4 describes the endowment of Tom, then Tom has a large amount of drinks but little food. In this case, he will value food more than drinks. If the indifference curve were a straight line, he would be indifferent between drinks and food ($\Delta x_2 = \Delta x_1$). Only a convex curve describes the fact that he would be willing to move from A by giving up a considerable amount of drinks to get some more food ($\Delta x_2 > \Delta x_1$). Figure 3.4 shows a plausible convex (left side) and an implausible straight indifference curve (right side) with an endowment point A as part of the argument.

Economists call the concept causing the convexity of indifference curves decreasing marginal utility. A marginal observation focuses on the last unit of a quantity consumed. We can observe what happens if we take one unit away from that quantity (subtractive

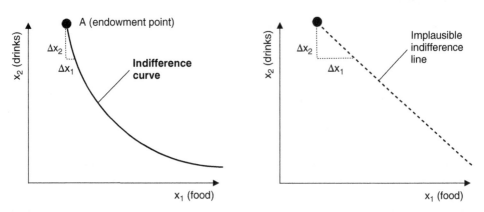

Figure 3.4 Indifference curve.

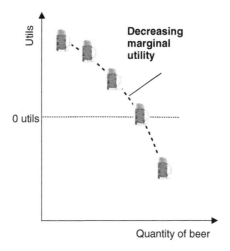

Figure 3.5 Decreasing marginal utility (beer).

marginality) or what happens when we add one unit to it (additive marginality). For a convex indifferent curve and a high quantity of good x_2 (q_2) we assume a willingness to give up quite a bit in exchange for good x_1. If someone has little of a good x_2, we assume that person to be willing to give up quite a bit from a good in abundance (x_1) to add just one unit of x_2. This is the first time we observe marginal behavior and I will make much more often use of the concept in later chapters.

Figure 3.5 provides an example of decreasing marginal utility that many people understand from experience. Tom likes beer and he can assign several utils (dimension of utility) to the beer he consumes. Tom orders the beer in litre mugs and is a strong drinker. He becomes indifferent only as he is drinking his fourth litre. The fifth one has negative utils and is a bad: It will give him raging headaches next day.

Each indifference curve represents a distinct level of preference. Thus, we can use indifference curves instead of preferences. Utility is a measure of the degree of satisfaction with the consumption of a good. The greater the utility, the greater the satisfaction. Utility can derive from egoism as well as from altruism. Tom could find a high utility in buying more beer; he could also find high utility from spending for charity (he could prefer to contribute a certain amount to charity).

Francis Edgeworth (1845–1926)

Edgeworth proposed the idea of indifference curves. He developed a mathematical approach to economic problems and contributed to statistics where he developed the Law of Error, the Law of Great Numbers, and the Law of Correlation.

Edgeworth (1881).

Source: Unknown Source/Wikimedia Commons/Public domain

3.2.3 Utility Maximization

If Tom wants to spend all his income as Susan did before, then he would have to reconcile his wishes in form of utility maximization with reality in the form of the budget line. If he would further make a rational decision, he would choose the bundle of goods where one of his indifference curves touches the budget line. This indifference curve represents the highest utility Tom can reach given his budget (Figure 3.6).

This utility maximising choice follows a subjective rationality. Subjective is Tom's preference order expressed by his indifference curve and utility maximization provides the rationality by reconciling wants and means.

We have discussed the convexity of indifference curves by looking at the exchange of one unit of a good for a quantity of another. When we think in units, we are referring to a discreet function. However, since preference orders are continuous, the same follows for the indifference curve representing a constant utility. The willingness to exchange an infinite amount of good x_1 for good x_2 is called the marginal rate of substitution. In a graph, this the slope of a tangent to the indifference curve (Figure 3.7).

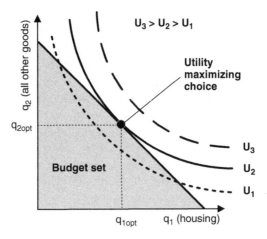

Figure 3.6 Utility maximising choice for a given budget.

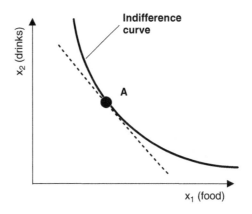

Figure 3.7 Marginal rate of substitution.

Instead of looking at two bundles of goods, we can also look at a large number of goods that Tom wants to buy. The total utility that he would obtain from a set of goods x can be expressed by the flowing function:

$$U = U\left(x_1, x_2, \ldots, x_n,\right) \tag{3.10}$$

Total utility always increases as there is no point of saturation (i.e. the first partial derivative is positive) and we call it marginal utility:

$$\partial U/\partial x_1 > 0 \tag{3.11}$$

The more we already have from a good, the smaller utility we derive from an infinitesimal increase of the same good (i.e. the second partial derivative is negative):

$$\partial^2 U/\partial^2 x_1 < 0 \tag{3.12}$$

The utility function and the marginal utility are shown in Figure 3.8. How can we describe rational behavior in case of an exchange of two goods? In such a case, the marginal utility of good x_1 divided by its price p_1 must equal the marginal utility of good x_2 divided by its price p_2:

$$\left(\partial U/\partial x_1\right)/p_1 = \left(\partial U/\partial x_2\right)/p_2 \tag{3.13}$$

Thus, for maximising utility, the slope of the budget line (p_2/p_1) must be equal to the tangent to the indifference curve:

$$p_2/p_1 = \left(\partial U/\partial x_2\right)/\left(\partial U/\partial x_1\right) \tag{3.14}$$

This is the case for the point of utility maximising choice in Figure 3.6.

In sum, a person can maximize utility by spending the complete budget (no savings) and by choosing the combination of goods for which the absolute value of the marginal rate of substitution equals the relation of the prices.

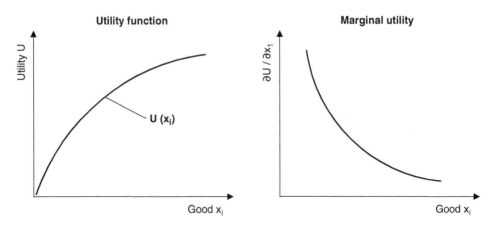

Figure 3.8 Utility function and marginal utility.

William Jevons (1835–1882)

"One of the most important axioms is that as the quantity of any commodity, for instance, plain food, which a man must consume, increases, so the utility or benefit derived from the last portion used decreases in degree. The decrease in enjoyment between the beginning and the end of a meal may be taken as an example".

Jevons (1886), pp. 151–152.

Jevons, Walras, and Menger developed simultaneously and independently the concept of marginal utility.

Source: Unknown Source/Wikimedia Commons/Public domain

3.3 Demand Curve

We have established that the quantity q demanded by an individual depends on the consumption c, the preferences and the prices of considered goods. In the case of two goods we have:

$$q_1 = f\left(p_1, p_2, c, \text{preferences}\right) \tag{3.15}$$

Starting with variations of the price in the utility maximising model with the budget constraint and the highest indifference curve, we can derive any number of price/quantity combinations. Such price/quantity combinations constitute the individual demand curve (Figure 3.9).

The budget line is based on the price of our two goods. We keep the price of good x_2 fixed in the upper part of Figure 3.9, while we vary the price of good x_1. Remember that we can buy more of a good if the price goes down and the income y remains the same. Next, we can draw three budget lines, one for each price. We will find the utility-maximising quantity where the highest possible indifference line touches a budget line. This provides us with a combination of price of good x_1 (from the budget line) and quantity (point of contact). At the point of contact, the slope of the budget line (p_2/p_1) is equal to the tangent to the indifference curve.

To build the market demand curve from different individual demand curves is easy: We only need to add up the individual quantities for given prices horizontally. Figure 3.10 shows this for three individual demand curves. These demand curves are drawn as straight lines for the sake of simplicity.

Variations in price cause the consumer to change the quantity demanded along the aggregate demand curve. This is a movement along the demand curve. Price and quantity are endogenous variables (i.e. they are part of the model). Changes of income, changes of savings (zero in the model), and changes of preferences are exogenous variable; they lead to shifts of the demand curve.

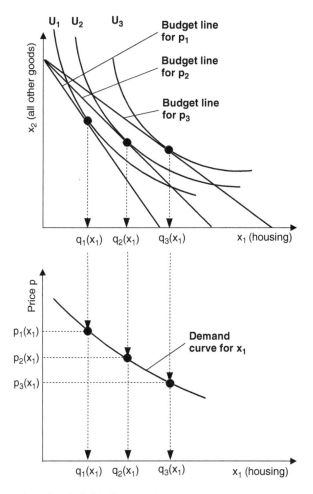

Figure 3.9 Construction of an individual demand curve.

The following box sums up the results of Chapter 3 on consumers:

SUMMARY

Perfectly competitive markets rely on several assumptions. The most important ones are:

- Buyers and sellers are price takers and quantity adjusters, the traded goods are homogenous, and information is complete.
- Consumers face a budget constraint.
- Consumers have defined preferences which can be transformed into a utility function.
- Marginal utility is decreasing.
- Consumers maximize their utility.
- The individual and market demand curves slope downward.

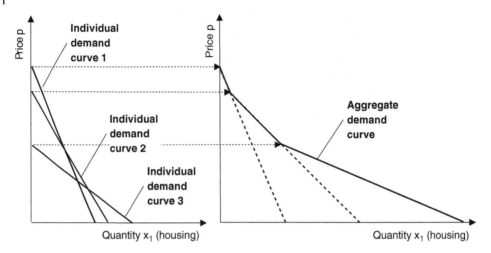

Figure 3.10 Aggregate demand curve.

3.4 Further Reading

The discussions above have shown that the development of the demand curve is based on the many assumptions used for modelling. The resulting demand curve is a neoclassical model. The human behavior in this model describes the homo economicus. Many authors have criticized some of the assumptions as unrealistic, and they have proposed different and more realistic ones. However, these other assumptions make modelling much more difficult.

New institutional economics (NIE) originated in the late 1930s. Simon (1955) became a Nobel laureate in 1978, for, among other findings, introducing the idea of bounded rationality. This means that individuals try to be rational when making decisions but at times fail to follow through; they are only intendedly rational. He also thought that individuals are not maximising their utility but instead satisficing their needs and wants – that is, they choose a solution that makes them happy but that is not necessarily maximal. The choice is not straightforward, since the individuals have no market transparency. They must make a strong effort when choosing, and they will stop this effort once they have exceeded (satisficed) their expectations.

Williamson (1985) developed the idea of transaction costs: Transactions in a market cost money. The total costs for the consumer are the production costs plus the transaction costs. The behavior of consumers and producers is opportunistic; people are self-interest seeking with guile (Williamson 1985, p. 47). Williamson became a Nobel laureate in 2009. Wallis and North (1986) measured the magnitude of transaction costs as percentage of the gross domestic product (GDP). In the hundred years from 1870 to 1970, transaction costs in the US increased from 20% to 40%. It is clear that transaction costs are not negligible. North is also a Nobel laureate (1993).

Kahneman (2011) combined psychology and economics in his research with Tversky. His description of heuristics used in decision-making and the biases influencing this differ strongly from the model of the home economicus. Kahneman influenced decisively the strand of economics called behavioral economics. You will not be surprised by now to learn that Kahneman became a Nobel laureate in 2002. Thaler (2015) differentiates between econs (the homo economicus) and humans. The observed behavior of humans differs from econs. Preferences are, for example, not fixed and not well developed. He received the Nobel Memorial Prize in Economic Sciences in 2017.

The number of Nobel Memorial Prize awards for the discussed economists signals the attention that these newer ideas receive in the world of economics. The evolution of economic thinking is fascinating. However, you should not conclude that research supporting the neoclassical model stopped many years ago. It received very strong support from the Chicago School of Economics. Contributors to the development of consumer behavior are Friedman (1957) (consumption analysis; Nobel laureate 1976) as well as Stigler and Becker (1977) (maximising, preferences; Nobel laureates 1982, respectively, 1992).

A fight for methodology accompanies the fight for the ideas and models. The described neoclassical model of consumer behavior was developed by deductive reasoning, although the assumptions were also based on observation. Contrary arguments (NIE, behavioral economics) make much stronger use of inductive reasoning.

References

Edgeworth, F. (1881). *Mathematical Psychics: An Essay on the Application of Mathematics to the Moral Sciences*. London: Kegan Paul.

Friedman, M. (1953). *Essays in Positive Economics*. Chicago: Chicago University Press.

Friedman, M. (1957). *A Theory of the Consumption Function*. Princeton: Princeton University Press.

Jevons, H. (ed.) (1886). *Letters and Journal of W. Stanley Jevons*. London: Macmillan.

Kahneman, D. (2011). *Thinking, Fast and Slow*. London: Penguin Books.

Popper, K. (1935, 2002). *The Logic of Scientific Discovery*. Abingdon: Routledge.

Simon, H. (1955). A behavioral model of rational choice. *Quarterly Journal of Economics* 69 (1): 99–118.

Stigler, G. and Becker, G. (1977). De gustibus non est disputandum. *The American Economic Review* 67 (2): 76–90.

Thaler, R. (2015). *Misbehaving*. New York: Norton.

Wallis, J. and North, D. (1986). Measuring the transaction sector in the American economy, 1870–1970. In: *Long-Term Factors in American Economic Growth* (ed. S. Engerman and R. Gallman), 95–161. Chicago: University of Chicago Press.

Walras, L. (1874). *Éléments d'économie politique pure, ou, Théorie de la richesse sociale*. Lausanne: Corbaz.

Williamson, O. (1985). *The Economic Institutions of Capitalism: Firms, Markets, Relational Contracting*. New York: Free Press.

4

Producers in Perfectly Competitive Markets

Case study   Thinking about Your Own Construction Company

In the last case study, you decided to buy a house instead of a used convertible. That was some years ago and your savings increased considerably even after buying a one-family home. A little later you were also able to replace the old, battered Toyota by a used convertible. On the job, the two of you have learned quite a bit about construction and another warm evening on the porch with a caipirinha in your hand you start discussing your dream of setting up your own construction company.

You wonder what you should produce, residential housing, industrial buildings, or infrastructure? What would the decision mean for construction, are there any general ideas that describe the transformation from input to output, i.e. the production process? Are there different ways for the transformation process?

You also have spent some time in the estimating department of your current employer, and you had to deal with estimates as project manager. Your experience confuses you a bit because what you see is that costs are calculated and in the end a markup for risk and profit is added. On the other hand, you learned in economics that decisions are made on the margin by adding one more unit (one more contract?). You have never seen that approach used and want to learn a bit more about it. Is the construction sector missing an important point, or is the concept of marginal pricing not applicable to construction projects?

If there were different transformation possibilities, what would they look like? You have noticed that in an area where you work, sewage systems, there are different technologies available such as open trenches and trenchless micro-tunnelling. Certainly, such different technologies impact the transformation process. During your last

---

(*continued*)

Construction Microeconomics, First Edition. Christian Brockmann.

submission, you saw competitors submit lower bids using open trenches while you were relying on trenchless technology. They won the bid with a less-advanced technology; how could that be? Of course, it had to do with the costs.

Thinking about costs, you wonder how different transformation processes and technologies affect costs. Is it possible to learn something about the costs not only for the project but for the company? In the end, as an owner you must start thinking about the company and not only one project as up to now. There is another puzzle: Are costs the same in the short and the long run? Construction time for a project is limited, but a company should stay in business for a long time.

Finally, you are curious what a supply curve can look like. By now you know already that demand curves are typically downward sloping and you can provide reasons for this, together with the assumptions that were the starting point for the theoretical deductions. Maybe it is possible to develop a supply curve in a similar way? As you ponder these questions, you remember that one assumption for perfectly competitive markets was the homogeneity of goods. Looking at construction output, you see diversity and not homogeneity.

There are so many questions that you decide to dig deeper into the theory of perfectly competitive markets and the role producers such as contractors play in them. You put the problem of homogeneity/heterogeneity aside for the moment, but you promise to return to it.

In Chapter 3 we have started with a short introduction of the supply and demand model and looked then closer at the role of the consumer in that model. This model is lacking an important component without producers, so it is time to analyse them more closely.

For deductive reasoning and model building, we need some general assumptions about the behavior of producers and, as it will turn out, this is profit maximising instead of utility maximising (Section 4.1). We will certainly find some criticism concerning this assumption. Next, we must conceptualize the transformation of inputs into outputs of higher value by describing the role of technology and production functions (Section 4.2). There are three different types of basic production functions, and they lead to different cost curves. The cost curves vary with time because producers have more options in the long run; they can sell assets that are fixed in the short run (Section 4.3). Finally, I will show how we can construct a supply curve from all this information (Section 4.4).

4.1 Producer Behavior

Producer behavior is modeled on assumptions similar to consumer behavior. The assumptions generalize how producers approach production decisions. They must therefore comply with the neoclassical model of supply and demand in the same way as consumer behavior.

1) Factors of production (inputs) are scarce.
2) Producers face investment constraints.
3) Producers maximize their profits.
4) Producers have complete information.
5) Producers decide rationally.

We can conceive quite a number of motives to become an entrepreneur by investing in a construction firm. Among them are the chance to be your own boss with the ability to take responsibility, the love for working as contractor, the status that comes along with it, the power the position accords, or for the profit one might make. Except for profit, the motives are more or less of a psychological or social-psychological nature. Profit is clearly an economic incentive, but it might provide satisfaction as well as money.

Profit is also a term that comes with a bag of negative connotations. To Marxists it is the part of value created by workers that entrepreneurs reserve for themselves by exploitation. To others, it might more generally be a word for greed.

Looking back at a natural experiment lasting from the October revolution in Russia (1917) to the collapse of the USSR (around 1990) and to economic changes in the People's Republic of China (1982), we can observe that market economies based on profit maximization outperformed planned economies based on production quotas regarding efficiency, i.e. the quantity of goods produced with scarce resources. Quality and availability of goods and services as well as average purchasing power were considerably better in market economies. The neoclassical model proposes that the reasons are efficiency of decentralized planning and profit as incentive.

In the supply and demand model, profit is the incentive for entrepreneurs to produce a good or to offer a service. We therefore need a clear definition of the term:

Profit = (def.) We call the difference between revenue and costs a profit if the difference is positive and a loss if it is negative. The value depends on the definition of the costs. There are two alternative concepts, accounting costs and economic costs. Accordingly, we can distinguish between accounting profits and economic profits. While accounting costs comprise only direct costs, economic costs include direct as well as opportunity costs.

Accounting costs do not include remuneration of entrepreneurship and interest on equity. However, these items are part of the microeconomic concept of economic costs in a construction firm (Table 4.1). Accordingly, a profit made by a partnership or by a corporation will also differ. Managers are typically employees in corporations and the owners are stockholders. The salaries of managers are accounting costs. However, the income of the owners in a partnership does not constitute accounting costs. A similar observation holds

Table 4.1 Profits in accounting and economics.

Accounting profits	Economic profits
Basis: Direct costs	Basis: Direct and opportunity costs
Revenue	Revenue
– Labor costs	– Labor costs
– Material costs	– Material costs
– Equipment costs	– Equipment costs
– Subcontractor costs	– Subcontractor costs
– Site installation costs	– Site installation costs
– Company overhead	– Company overhead
	– Interest on equity
	– Owners' salary

for equity. A higher accounting profit in a partnership might end up lower than that in a corporation once the owner's salary and interest on equity are subtracted.

Economic profits are always lower than accounting profits.

The long-term result in perfectly competitive markets is an economic profit of zero. The producers are indifferent between staying in the market or investing in another market. They earn the opportunity costs of doing business but no more. This is an attractive and interesting result. How it comes about will require further explanations in the coming chapters.

Zero profit is not a strong incentive. However, we should be clear about the fact that the average outcome is zero. As there are always some winners and some losers, the winners take home more money while the losers, well, lose some money. Losing money for a longer time leads to bankruptcy. Profits have the double function of carrot and stick as incentive: a reward for those who do well and punishment for those who do not. Smith (1776) famously claimed that the pursuit of self-interest serves society more than benevolence of the producers (but only in perfectly competitive markets).

In sum, profits are necessary and can be beneficial to all. Not beneficial is market power, the possibility to reap higher profits than in the system of free supply and demand in perfectly competitive markets. If you look at the market assumptions for the neoclassical model, then you find competition (the opposite of market power) as being one. It takes the form of many consumers and producers acting as price takers and quantity adjusters.

Entrepreneurs must decide what and how to produce, i.e. which production technology to employ. They must also consider the costs of the factors of production. Production has a technical touch while costs are at the core of microeconomics.

4.2 Production Theory

Classical factors of production are land, labor, and capital. In this context, land provides for agricultural as well as forestry products, and it is a source for mining. In construction, land is primarily the base on which contractors build. The structure interacts with the soil and the stability, and usability of the structure depends on its properties. In this way, land is a direct factor of production in construction and inseparable from the end product. The extracted goods from mining (sand, stone, minerals, fuels, etc.) are used as materials in construction. Wood from forests is also an important input.

Labor is measured in time units, most often in hours and the costs of labor are wages. The total labor costs C_L for a period (or a project) can be calculated from the total hours n and the average wage w:

$$C_L\left[\text{USD}\right] = n\left[\text{h}\right] \cdot w\left[\text{USD/h}\right] \tag{4.1}$$

Capital as factor of production includes material, equipment, fixed assets (offices, factories, stores, yards, etc.) and working capital in the form of money. However, material costs differ in nature from other capital; they vary with the production volume. On the other hand, equipment, fixed assets, and financing varies much less with production volume; these items are for a certain amount of time fix. Contractors must pay for them even when they have no job at hand. Yet, without a job, contractors will not buy material. For this reason, I will treat material costs as a separate factor of production and not as part of capital.

Material as a factor is purchased only for specific projects. Contractors will need a bill of quantities to determine the required amounts. Important items are steel, concrete, bricks, and wood for structural works. A list of items required for a turnkey project is very long.

Other factors that are not part of the neoclassical model are skills of workers, engineers, and managers (human capital), organizational capital or organizational process assets of the firm (Project Management Institute 2017), institutional capital of the firm, and the environment as a source and a sink.

Human capital relies on education, training, and on-the-job learning. Firms develop organizational process assets over time, and they include standardized procedures and processes. Institutional capital describes the economic institutions of a society. Acemoglu and Robinson (2012) attribute the success of different nations mainly to this factor. It is most important as a national concept, but it can also have a regional impact, especially in federally organized countries such as the USA or Germany. Institutions rest on regulative, normative and cognitive-cultural pillars (Scott 2001). We can also find such pillars in firms. The environment provides not only exhaustible resources such as sand, stone, and other minerals but also renewable energy from sun and wind. It is the source of water and the sink for emissions.

In the following discussions, labor and capital are of central importance.

 Factors of production = (def.) Factors of production include (i) land, (ii) labor, (iii) capital, (iv) materials, (v) human capital, (vi) institutional capital, and (vii) environment.

4.2.1 Technology

The possible maximum output with given resources is the most important question in production. Assume, a single large firm (microeconomics) or a sector (macroeconomics) has the capability of producing housing and bridges with given factors of production, notably

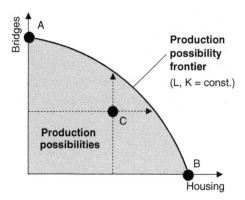

Figure 4.1 Production possibility frontier.

labor (L) and capital (K). The maximum output is on a line, called the production possibility frontier (Figure 4.1).

The large firm can decide to use the given factors (L, K = const.) of production only for building bridges (point A) or only for housing (point B). Production on the possibility frontier is, by definition, efficient (i.e. there is no waste). Accordingly, point A and B show efficient use of the factors of production. Point C is an inefficient production choice. To the right of point C, there are possibilities to produce more housing and above point C more bridges using the same factors of production.

We can explain the concave shape of the curve by the specificity of factors of production. Specificity describes to what degree an input is adapted to a certain use. It is plausible that some workers (L) are more skilled to build bridges and others to build housing. The same holds true for equipment and reusable material such as formwork or sheet piles (K). Employing the factors of production for different outputs allows to benefit from specialized skills and use. Thus, starting at point A and taking some workers from bridges to housing (the least able to build bridges and the most able to build housing) will decrease the productivity in bridge building just a bit but increase the productivity in housing a lot: The effect is overproportionate, i.e. production is above a straight-line connecting points A and B. From this discussion, we can arrive at a definition for technology.

 Technology = (def.) A technology represents the set of production activities known by a firm. In microeconomics, these activities are all the technical possibilities of a firm.

Assumptions for technology:

1) Input without output is possible (waste or elimination of factors of production).
2) There exists production with a positive output.
3) Production is not reversible.
4) The production possibility set is closed, i.e. there exists a production possibility frontier.
5) The firm produces a homogenous good with the use of factors of production. The homogenous good and the factors are divisible at will.

For modeling technology, we need some assumptions (Fandel 1991).

Every input/output combination in a production process can be represented by a vector v of goods with k inputs or outputs ($k = 1, 2, \ldots, n$)

$$v = \left\{ \begin{array}{c} v_1 \\ \vdots \\ v_n \end{array} \right\}$$

(4.2)

Inputs are defined as negative (consumption):

$$v_k < 0$$

(4.3)

Outputs are defined as positive (production):

$$v_k > 0$$

(4.4)

There exist goods (input or output) without impact:

$$v_k = 0$$

(4.5)

Every vector $v \in \mathbb{R}^n$ with the above characteristics is a called an activity in production. Figure 4.2 gives a graphical presentation of these points for $\mathbb{R}^2$. There are three production activities shown (v^1, v^2, and v^3). Remember the convention that inputs are defined as negative ($v_k < 0$) (Jehle and Reny 2011).

Input without output means that the production result turns up in the negative orthant of graph of Figure 4.2 ($\mathbb{R}^2_-$). The space of real numbers $\mathbb{R}$ is applicable because we assume the infinite divisibility of inputs and outputs.

Technologies are a subset of the space $\mathbb{R}$ ($T \subset \mathbb{R}^n$). This is evident when looking at Figure 4.2. All technologies are characterized by at least one activity with a positive outcome:

$$v \in T \text{ with } v_k > 0 \text{ for at least one } k, k = 1, 2, \ldots, n$$

(4.6)

Thus, elimination of inputs and idleness are not technologies.

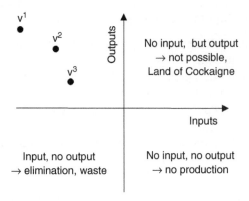

Figure 4.2 Two-dimensional vector space for production.

The nonreversibility of production means that we cannot reproduce the inputs from the output. This is most evident for labor as an input. It does not exclude reusability of some inputs by recycling.

We have already seen that production is only efficient on the border of the technology space. This is the importance of the production possibility frontier:

- An activity $v \in T$ is efficient if it maximizes output with a given input and no waste (maximum principle).
- An activity $v \in T$ is efficient if it realizes a given output with a minimum of input (minimum principle). This typically is the case in construction when the owner provides the design (i.e. fixes the output).
- An activity $v \in T$ is efficient if there is no other activity $w \in T$ with $w \neq v$ with which a firm can produce the same or a larger output by using the same or less inputs (optimum principle). This plays a role in construction during tendering. It is the principle task of contractors to determine the optimum activity among the group of bidders.

We can differentiate between three important forms of technologies: (i) increasing returns to scale, (ii) decreasing returns to scale, and (iii) constant returns to scale (Figure 4.3):

- If all inputs are increased by a factor λ and output increases by more than λ, then we face increasing returns to scale.
- If all inputs are increased by a factor λ and output increases by less than λ, then we face decreasing returns to scale.
- If all inputs are increased by a factor λ and output increases exactly by λ, then we face constant returns to scale.

Innovation changes the production possibility set by expanding the production possibility frontier. It is possible to imagine the introduction of a new bridge-building technology. This is shown on the left side of Figure 4.4. On the right side you can observe a technology change that affects both outputs. This could be the introduction of a new management approach such as lean construction.

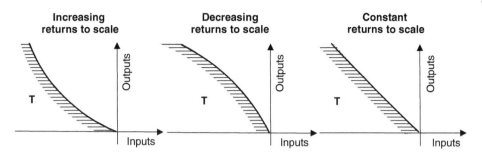

Figure 4.3 Important forms of technologies.

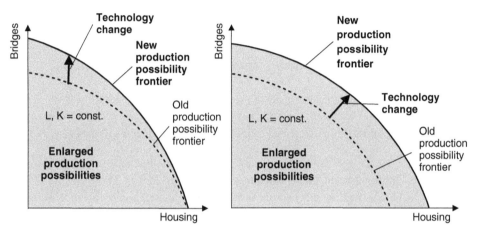

Figure 4.4 Impact of technology changes.

4.2.2 Production Functions

As we have seen above, technology is the set of production activities known to the firm. The production function is nothing but a nominal definition of the efficient border of a technology, i.e. of the production possibility frontier. While we have seen waste in a technology (point C in Figure 4.1), we concentrate in a production function on efficiency. Rational behavior of producers excludes the acceptance of waste if they can avoid it. The important point is that technologies and the derived production functions include only activities that are known to a firm. An enlargement of technology allows us to detect waste that was hitherto unknown and thus to increase efficiency.

A production function describes for each vector of inputs x the amount of efficient output y:

$$y = f(x) \tag{4.7}$$

When working with production functions, we need a few definitions. As there are numerous inputs, this requires at times an analysis based on partial derivatives.

Economists define productivity as an input–output relation:

$$\text{Productivity} = \frac{y}{x_i} \qquad (4.8)$$

Most important is labor productivity (y/x_{Labor}). Therefore, the word productivity refers most likely to labor productivity when it is not further qualified.

The production coefficient is the inverse of productivity:

$$a_i = \frac{x_i}{y} \qquad (4.9)$$

The partial marginal derivative between input and output equals:

$$\frac{\partial y}{\partial x_i} \qquad (4.10)$$

$\frac{\partial y}{\partial x_i} > 0$: An increase (decrease) of input x_i increases (decreases) output y.

$\frac{\partial y}{\partial x_i} = 0$: An increase (decrease) of input x_i does not change output y.

$\frac{\partial y}{\partial x_i} < 0$: An increase (decrease) of input x_i decreases (increases) output y.

Factor combinations that lead to the same output are called isoquants and they are the equivalent to indifference curves in consumer theory (Section 3.2). The equivalent to the marginal rate of substitution in consumer theory is the marginal rate of technical substitution (MRTS). Figure 4.5 gives a graphical depiction of the MRTS for a two-dimensional case; the slope in point A represents the MRTS. It tells us by how much we must increase factor x_2 if we decrease factor x_1 by an infinitesimal amount.

The reasons for the concave shape of the isoquant are similar to those given in Section 3.2 for indifference curves. In mathematical terms, the MRTS equals:

$$MRTS = \frac{\partial y / \partial x_1}{\partial y / \partial x_2} \qquad (4.11)$$

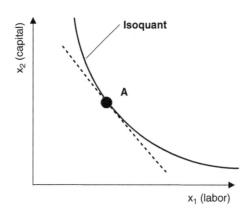

Figure 4.5 Marginal rate of technical substitution.

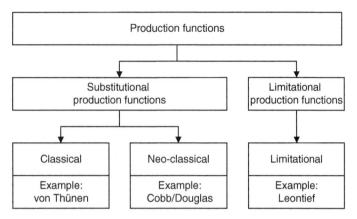

Figure 4.6 Classification of production functions.

Different scholars have proposed numerous production functions; however, the most important ones are shown in Figure 4.6. Substitutional production functions allow infinite small substitutions of one factor by another. On the other hand, a fixed ratio of inputs characterizes limitational production function. I will argue in Chapter 12 (contractor theory) that substitutional production functions are typical for construction. However, lean construction tries to establish fixed ratios of inputs and the same is often advantageous for megaprojects (Brockmann 2021).

4.2.2.1 Classical Production Function

In 1768, Turgot was the first formulate ideas about production functions (Schumpeter 1954) and almost a century later in 1842, von Thünen formalized the idea in mathematical terms. Basically, the production function assumes substitutional factors of production and in the beginning increasing ($\partial y / \partial x_i > 0$) and then decreasing marginal rates ($\partial y / \partial x_i < 0$).

Anne Robert Jacques Turgot (1727–1781)

He was the first to formulate the law of diminishing returns in production. Turgot was not only interested in theoretical economic reasoning but he served also as a minister under Louis XVI in France.

Source: Antoine Graincourt/Wikimedia Commons/Public domain

Figure 4.7 depicts a classical production function and the corresponding average and marginal labor productivity. A production function shows total output, varying with increasing inputs. Labor L is a variable in the figure and capital K is constant. The underlying production function has accordingly only two inputs, L and K. Its mathematical form is $y = f(K, L)$.

All production functions show an output per period (day, week, month, year. . .). The total sum of output always increases. If you look at construction, we always add to a building until the building is finished; then production ceases. Only when you compare between periods, you can find a decrease: period $n + 1$ might show a smaller output then period n.

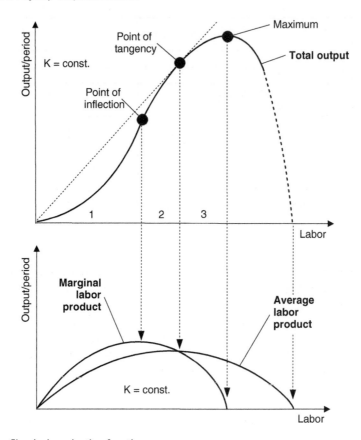

Figure 4.7 Classical production function.

You should also keep in mind the ceteris paribus clause (K = const.) when you interpret the graph. This means in a construction context that a site (land), design (materials) and site installation (capital) are given and only the amount of labor varies. Based on a classical production function, we should plan with the amount of labor that is equivalent to the maximum average labor productivity. If we move beyond the maximum of total output, then the output per period decreases, although there are more workers on site. This is quite often the case when the schedule slips at the beginning and the contractor has to install a crash program at the end. The important point is that we can at least in theory define an optimum crew size for a given design, site, and capital.

In the lower part of Figure 4.7, we find the marginal and the average labor product. The tangent to the total output curve is the graphical equivalent to the marginal labor product and it measures the input of an additional worker (or more precise the infinitely small increase in work hours). The slope increases from the origin to the inflexion point and then decreases to the maximum. At the maximum, the tangent is horizontal and the slope equals 0. The marginal labor product is negative beyond this point.

The slope of a vector from the origin to any point of the total output curve gives us the amount of average labor product. It keeps increasing to the point of tangency and then

Table 4.2 Summary of the classical production function.

	Total product (y)	Marginal product ($\partial y/\partial x$)	Average product (y/x)
Phase 1	Progressively increasing	Degressively increasing ($\partial y/\partial x > y/x$)	Degressively increasing ($y/x < \partial y/\partial x$)
Phase 2	Degressively increasing	Progressively decreasing ($\partial y/\partial x > y/x$)	Degressively increasing ($y/x < \partial y/\partial x$)
Phase 3	Degressively increasing	Progressively decreasing ($\partial y/\partial x < y/x$)	Progressively decreasing ($y/x > \partial y/\partial x$)

decreases to become 0 where the dashed line intersects with the horizontal axis. This point signifies inputs without output and is total waste.

As more and more labor hours are added to the other fixed inputs, the classical production function goes through three major phases (Figure 4.7). The results of this production function are summarized in Table 4.2. Total output is also called total product, marginal labor product more generally marginal product and average labor product average product.

4.2.2.2 Neoclassical Production Function

An example of a neoclassical production function is the Cobb–Douglas production function, proposed by the authors in 1928. It takes the following mathematical form:

$$y = a_0 x_1^{\alpha_1} \cdot x_2^{\alpha_2} \ldots x_n^{\alpha_n} = a_0 \prod_{i=1}^{n} x_i^{\alpha_i}$$

(4.12)

with $0 < a_0 = \text{const.}, 0 \leq \alpha_i = \text{const.} < 1, i \in \{1, \ldots, n\}$

The input factors x_i are substitutional since the form is multiplicative but no input can be reduced to zero. In that case, we would face complete waste, i.e. input of the other factors but no output.

Often, we find a simplified form with just two inputs, labor L and capital K.

$$y = f(L, K) = L^{\alpha} \cdot K^{\beta} \text{ with } \alpha + \beta = 1$$

(4.13)

$$\partial y / \partial L > 0 \text{ and } \partial y / \partial K > 0$$

(4.14)

$$\partial^2 y / \partial^2 L < 0 \text{ and } \partial^2 y / \partial^2 K < 0$$

(4.15)

The partial derivatives tell us that output always increases with more input (Eq. 4.14) but that the increases are getting smaller (Eq. 4.15), i.e. we find diminishing returns.

The value α is the percentage of labor regarding all inputs and β the percentage of capital. We find often in high-wage countries that 60% of inputs in buildings is labor and if $\alpha + \beta = 1$, it follows that capital amounts to 40%. The corresponding Cobb–Douglas production function would then become: $y = L^{0.6} \cdot K^{0.4}$.

The production elasticity determines the relation between the increase of output in percent for a given increase of one factor of production in percent:

$$E_{x_i,y} = \left(\frac{\Delta y}{y}\right) / \left(\frac{\Delta x_i}{x_i}\right) \tag{4.16}$$

It can be shown that this equals the coefficient for a Cobb–Douglas production function in the form of:

$$y = x_1^{\alpha_1} \cdot x_2^{\alpha_2} \rightarrow \frac{\partial y}{\partial x_1} = \alpha_1 \cdot x_1^{\alpha_1 - 1} \cdot x_2^{\alpha_2} \tag{4.17}$$

For infinitesimal small changes in Eq. (4.16) we get:

$$E_{x_i,y} = \left(\frac{\partial y}{y}\right) / \left(\frac{\partial x_i}{x_i}\right) = \frac{\partial y}{\partial x_i} \cdot \frac{x_i}{y} \tag{4.18}$$

By inserting this into Eq. (4.17) and substituting $y = x_1^{\alpha_1} \cdot x_2^{\alpha_2}$, we get:

$$E_{x_i,y} = \frac{\partial y}{\partial x_1} \cdot \frac{x_1}{y} = \alpha_1 \cdot x_1^{\alpha_1 - 1} \cdot x_2^{\alpha_2} \cdot \frac{x_1}{y} \rightarrow E_{x_1,y} = \alpha_1 \tag{4.19}$$

The coefficients in the Cobb–Douglas production function thus equal the production elasticities.

Paul Douglas (1892–1976), Charles Cobb: no picture (1875–1949)

Douglas was an economist and senator in the USA. Cobb was a mathematician and economist. Together they developed the influential Cobb–Douglas production function.

Source: Mysticgames.com

Partial factor variation shows that the production function is degressively increasing from the beginning (Figure 4.8). The marginal product (slope of the tangent to the curve) decreases continuously as does the average product (the vector from the origin to any point on the curve). The average product is always greater than the marginal product.

A Cobb–Douglas production function assumes that we can replace labor by capital in any conceivable way (except by reducing it to zero), we can substitute one input by the other. This again presupposes the infinite divisibility of all inputs.

We have already seen that average and marginal products increase degressively. Economists call this the law of diminishing returns. It is called a law, i.e. a regularity, because we can observe it many times. Think of your own workday of 10 hours to which you must add another 2 hours. Most likely, the productivity during the day was greater than at the end.

There is a single variable input in Figure 4.9, labor, and you can see that the average productivity decreases from point A to point B ($\Delta_{o1}/\Delta_{i1} > \Delta_{o2}/\Delta_{i2}$). Time and space provide a reasoning for this phenomenon in construction. When you let a crew work longer hours, productivity in the late evening will be lower than in the morning. When you add more

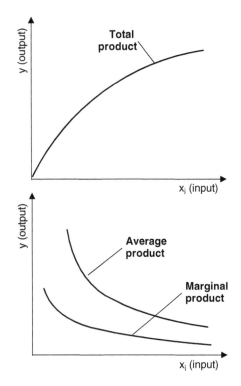

Figure 4.8 Cobb–Douglas production function.

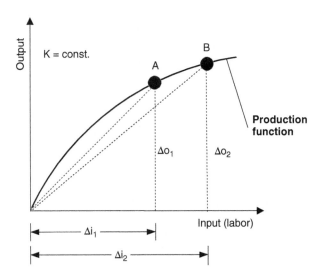

Figure 4.9 Diminishing returns in a Cobb–Douglas production function.

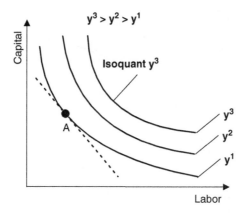

Figure 4.10 Isoquants for a neoclassical production function.

workers to a construction site with limited space, they will start impeding each other: Adding more workers can decrease average and marginal product. As we have seen, marginal product is the tangent to the production function, and you can see that the slope of the tangent decreases as you move along the production function. The argument for labor also holds true for equipment and reusable material. A higher use will result in more repair work, which must be deducted from working hours.

Isoquants describe constant output levels for different combinations of inputs. Figure 4.10 shows three isoquants for a neo-classical production function. The tangent to an isoquant is the MRTS at this point. This is shown for point A.

We saw in Figure 4.8 what happens if we only vary one factor of production (x_i) and keep the others constant. This is a partial factor variation. In a total factor variation or variation of scales, all factors of production are simultaneously increased by a common factor λ. The corresponding measure is called elasticity of scales $E\lambda_y$. It is defined as the derivative of output change over the derivative of factor change:

$$E_{\lambda,y} = \frac{dy}{y} \Big/ \frac{d\lambda}{\lambda} = \frac{dy}{d\lambda} \cdot \frac{\lambda}{y} \tag{4.20}$$

An elasticity of scales, or output elasticity, greater than 1 ($E\lambda_y > 1$) indicates increasing returns to scale, one that equals 1 ($E\lambda_y = 1$) constant returns to scale and one that is smaller than 1 ($E\lambda, y < 1$) decreasing returns to scale (Figure 4.11).

The returns to scale are constant for a Cobb–Douglas production function with $a + b = 1$, this is the typical case. With $a + b > 1$, we find increasing returns to scale and for $a + b < 1$ decreasing ones. A reason for increasing returns can be seen in synergies, where the whole is greater than the sum of the parts, and decreasing one factor thus results in a need for a disproportionate increase in organizational efforts.

4.2.2.3 Limitational Production Function

We are facing a limitational production function if we cannot substitute the factors of production at will for each other. The ratio of inputs is fixed in such a case. If a table consists, for example, of four legs and one tabletop, then the input relationship is 4 : 1. Adding three more legs as input does not allow the construction of another table. An example of a

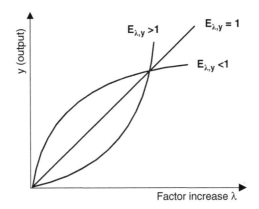

Figure 4.11 Total factor variation, elasticity of scales.

limitational production function is the Leontief function (1966). This function is called linear-limitational since doubling all inputs leads to a doubling of outputs.

The mathematical form of a Leontief production function is:

$$y = {x_n}/{a_n} \text{ with } {x_1}/{a_1} = {x_n}/{a_n} \text{ for all } n \in \{1, ..., n\} \tag{4.21}$$

The values a_i are the production coefficients for the different inputs. If the inputs x_i are available at such a ratio that $x_1/a_1 \neq x_n/a_n$, then production function has a limit:

$$y = \min\{{x_n}/{a_n}, n = 1, ..., n\} \tag{4.22}$$

For the example of the table, we might have 17 legs (x_L) and 5 tabletops (x_T). The production coefficients are $a_L = 4$ and $a_T = 1$. $17/4 = 4.25$ and $5/1 = 5$. Since we can use only full sets of legs, we can produce four tables with the legs and five with the tabletops. The minimum value is 4, and this is the maximum output y. One leg and one tabletop are waste in this process. If we double the inputs, we get $34/4 = 8.5$ and $10/1 = 10$. i.e. the maximum output is 8, which shows the linearity: double inputs bring double outputs.

Figure 4.12 shows the isoquants of a Leontief production function with labor and capital as inputs. Efficient production is only possible in the corner of the isoquants.

4.2.2.4 Technological Change (Innovation) and Learning

Technological change works like a moderator variable c in the production function:

$$y = c \cdot f(L, K) \text{ with } c > 1 \tag{4.23}$$

Figure 4.13 shows the result graphically for a Cobb–Douglas production function. The same inputs allow for a higher output because of technological change. When Malthus (1798) published his ideas about population growth, he was not aware of the influence of technological growth. He predicted strong population growth and smaller increases in food production; this would have led to an ever-starving population and later earned economics

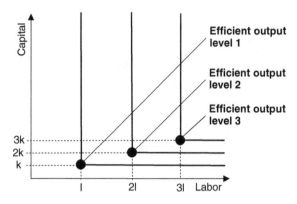

Figure 4.12 Linear-limitational production function.

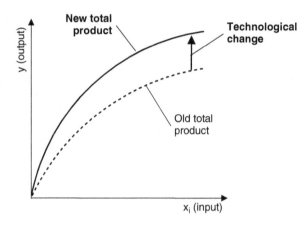

Figure 4.13 Effect of technological change.

the label of a dismal science. Thanks to innovation, we are in many parts of the world able to keep food production ahead of the needs of the population.

Learning allows us to do things better than before. We formulate this idea in economics by progress curves, which state that it is possible to reduce inputs by a certain percentage when doubling cumulated output using the same technology. With learning, we might need less labor, less equipment, or fewer fixed assets per unit. Costs allow us to summarize these different inputs and progress curve theory states that it is possible to reduce costs by 20–30% when doubling cumulative output. Assume you produce 100 units of output at a value of 100, then it could be possible to produce the unit 200 at a value of 80, given a 20% cost reduction. Notice that there is no change in technology. Figure 4.14 depicts the results for a classical production function.

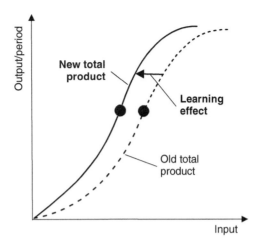

Figure 4.14 Learning effect for a classical production function.

4.3 Cost Theory

Costs play a very important role in markets. Many facets of historic economic thinking centered on the right, real, or natural value of a good. Ricardo (the capitalist) and Marx (the communist) both thought that such a value depends on the amount of labor required for the production of any good. While we think today that supply and demand simultaneously determine the price (and not costs alone); costs remain an important part of the market price.

In the previous chapter on production theory, output was variable, at least most of the time, with the exception of isoquants. By focusing on fixed outputs and variable inputs, we can move on toward cost theory. A cost function is the priced (monetarily evaluated) inverse of a production function. This remains true only as long as input prices remain constant.

Fixed costs do not vary with output, while variable costs change depending on the output level. However, what factors of production a producer cannot sell today – and what therefore are fixed assets – can in fact be sold eventually. A firm's headquarters cannot be sold the next day, but with time a buyer can be found. In the long run, there are no fixed costs. If we look at changes over time, then we can also distinguish between time-independent and time-dependent costs. Total costs (C_t) are the sum of fixed (C_f) and variable costs (C_v):

$$\left(C_t\right) = \left(C_f\right) + \left(C_v\right) \tag{4.24}$$

Marginal costs are the costs of producing one more unit of output y, i.e. $\partial C_t/\partial y$. We can define a few different cost curves based on cost functions, i.e. marginal costs ($\partial C_t/\partial y$), average total costs (C_t/y), average fixed costs (C_f/y), and average variable costs (C_v/y).

Technical efficiency tells us whether a production is wasting resources. It does not tell us which of the many possible efficient combinations of factors of production to prefer. With the introduction of costs, we can choose the minimum cost combination, which is what we are looking for if we are trying to maximize profits. Structural engineers are trained to reduce quantities (material inputs). Some structural solutions minimize material quantities at the same time as they increase labor input. Such problems are addressed by the term buildability. The knowledge of costs helps us to find optimal solutions in construction. It is not the only criterion – but it is a very important one. Others are aesthetics, construction time, quality, sustainability, and safety.

4.3.1 Cost Curves for Classical Production Functions

Figure 4.15 shows the total product of the classical production function and to the left the corresponding cost curves as monetarily weighted inverse with the addition of a socle of fixed costs.

Variable costs increase degressively at low levels of output and progressively at high levels. Marginal costs $\partial C_t/\partial y$ decrease from zero to the point of inflexion to increase again afterwards; remember, this is the tangent to the cost function (Figure 4.16). The minimum of marginal costs is exactly at point A. The curve of average variable costs follows the vector from the origin to different points of the cost function. It decreases until point B and then increases again. At point B, the marginal cost curve and the average variable cost curve must intersect because the slopes of the tangents are the same. The average total cost curve is the vertical sum of the average variable and average fixed costs. The average fixed costs must continually decline as the output y grows. It is most important to note that the marginal and the average total cost curves intersect. In point C, we find that marginal costs equal average total costs.

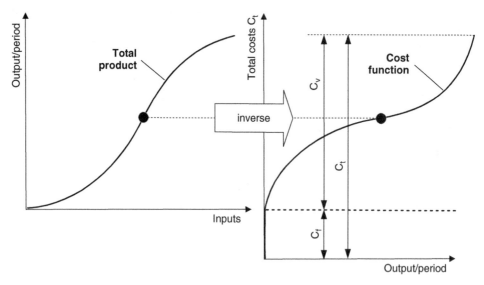

Figure 4.15 Cost function based on a classical production function.

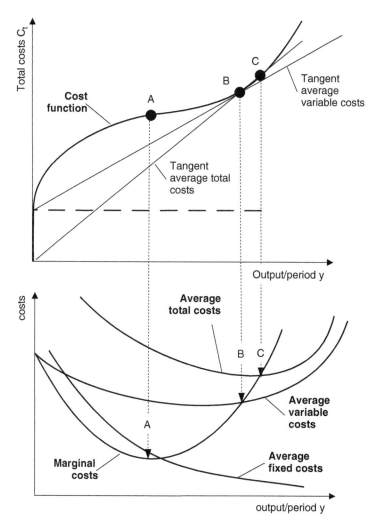

Figure 4.16 Cost curves for a classical production function.

Up to now, we have been looking at a partial factor variation because we assume some factors to be fixed. This is only reasonable in the short run. In the long run, we must assume a variation of all factors of production, i.e. in a simplified production function, the variation of labor L and capital K.

The cost minimum (C_{min}) with prices for labor (p_L) and capital (p_K) must satisfy:

$$C_{min} = p_L \cdot L + p_K \cdot K \rightarrow min \tag{4.25}$$

Partial differentiation brings the result that the ratio of the marginal products must be equal to the ratio of the prices:

$$\left(\frac{\partial y}{\partial L} \right) \bigg/ \left(\frac{\partial y}{\partial K} \right) = \frac{p_L}{p_k} \tag{4.26}$$

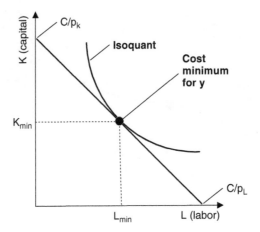

Figure 4.17 Minimal cost combination.

We can construct a cost line in analogy to the budget line of consumer theory for a graphic presentation of this formula. This line has two distinct endpoints on the y- and x-intercept, which we can calculate by putting one input value to zero, e.g. $C = p_L \cdot L + p_K \cdot 0 \rightarrow L = C/p_L$. We find the cost minimal combination at the point of tangency by the maximum isoquant and the cost line. Thus, we can determine the minimum cost for any chosen output level y (Figure 4.17).

For the cost minimum in Figure 4.17, we find the slope of the cost line $(-p_1/p_2)$ to equal the slope of the isoquant $(\partial y/\partial L) / (\partial y/\partial K)$.

4.3.2 Cost Curves for Neoclassical Production Functions

We can formulate the cost function for a neoclassical production function in analogy to the classical one: It is again the inverse (Figure 4.18).

From the cost function we can deduct qualitative cost curves. The average fixed costs decrease with increasing output, the average total costs first decrease and then increase again; average variable and average marginal costs increase (Figure 4.19).

The average total costs have a minimum at x^1. This is the point where a vector from the origin touches the total cost curve. At this point, the slope of the vector is the same as the tangent to the curve (marginal costs). Thus, average total costs equal marginal costs.

The result for a cost minimum of a Cobb–Douglas production function shows that the ratios of inputs must be equal to the inverse of the price ratios weighted with the production elasticities:

$$y = L^\alpha \cdot K^\beta \tag{4.27}$$

$$L/K = \frac{\alpha p_K}{\beta p_L} \tag{4.28}$$

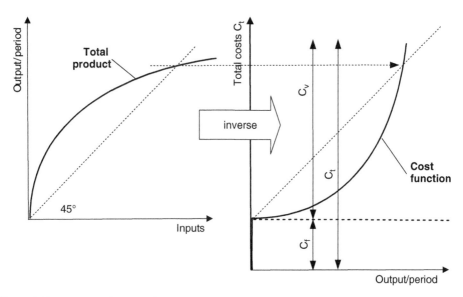

Figure 4.18 Cost function based on a neoclassical production function.

4.3.3 Cost Curves for Limitational Production Functions

We have seen in Figure 4.12 that a line from the origin passes through all efficient production points in a Leontief production function. If we put labor and capital input at 0, we have no costs, i.e. this production function has no fixed costs. As the factors of production have a fixed relationship with each other, a partial variation of one factor does not make much sense.

It is, however, possible that several efficient production processes can be mixed with the result of an efficient new process. For the purposes here, I will limit the discussion by showing the result. The total product is linear with kinks at the points where there is a change in the process combination. The process combination becomes less productive with every step, the slope decreases. The cost function is again the inverse, mirrored around the 45° line (Figure 4.20).

As the cost function is linear, the marginal costs as derivatives must be constant from one kink to the next.

4.3.4 Simplified Cost Function with Constantly Increasing Variable Costs

To many students all the above production and cost functions as well as the costs curves might look new and somewhat bewildering. So, it is time for a cost function that you should have encountered in construction management (Figure 4.21).

$$C(y) = ay + C_f \, a > 0, \text{const.} \tag{4.29}$$

Not only the average variable costs but also the marginal costs are constant.

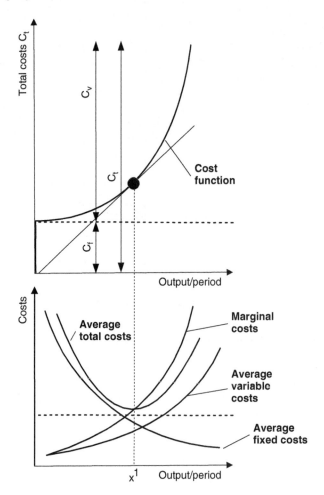

Figure 4.19   Cost curves for a neoclassical production function.

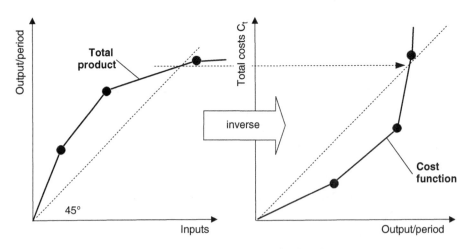

Figure 4.20   Cost function based on a limitational production function.

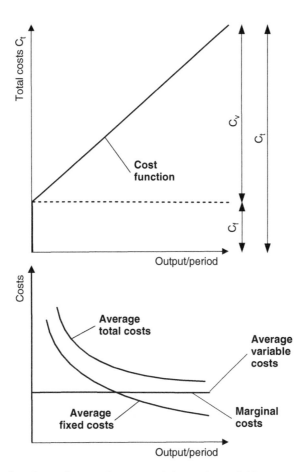

Figure 4.21 Cost function and curves for constantly increasing variable costs.

4.3.5 Long-Run Cost Curves

In the long run, all fixed costs become variable. Instead of analysing partial factor variation, we now focus on total factor variation.

For a Cobb–Douglas production function with constant returns to scale we will have constant variable costs in the case of total factor variation. Once a firm has found the optimal input, it just needs to extent the production. This applies, of course, only to the continuous production of the same good. If it would cost 1 USD to produce one pencil, 100 pencils would cost 100 USD in such a case.

For a classical production function, the long-run cost curve is the envelope of short-run cost curves and has the same form: first average costs decrease and later increase again (Nicholson and Snyder 2014). These authors also provide the conclusion from empirical (not theoretical) studies and in this case, the long-run average cost curve has an L-shape (Figure 4.22).

Especially the data from the empirical studies show that there exists some uncertainty about the shape of long-run cost curves.

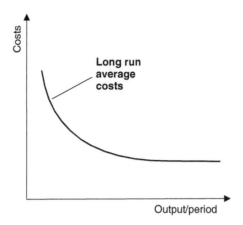

Figure 4.22 Long-run average cost curve.

4.4 Supply Curve

The production function together with the cost curves lead to the supply of a single firm and then to the aggregated supply of an industry.

4.4.1 Short-Run Supply Curve of a Firm

We have modeled the behavior of a producer as profit maximising. Profit (π) is the difference between revenue (R) and costs (C), both depend on the output y:

$$\pi(y) = R(y) - C(y) = p \cdot y - C(y) \tag{4.30}$$

For the maximum of the function, we need the first derivative:

$$\frac{dP}{dy} = p - \frac{dC}{dy} = 0 \rightarrow p = \frac{dC}{dy} \tag{4.31}$$

This means that a producer who wants to maximize his profits must set the price to equal marginal costs; this is true since the second derivative is positive (i.e. there is a maximum):

$$\frac{dP^2}{d^2 y} = -\frac{dC^2}{d^2 y} < 0 \rightarrow \frac{dC^2}{d^2 y} > 0 \tag{4.32}$$

This result also makes sense intuitively: if the costs of an additional unit of input is below market price, then the producer can increase his profit by adding the unit. Once the last unit of input equals price, the profit maximising producer is indifferent about it. When the costs of the last unit are higher than the price, then the producer makes a loss.

Figure 4.23 shows the short-run supply curve for a profit-maximising producer for neoclassical production function. The producer cannot cover his average variable costs for any price between 0 and p_1. He would make a loss; this certainly is not rational. If he does not produce at all, the maximum loss is equivalent to his fixed costs. In the price range between

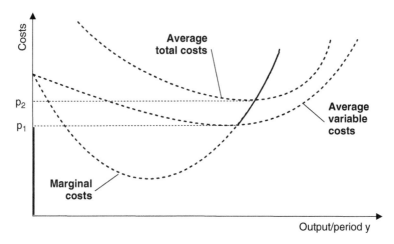

Figure 4.23 Short-run supply curve.

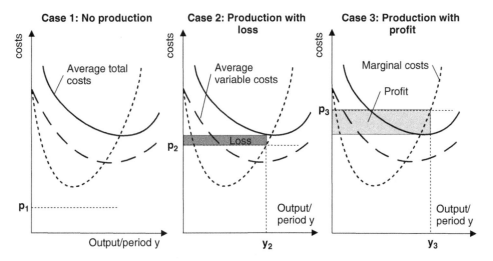

Figure 4.24 Influence of price levels on decision-making by producers.

p_1 and p_2, the producer will make a loss but can at least cover some of his fixed costs. This makes sense in the short but not in the long run. The producer makes an economic profit only if the price exceeds p_2. Only the upward-sloping part is shown in the model of supply and demand because a supply of zero (price below p_1) is irrelevant.

Figure 4.24 shows the economic result for three distinct price levels p_1, p_2, and p_3. In the right box, the producer makes a loss; in the middle one, he will cover some fixed costs and in the right one, he makes a profit.

The producer in case 1 is not even covering his average variable costs; for each additional output, he will lose money. In case 2, he can cover all his average variable costs and some of his fixed costs, In the short run this makes sense; however in the long run, it will lead to bankruptcy. Only in case 3 will the producer make an economic profit. The decision-making

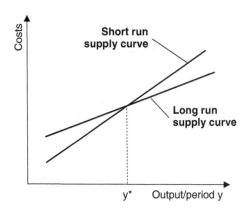

Figure 4.25 Long- and short-run supply curve by a firm.

is based in case 2 as well as in case 3 on $MC = p$. To avoid a loss, the producer must supply quantities to the right of the minimum of average variable costs (AVC).

The thoughts allow formulating the shutdown condition:

$$AVC(y) = \frac{c_v(y)}{y} > p \qquad (4.33)$$

4.4.2 Long-Run Supply Curve of a Firm

If we denote a fixed factor of production by k (e.g. the factory size) then we can formulate for profit maximization in the short run by using MC for marginal costs:

$$p = MC(y, k) \qquad (4.34)$$

In the long run, k depends on output, and we get:

$$p = MC(y, k(y)) \qquad (4.35)$$

The quantity y^* for which k is optimal in the short run coincides with the long run; this point (p^*, y^*) is common to both curves. Since the producer can adjust fixed costs in the long run, the corresponding cost curve will be more elastic, i.e. less reactive to price changes (Figure 4.25).

In the case of constant average costs, the long-run supply curve will be horizontal (Figure 4.26).

4.4.3 Market Supply Curve

The short-run market supply curve is the horizontal sum of all individual supply curves: We must add the quantities of all producers who are willing to supply a quantity for a given price (Figure 4.27).

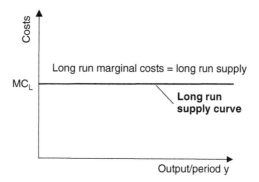

Figure 4.26   Long-run supply curve for constant average costs.

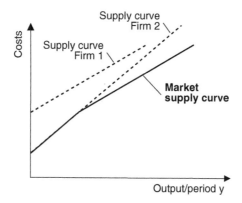

Figure 4.27   Short-run market supply curve.

References

Acemoglu, D. and Robinson, J. (2012). *Why Nations Fail: The Origins of Power, Prosperity, and Poverty*. New York: Crown Business.

Brockmann, C. (2021). *Advanced Construction Project Management: The Complexity of Megaprojects*. Hoboken: Wiley Blackwell.

Cobb, C. and Douglas, P. (1928). A theory of production. *American Economic Review* 18 (1 Supplement): 139–165.

Fandel, G. (1991). *Theory of Production and Cost*. Berlin: Springer.

Jehle, G. and Reny, P. (2011). *Advanced Microeconomic Theory*. Harlow: Pearson.

Leontief, W. (1966). Input-output analysis. In: *Input-Output Economics* (ed. W. Leontief), 134–155. New York: Oxford University Press.

Malthus, T. (1798, 2015). *An Essay on the Principle of Population and Other Writings*. London: Penguin Classics.

Nicholson, W. and Snyder, C. (2014). *Intermediate Microeconomics and its Application*. Mason: Cengage Learning.

Project Management Institute (ed.) (2017). *A Guide to the Project Management Body of Knowledge*. Newtown Square: Project Management Institute.

Schumpeter, J. (1954). *History of Economic Analysis*. London: Allen & Unwin.

Scott, R. (2001). *Institutions and Organizations*. Thousand Oaks: Sage.

Smith, A. (1776). *An Inquiry into the Nature and Causes of the Wealth of Nations*. Ware: Wordsworth Editions (2012).

5

Interaction in Perfectly Competitive Markets

Case Study   Your Own Company, a Precast Concrete Factory

Some years have passed and you run your own precast concrete factory. This is what you decided after looking at the chances in your regional market. As you live in an area with several mid-size cities in close distance to each other, you noticed that the manufacturing and logistic industries are building all kinds of factories, warehouses, and distribution centers. These are structures that are often designed using prefabrication, especially with prefabricated columns and beams. There are only two competitors but there is also a close substitute, steel prefabrication. Luckily, the input prices for concrete elements are cheaper than for steel construction; concrete has a cost advantage over steel in your area.

The three competing precast concrete factories fight for their market shares, so competition is lively and demand is rather strong. If the market prices of concrete elements go up, there is always the threat of substitution through steel. You can't dictate a price to an owner; rather it is true that you have to adjust your capacity to demand. Unfortunately, you do not know your marginal costs since you produce most elements in small batches and all batches are different from each other. So, you just try to cover your estimated costs plus a margin.

Whenever there is a request for a quote and one of your competitors has a lower price, he gets the contract. Sometimes good customers offer you to sell at your competitor's low price – they give you the chance of a last offer. Anyway, you have no chance to sell above the lowest price.

There seems to be no general market equilibrium if you do not consider each request as a market. This is quite different from what you know about mass production, where different producers provide close substitutes to the market and they all sell more or less at the same price. Once you start thinking about it, you wonder what constitutes a market.

In the moment, you do not worry too much. Demand is high and you can procure good contracts. Last year you made a profit of 8% on turnover and it has been similar all the three

---

(continued)

Construction Microeconomics, First Edition. Christian Brockmann.
© 2023 John Wiley & Sons Ltd. Published 2023 by John Wiley & Sons Ltd.

years that you ran your company. In the end, this is a result close to your own initial analysis of market chances and it persuaded you to set up your own firm. Sometimes you worry about further competitors who certainly would reduce your profits, but somehow your area does not attract that much attention and you don't know of any new startups.

While you feel comfortable with the situation, you also wonder what it means to society. Are your profits taking away benefits from others? Whichever way your thoughts are turning, you have no idea how to conceptualize economic benefits in a society. It seems to be a fuzzy concept.

At an industry meeting last week you met one of your competitors who is already more than 20 years in business. When you were happily chatting about the good times, he reminded you of a recession some 10 years ago, which lasted 6 years. He lost a lot of money during that time and barely survived. You make a mental note to put some money away for harsher times.

The demand of consumers and the supply of producers meet in a market. The market price lies at the intersection between the demand and supply curve. This intersection also determines the market quantity. The curves are presented in Figure 5.1 as straight lines for reasons of simplicity only.

It took mankind approximately 2500 years to develop this graph that looks so very self-evident today. Please, do not forget to remember all the assumptions that are required to reach the result (Chapters 3 and 4). Any parrot can pronounce supply and demand but that does not mean that it understands the functioning of markets. If you listen closely, there are still today many people who claim that a price is not correct (too high or too low). Maybe, the argument goes, the price does not reflect the amount of labor required by a

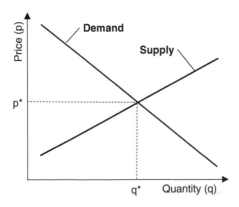

Figure 5.1 Market equilibrium in perfectly competitive markets.

craftsman for manufacturing. The idea of a right price determined by some inherent quality of the good has been prevalent in economics for all but the last 150 years. There are many theories about price determination besides supply and demand, among them intrinsic theory of value, labor theory of value, exchange theory of value, monetary theory of value, and power theory of value. Power, for example, plays no role in the supply and demand model. In perfectly competitive markets this holds true because of the assumption of many consumers and producers acting as price takers: It is the consequence of an assumption.

5.1 Equilibrium Price and Quantity

An important question is whether a perfectly competitive market can reach an equilibrium. The idea of an equilibrium is well known to civil engineers as they have to find equilibrium conditions in structural analysis to guarantee stability. A typical way of conveying the idea can be found in Figure 5.2. The ball on the left side will come to rest in its initial position after an interference by a force; this is an equilibrium position. The ball to the right will never regain its initial position once disturbed; this is an instable position.

To find out how perfectly competitive markets react, a thought experiment is helpful: Let all producers raise the price from p^* to p^+ (Figure 5.3). The producers can sell the quantity demanded q^* at price p^* but they are unable to sell the quantity supplied at p^+, there is a surplus. Nobody in the market is willing to by the surplus and storage costs pile up every day. Sooner than later the first producer will lower the price and the others must follow in order to empty the stores. In the end, the market will return to the equilibrium position with price p^* and quantity q^*.

Something similar happens when the price p^- is below the equilibrium price p^* (Figure 5.4). Producers are only willing to supply the quantity q^- at price p^- while the consumers want the larger quantity demanded at p^-. This causes a shortage in the market. Either some consumers raise their offer or the producers understand the situation and raise

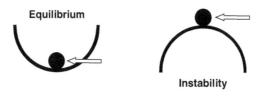

Figure 5.2 Equilibrium and instability.

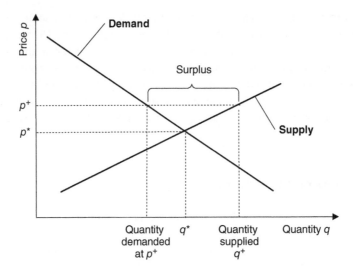

Figure 5.3 Market surplus.

the price by themselves. In both cases the market will end up at the initial equilibrium with $p*$ and $q*$.

Thus, the market tends to return to an equilibrium after being disturbed. In the case of a surplus the cause are frustrated producers and in the case of a shortage frustrated consumers. Only at $p*$, $q*$ the wishes of consumers and producers find a balance. Changes in price always

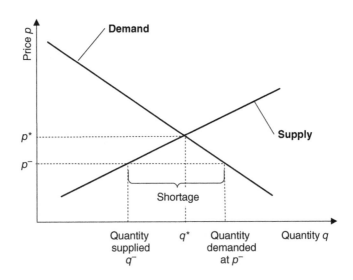

Figure 5.4 Market shortage.

call for changes in quantities and vice versa. Price and quantity are endogenous variables. These are part of the model and a result of the interaction.

5.2 Comparative Statics

There are other market changes possible by exogenous variables; these are not part of the model and they are inputs. If the income of the consumers decreases, for example, the demand curve shifts to the left. At each price, consumers can only afford to buy a smaller quantity (Figure 5.5).

More generally speaking, any contraction of demand will cause the demand curve to shift to the left. At any price, the consumers will demand less. This market disturbance causes the quantity q^* to decrease to q^{**} and the price p^* to decrease to p^{**}. An increase in income would shift the demand curve to the right and accordingly the equilibrium price and quantity would rise.

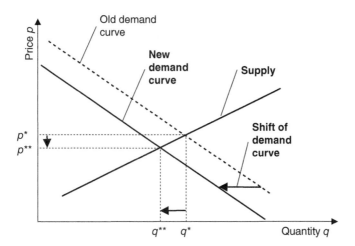

Figure 5.5   Contraction of demand.

Other factors (exogenous variables) that decrease demand are negative changes in taste, an expected price decrease, or a decrease of the number of consumers. Other factors that increase demand are positive changes in taste, an expected price increase, or an increase in the number of consumers (Krugman and Wells 2018).

The supply curve can also shift, an exemplary cause is an improved technology. This allows the producer to produce more while keeping costs constant. At any price level, the producer can offer a higher quantity. This leads to an increase in quantity from q^* to q^{**} and a decrease in price from p^* to p^{**} (Figure 5.6). Other factors shifting the supply curve outwards are decreases of input prices, expected price falls and an increase of the number of producers. Supply decreases if prices of inputs rise, when cost-saving technologies become unavailable, for expected price rises and when the number of producers falls.

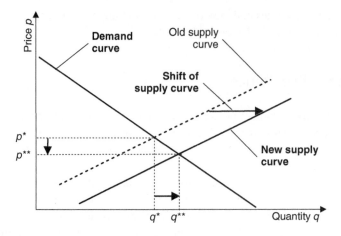

Figure 5.6 Expansion of supply.

5.3 Elasticities of Demand and Supply

We can describe the shape of the demand curve by using the concept of price elasticity. A demand (supply) is elastic when a small change in price leads to large changes in quantity. It is inelastic when a large change in price leads to small changes in quantity (Figure 5.7). There are two extremes: perfectly elastic demand (supply) and perfectly inelastic demand (supply). It might seem that, for example, a perfectly inelastic supply is a nonexistent extreme. This is not true, if you think of the short-term additional supply in housing. It takes time to develop property, acquire permits, design the housing, procure contractors, and erect the buildings. In the short run (maybe two years), there might be no additional supply on the local market.

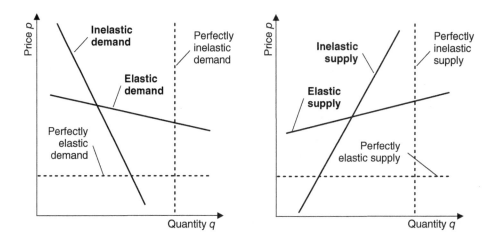

Figure 5.7 Demand and supply elasticities.

There are two very interesting elasticities: price and income elasticity. The definition of price elasticity is the change in quantity $\Delta Q/Q$ [%] divided by the change in price $\Delta P/P$ [%]. We measure prices in currency (e.g. USD), thus by dividing two prices, we get as dimension percent (%); the same result applies to dividing quantities. Thus, price elasticity $e_{P,D}$ of demand is defined as:

$$e_{P,D} = \Delta Q_D/Q_D\,[\%]\,/\,\Delta P/P\,[\%] \tag{5.1}$$

As the demand curve slopes downwards, price elasticity of demand is always negative. If the change in quantity for a unit change in price is large, we call the corresponding price elasticity of demand elastic ($e_P < -1$); such a curve is trending toward a horizontal line (Figure 5.7). If the changes in quantity and price are the same, we call the elasticity to be unit elastic ($e_P = -1$). Finally, price elasticity is inelastic when quantity changes are small for a unit change in price; such a curve is trending toward a vertical line. A horizontal line represents a perfectly elastic demand ($e_P = \infty$) and a vertical line a perfectly inelastic demand ($e_P = 0$).

Price elasticity of demand is not constant. Figure 5.8 shows a straight-line demand curve. The calculation for point A provides an elastic demand ($e_P = -4 < -1$), for point B a unit-elastic demand ($e_P = -1$) and for point C, an inelastic demand ($e_P = -0.25 > -1$) (Table 5.1).

Elasticity is also connected total revenue (P·Q) (Table 5.2).

We can check these conclusions by calculating the total revenue with the data from Figure 5.8 (Table 5.3).

Total revenue changes when the price rises or falls. This is demonstrated in Table 5.3. A specific point on the demand curve allows to calculate total revenue (P·Q) because price P and quantity Q are given for this point. Figure 5.9 shows an example where a fall in prices from 40 (A) to 25 (B) increases total revenue. As point A lies in the elastic part of the demand curve, this comes as no surprise.

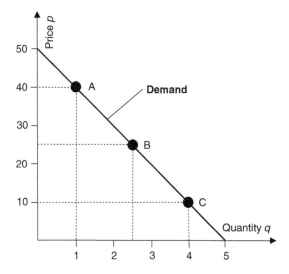

Figure 5.8 Elasticities of a straight-line demand curve.

Table 5.1 Elasticity of a straight-line demand curve.

Point/ movement	Initial price P	Change in price ΔP	% change in price ΔP/P	Initial quantity Q	Change in quantity ΔQ	% change in quantity ΔQ/Q	Price elasticity e_p	Description
A	40	—	—	1.0	—	—	—	
B	25	—	—	2.5	—	—	—	
C	10	—	—	4.0	—	—	—	
A→B	40	−15	−0.375	1.0	1.5	1.500	−4.00 (A)	Elastic
B→A	25	15	−0.600	2.5	−1.5	−0.600	−1.00 (B)	Unit-elastic
B→C	25	−15	−0.600	2.5	1.5	0.600	−1.00 (B)	Unit-elastic
C→B	10	15	1.500	4.0	−1.5	−0.375	−0.25 (B)	Inelastic

Table 5.2 Impact of price elasticity and price changes on total revenue.

Demand	If the price increases, P.Q will	If the price decreases, P.Q will
Elastic	Fall	Rise
Unit-elastic	Stay unchanged	Stay unchanged
Inelastic	Rise	Fall

Table 5.3 Price decreases, price elasticity and total revenue.

Point	Description	Price		Quantity	P·Q	
		50	Price decrease	0.0	0	Rise
A	Elastic	40		1.0	40.0	
		30		2.0	60.0	
B	Unit-elastic	25		2.5	62,5	Stable
		20		3.0	60.0	Fall
C	Inelastic	10		4.0	40.0	
		0		5.0	0	

The other important elasticity measure is the income elasticity e_I, which shows the effect of income (I) changes on demand. In analogy to price elasticity it is defined as:

$$e_I = \Delta Q/Q[\%] / \Delta I/I[\%] \tag{5.2}$$

The value is positive for normal goods, i.e. income increases lead to increases in quantity bought. However, the increase in quantity bought varies from good to good.

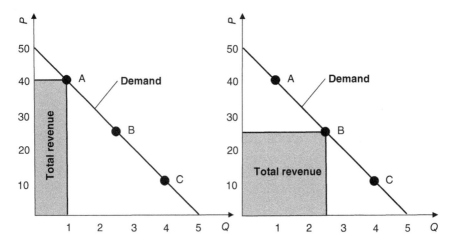

Figure 5.9 Price changes and changes of total revenue.

Table 5.4 provides some examples of income and price elasticity. Unfortunately, many of the data are quite old, dating from the 1970s (Nicholson 1990, p. 134).

Krugman and Wells (2018, p. 162) also provide some more recent data and the most interesting one is owner-occupied housing with a price elasticity of 1.2, the same as the value in Table 5.4. Quantities demanded for necessities such has food, medical services, and rental housing are not very responsive to price changes; we always need some quantity of such items. It is easier to renounce buying something more luxurious such as owner-occupied housing or automobiles.

Supply elasticities are similar to demand elasticities. The difference is that we are no longer looking at the quantity demanded but at the quantity supplied; the same holds true for the prices (requested instead of offered). Thus, the price elasticity of supply is defined as:

$$e_{P,S} = \Delta Q_S / Q_S [\%] / \Delta P / P [\%] \tag{5.3}$$

Table 5.4 Examples of income and price elasticity.

Item	Income elasticity	Price elasticity	
Coffee	0.51	−0.16	Inelastic
Housing rental	1.00	−0.18	Inelastic
Medical services	0.22	−0.20	Inelastic
Food	0.28	−0.21	Inelastic
Gasoline	1.06	−0.54	Inelastic
Beer	0.93	−1.13	Elastic
Electricity	0.61	−1.14	Elastic
Housing owner-occupied	1.20	−1.20	Elastic
Automobiles	3.00	−1.20	Elastic

Instead of looking at movements along the demand curve, we now look at movements along the supply curve. A vertical line represents a perfectly price inelastic supply and a horizontal line a perfectly price elastic supply. The higher the availability of inputs, the higher the price elasticity of supply. Also, the more time producers have to adjust production, the higher the price elasticity of supply.

5.4 Consumer and Producer Surplus

The aggregated demand curve (the market demand curve) informs us about the consumers' willingness to pay. If we have an equilibrium price in a market, then some people would be willing to pay more than the market price. Thus, the market price makes this group of consumers better off. We call the difference between willingness to pay and market price consumer surplus. A similar argument applies to producers: Some are willing to supply the good at a lower than equilibrium price. They are also better off and profit from the producer surplus. Figure 5.10 clearly shows the consumer and producer surplus.

The size of the surplus depends on elasticities. In Figure 5.11 you can see the surplus for inelastic and elastic demand curves; the size of the surplus varies considerably.

We should never forget that the willingness to pay depends on the ability to pay. A family who would like to buy a one-family home and does not have the necessary funds will not be able to bring its demand to the market. A similar observation applies to producers with average total costs above market price.

A decrease in price will increase the consumer surplus as long as there are no changes to the demand curve. What happens to the producer surplus depends on the elasticities. In Figure 5.12, the producer surplus also increases because the increase in quantity offsets the decrease in price. What is important is the idea that price changes lead to changes of the surpluses.

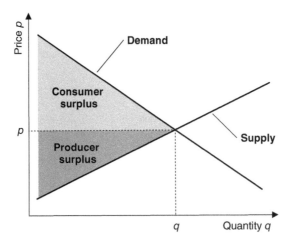

Figure 5.10   Consumer and producer surplus.

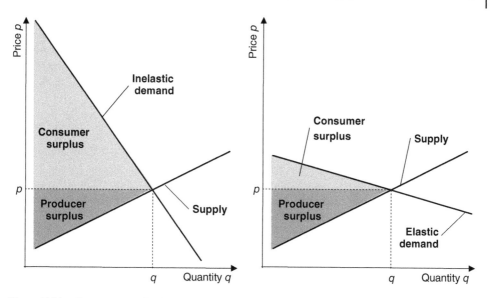

Figure 5.11 Consumer surplus for elastic and inelastic demand.

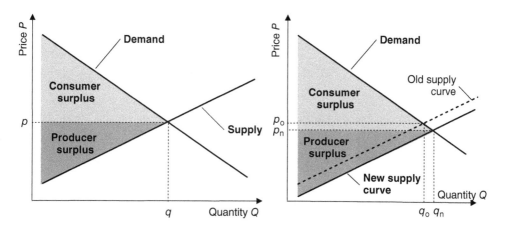

Figure 5.12 Changes to the surpluses by a shift in supply.

5.5 Time-Dependent Supply Curves and Market Outcomes

As discussed in Chapter 4, we can distinguish different supply curves which reflect the reaction possibilities of the producers. This reaction depends on the time available to the producers when demand increases. In the very short run, producers cannot react at all. In the short run, they can adjust their resources to supply larger quantities and in the long run, additional producers can enter or leave the market to increase capacity.

5.5.1 Very-Short-Run Supply Curve

Between a sudden increase in market demand for housing and additional housing supply, there is a time lag. Land needs to be appropriated for a building zone, utilities and streets must be provided, the land must be allocated to parties, designs need to be established, building permits must be given, and construction must proceed until handover. If such a process is going fast track, it might take two years. This period denotes a very short run in providing new housing to the market. During this time, the supply is completely inelastic. A shift of demand in such a situation will only raise the price while quantity remains fixed (Figure 5.13).

The shift of the old to the new demand curve raises the price p_o to p_n while the quantity supplied q remains fixed, since the contractors have no chance to bring a higher quantity to the market. The contractor (producer) surplus increases, as the dark grey quadrangle is added to the light grey one. The consumer surplus remains unchanged if the new demand curve is exactly parallel to the old one; the triangle of the surplus just moves up.

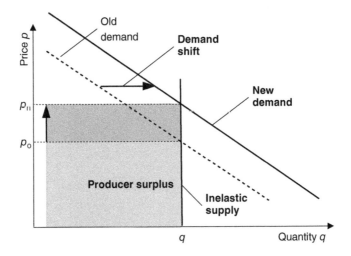

Figure 5.13 Very short run supply curve.

5.5.2 Short-Run Supply Curve

The number of producers is fixed in the short run. There is no time for companies to enter or exit the market. The lag between the short and the long run depends on information about the market and the preparation for entry or exit. In comparison to the very short run, producers can extend their supply along the supply curve: Changing prices will encourage producers to adjust the supply. The prices of the factors of production and, accordingly, the cost curves remain unchanged in this model.

Figure 5.14 shows the results of a demand change for the individual producer, the individual consumer, and the market. The initial equilibrium price and quantity in the market are at P_1 and Q_1 (or q_{1P} for the producer, q_{1C} for the consumer). The individual consumer (and others like her) then extends his demand to q_{2C}. This causes the market demand curve

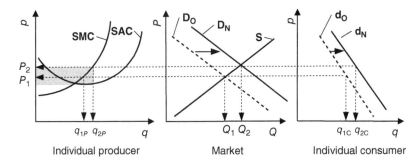

Figure 5.14 Short run supply curve.

to shift to the right from D_O to D_N. The intersection between the new demand curve D_N and the supply curve determines a higher price P_2 and a higher quantity Q_2. The short-run marginal costs (SMC) and the short-run average costs (SAC) remain unchanged. However, the profit of the individual producer increases to the shaded area in Figure 5.14 (minimum SAC, P_2, q_{2P} instead of minimum SAC, P_1, and q_{1P}).

5.5.3 Long-Run Supply Curve

In the long run, producers cannot only supply larger quantities, they also must expect that the additional profits attract others to set up new production facilities for the market. Transaction costs for entering the market are neglected in the model and the assumption is that all producers face the same unchanged cost curves. In this market, a producer can adjust his factors of production in such a way that he can produce with minimum costs.

At the beginning of the following explanation, we assume a producer who has minimized his costs in such a way that he realizes an economic profit. The same applies to the other producers in the market. The profits attract new entrants. The new entrants increase supply and the supply curve shifts outward. The market price falls since demand is unchanged. We can spin the story: Some producers leave the market; the supply curve shifts back with the corresponding increase in price and profits for the remaining producers. Here, we return to the beginning of the story. The long-run supply curve thus becomes horizontal with fluctuations around the zero-profit line: The long-run supply curve allows no economic profits to the producer (Figure 5.16).

We saw already that economic profits differ from the definition of accounting or estimating profits. Economic profits are the difference between total revenues and total opportunity costs. The opportunity costs include adequate pay for the owners and interest on equity. If the owners would decide to invest their money in some other form (working as a manager for someone else and investing the equity in the stock market), the likely gains are the owners' opportunity costs of running his own firm. An owner who realizes zero economic profit has no reason to leave the market, there is no better way (opportunity) to employ his own resources.

Figure 5.15 repeats some of the information from Figure 5.16: Point A describes the initial situation. The producer supplies the quantity q_{1P} at the market equilibrium price P_1. The producer provides the quantity where the market price P_1 equals marginal costs.

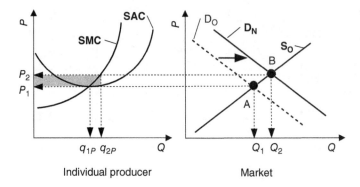

Figure 5.15 Short run curve with increase in demand.

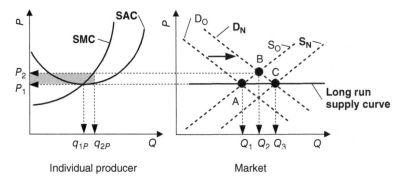

Figure 5.16 Long-run supply curve.

The price equals the minimum of short-run average costs. Hence, economic profit is zero. Next, the demand curve shifts outward for some reason to point B. The producer (and all others like him) increase output without changes to the cost curves. The price rises to P_2 and the quantity of the individual producer to q_{2P} and the market quantity to Q_2. The producer realizes now an economic profit since the price is higher than the average total costs for the quantity provided. This is the grey area in the left panel.

The economic profit of the producers does not remain unnoticed; it attracts others who will enter the market. Remember that there are no transaction costs to the new entrants and they have the same cost curves as all others. The new entrants expand supply from Q_2 to Q_3 by shifting the supply curves outward from S_O to S_N. The price adjusts to the new intersection of S_N and D_N (point C). The long-run supply curve runs through point A and Point C. The faster the market adjustments, the smaller the fluctuations around the horizontal long run supply curve. Thus, in the long run, economic profits are close to zero.

5.6 Welfare

The market welfare model postulates that the market equilibrium maximizes the net benefits to society from production and consumption (Dorman 2014). It rests on three assumptions.

Assumptions for Market Welfare

1) The demand curve represents the marginal benefits to society.
2) The supply curve represents marginal costs.
3) The supply and demand curve have a single and stable equilibrium.

The most problematic assumption is that for the demand curve as it rests on further rather restrictive conditions:

- Societal benefit is the sum of individual benefits.
- Each point on the demand curve represents the willingness to pay of some consumer.
- Willingness to pay represents an amount of utility.
- The marginal utility of money is the same for all consumers.
- There are no other influences than the willingness to pay.

The most controversial point is the societal benefit. As a sum of individual benefits (utility), we can write it as mathematical equation (Varian 2014):

$$W(u_1,\ldots,u_n) = \sum_{i=1}^{n} u_i \tag{5.4}$$

This idea is based on the utilitarian ethics of Bentham.

 Jeremy Bentham (1748–1832)

Bentham is the founder of modern utilitarianism. His philosophy has a strong impact on ethics. Ethics affect normative economics, and here utilitarianism is very influential.
  "*It is the greatest happiness of the greatest number that is the measure of right and wrong*".

Bentham (1776, 1977, p. 310)

Source: National Portrait Gallery/Wikimedia Commons/Public domain

---

Rawls (1971) published a different idea in his book *Theory of Justice*. His welfare function is a minimax solution:

$$W(u_1,\ldots,u_n) = \min\{u_1,\ldots,u_n\} \tag{5.5}$$

While Bentham concentrates on maximizing the welfare of a group, Rawls focuses on individuals by asking to maximize the welfare of the person with the minimal utility.

It should be evident that economics is not modelling in a vacuum and that ethics inform economic decision-making to a large degree. As this is not a book on ethics or justice, I will leave the topic now, but I believe it to be worthwhile mentioning and hope that you will form your own ideas.

5.7 Efficiency and Equity

We can measure the equality of the economic outcome of a national economic system with the Gini coefficient. This coefficient is a measure to evaluate the inequality in a set of data. If we focus on income as data set, then a value of zero would mean that all people have the same income; a value of one would represent an income distribution where all income is in the hands of one person.

The Gini coefficient is calculated with the help of the Lorenz curve which plots the cumulated proportion of the poorest percentile of a population (5%, 10%, 15%...) on the *x*-axis versus the cumulated share of the total income (*y*-axis). Accordingly, a 45° line would represent perfect income equality (Figure 5.17).

The calculation of the Gini coefficient based on Figure 5.17 is

$$G = A/(A + B) \tag{5.6}$$

Figure 5.19 illustrates that different nations accept different degrees of inequality.

Some Western countries with highly efficient economies exhibit rather uneven income distributions (darkly shaded areas in Figure 5.18). Such markets might be Pareto efficient, (i.e. there is no way to make someone better off without making someone else worse off but they are not equitable). An equitable allocation is one where nobody prefers someone else's bundle of goods over his own. If the definition of a fair allocation is one that is at the same time Pareto efficient and equitable, then it seems that economics has still a lot of work to do.

The discussions in Section 5.6 and 5.7 have often focused more on macro- than microeconomics. I feel this is necessary in order to place the argument of efficient markets into a wider context.

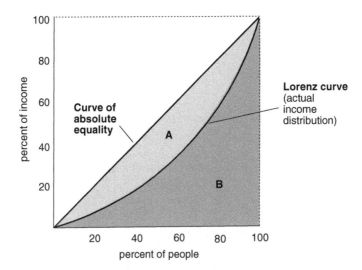

Figure 5.17 Lorenz curve and Gini coefficient.

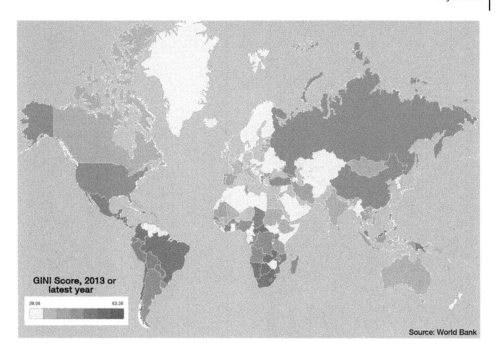

Figure 5.18 Global data of the Gini coefficient. *Source:* Joe Myers (2015). 5 maps on the state of global inequality, World Economic Forum. Retrieved from: https://www.weforum.org/agenda/2015/11/5-maps-on-the-state-of-global-inequality/.

References

Bentham, J. (1776). A fragment on government. In: *The Collected Works of Jeremy Bentham: A Comment on the Commentaries and A Fragment on Government* (ed. J. Burns and H. Hart), 391–551. Oxford: Oxford University Press (1977).

Dorman, P. (2014). *Microeconomics: A Fresh Start*. Heidelberg: Springer.

Krugman, P. and Wells, R. (2018). *Microeconomics*. New York: Worth.

Nicholson, W. (1990). *Intermediate Microeconomics and Its Application*. Chicago: Dryden Press.

Rawls, J. (1971). *A Theory of Justice*. Cambridge: Harvard University Press.

Varian, H. (2014). *Intermediate Microeconomics with Calculus: A Modern Approach*. New York: Norton.

6

Imperfect Markets

Case Study   Why Do You Always Have to Deal with Competition?

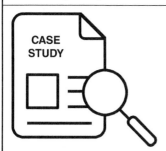

Your firm is doing fine. However, in the evening on the porch while watching the sunset, you surf the web and look up the profit margins for Apple. You find information on statista.com (https://www.statista.com/statistics/263436/apples-gross-margin-since-2005).

The gross profit margin as percentage of revenue is around 40%. You are not quite sure about the definition of gross profit margin, but 40% is far away from what you achieve in your precast element firm. You earned last year an overall margin of 12%, and after subtracting your business costs, your net margin turned out to be 5% of revenues. You assume that your competitors are the reason for the much lower profits in comparison to Apple. At the same time, you observe that you firm has no brand name. It is well respected among owners, but nobody brags about having precast elements from your company. There is not even an imprint of your firm's name on the elements. Your customers did not like the idea when you tried it a while ago. Surely you would like to be perceived by your customers as the apple of their eyes, but you do not find a way of achieving this. All the praise you get does not help you sign a contract when your price is higher than your competitors'.

There is a chance for the future, as one of your direct competitors wants to sell her company as she approaches the age of 70. You are not sure if that will happen and if there is someone interested in buying. Anyway, you start dreaming that both your direct competitors leave the market. Then you would just face the indirect substitutes by steel components. This would certainly allow you to raise your prices – but by how much? You know that you would lose contracts if the prices were too high.

As the situation happens to be in the moment, you still have to deal with your competitors. There are only two of them, and you could easily set up a cartel where you sit down together and determine who takes what contracts at what price. Instead of being competitors, you would be collaborators. You know that such cartels are illegal and that the practice of price rigging could earn you a prison sentence. Looking at the

(continued)

Construction Microeconomics, First Edition. Christian Brockmann.
© 2023 John Wiley & Sons Ltd. Published 2023 by John Wiley & Sons Ltd.

> sunset with a drink in your hand, you know what you would be missing in jail. You also
> remember you mom's advice: *"Make only such business during the day that allows you to
> sleep well at night."* You decide that this is very good advice.
>   Still, the size of Apple's profits keeps bugging you...

Perfectly competitive markets are only one type of many possible market structures. Table 6.1 shows a typology of market structures based on the number of buyers and sellers. We will find out later in Chapter 14 that the number of participants is not good enough a measure to determine real-world market structures; it is just a first orientation.

In the general literature on microeconomics, we find extensive discussions of monopolies and oligopolies. Here, the market power of producers influences market outcomes. We have looked at homogenous products in a market up to now. In the real world, however, we find many markets where the products have brand names. By branding, producers try to differentiate products and thereby decrease competition. Some consumers strongly prefer T-shirts from Lacoste over those from Hilfiger. Not so many see a difference in gas provided by Shell or Exxon. Branding, seemingly, works in some instances better than in others. Table 6.2 illustrates market structures based on the number of sellers and the possibility of product differentiation with many buyers in all cases. This typology allows introducing the important market of monopolistic competition.

In many real markets we find monopolies, oligopolies, and monopolistic competition. Perfectly competitive markets are rather an exception and the treatment in Chapter 5 serves mostly to introduce microeconomic concepts and to provide a clear benchmark. In this chapter follows the discussion of the three widespread imperfect market structures (monopoly, monopolistic competition, and oligopoly) and one market structure that is less often discussed: A monopsony. The treatment will be rather cursory. Therefore; I recommend consulting Tirole (2000) if you are interested in a more thorough treatment of the topic.

Table 6.1 Typology of market structures based on numbers of buyers and sellers.

	Many buyers	**Few buyers**	**One buyer**
Many sellers	Perfectly competitive market	Oligopsony	Monopsony
Few sellers	Oligopoly	Two-sided oligopoly	Limited monopsony
One seller	Monopoly	Limited monopoly	Two-sided monopoly

Table 6.2 Typology of market structures with differentiated products.

	Many buyers	
	Products are not differentiated	**Products are differentiated**
Many sellers	Perfect competition	Monopolistic competition
Few sellers	Oligopoly	
One seller	Monopoly	

In Chapter 14 on construction markets, we will see what structures determine the market outcomes in the construction sector.

6.1 Monopoly

In a monopoly, only one producer provides a good to a large number of consumers. It is plausible to expect that the monopolist wields market power. The important question is whether the market outcome is still efficient and whether it fulfils the conditions of the welfare functions in Chapter 5. To sum up the result, a monopoly does not maximize the sum of individual benefits (Bentham), nor does it maximize the benefits of the least lucky individual (Rawls).

There are a number of reasons for the existence of monopolies:

- Control of a scarce resource
- Natural monopolies
- Technical superiority
- Network externalities
- Government barriers

Whoever has control of a scarce resource used for direct sales or as a factor of production has power in the corresponding markets. Oftentimes this control has regional limits. A worldwide monopoly was the supply of diamonds by de Beers from South Africa. The company controlled almost 85% of the global diamond supply until 1990. One example of existing regional monopolies in construction are in some places asphalt production facilities.

Natural monopolies exist because of the ownership of costly infrastructure such as utility or railroad lines. Once a company has established a network of utility lines (fixed costs), the average costs decrease with each quantity delivered. It just does not pay for a competitor to build another network. In some countries, such monopolies are broken up by the government. Governments can force a natural monopolist to open the network for transportation of utilities (power, gas, telephone, etc.) by competitors through antitrust legislation.

Technical superiority includes design (Apple) or service superiority. In the 1950s, the Dywidag bar held an advantage in post-tensioning of structures. Later, Intel became well known for its computer chips. The last two are examples of product superiority. There are many more examples, but most provide monopolistic advantage only for a limited time.

A network externality allows an individual to profit from a service better if many others use the same. Examples are Microsoft Windows, Google, or Facebook.

Governments create barriers to incentivize innovation by granting patents and copyrights. A patent protects an innovation for a number of years.

In all these cases, the producers can wield some form of power and they are likely to exercise it to their own advantage. Historical examples are the American robber barons (Rockefeller, Carnegie, Hearst, Morgan, Vanderbilt, etc.). To counter the negative effects of their economic behavior, the US legislative introduced antitrust legislation (Sherman Act, passed in 1890).

Natural monopolies typically have decreasing average costs (i.e., the more the monopolists produce, the lower the average costs). The other types of monopolies face increasing marginal and accordingly also increasing average costs. Natural monopolies can increase welfare; normal monopolies have no incentive to do this.

6.1.1 Normal Monopolies

The monopolist has possibilities to set the price (instead of accepting the equilibrium price) because there are no alternatives to its offer. The monopolist can observe the market demand curve and can determine at which price profits can be maximized; the profit π depends on price p, quantity q, and costs c:

$$\pi(q) = p(q) \cdot q - c(q) \tag{6.1}$$

The derivative of this function provides the profit maximum:

$$\frac{d\pi}{dq} = \frac{dp}{dq} \cdot q + p(q) - \frac{dc}{dq} = 0 \tag{6.2}$$

$$\rightarrow \frac{dp}{dq} \cdot q + p(q) = \frac{dc}{dq} \tag{6.3}$$

In perfectly competitive markets the profit maximization demands that the price equals marginal costs ($p = dc/dq$), the equilibrium price does not depend on quantity. The producer in a perfectly competitive market gets the same price for each additional unit. A monopolist who increases his quantity will have to face a lower price applicable to all goods supplied because of the downward-sloping demand curve. In all market configurations marginal costs equal marginal revenue; however, marginal revenue is different for a monopolist.

Figure 6.1 shows the price mechanism in a monopoly for a classical production function and a linear demand curve. The marginal revenue curve is exactly half of the demand curve

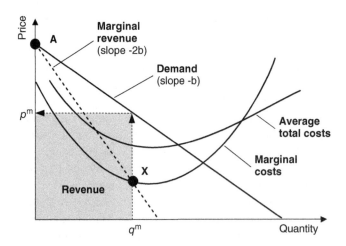

Figure 6.1 Profit maximization in a normal monopoly.

for each price level. We find for a linear demand equation with a y-axis intercept a and a slope of $-b$:

$$p = a - b \cdot q \tag{6.4}$$

Therefore, revenue r equals:

$$r = p(q) \cdot q = (a - b \cdot q) \cdot q = a \cdot q - b \cdot q^2 \tag{6.5}$$

We get the marginal revenue by calculating the first derivative:

$$\frac{dr}{dq} = a - 2bq \tag{6.6}$$

The demand function and the marginal revenue have the same y-intercept at point A but the slope of the linear marginal revenue curve is twice that of the linear demand curve. Point X in Figure 6.1 denotes the point where marginal costs equals marginal revenue, and this determines price and quantity of the monopolist. The quantity q^m of the monopolist maximizing revenue is vertically underneath point X and the price p^m is to the left of the intersection of the vertical line through X and the demand curve. Revenue R is shaded grey and equals p^m times q^m.

If we look at the welfare result from a monopoly, we find that there is a decrease in total welfare (deadweight loss) and consumer surplus but an increase in producer surplus. Compared with perfect competition, the monopolist offers a reduced quantity and demands a higher price. The quantity reduction leads to the deadweight loss, there are fewer goods available. In addition, the price increase takes some of the consumer surplus away and shifts it to increase the monopolist surplus (Figure 6.2).

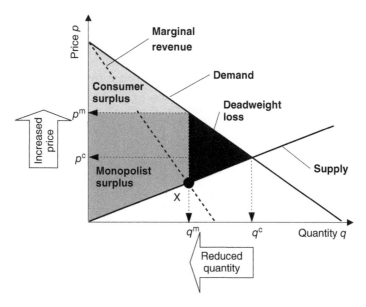

Figure 6.2 Consumer and monopolist surplus.

The profit maximizing monopoly price is also called the Cournot price because of Cournot's contribution to the model above.

Antoine Augustin Cournot (1801–1877)

Cournot developed the model for profit maximization. He made extensive use of mathematics in his works. Even better known are his contributions to the theory of oligopolies. He defined the Cournot equilibrium for duopolies.

"*Anyone who understands algebraic notation, reads at a glance in an equation results reached arithmetically only with great labour and pains*".

Cournot (1838, 1897, p. 4)

Source: Unknown/Wikimedia Commons/Public domain

In sum, a normal monopoly does not lead to an efficient market result. It decreases welfare and benefits the producer at the expense of the consumers.

6.1.2 Natural Monopolies

The main characteristic of a natural monopoly such as utility provision is that the average costs approach the marginal costs because the high fix costs of the utility network spread over more and more units. Marginal costs are in addition constant (Figure 6.3).

Although average total costs constantly decline, the natural monopolist has no incentive to give up profit maximizing. If free to do so, the natural monopolist will produce the same quantity q^m at the same price p^m as the normal monopolist, following the rule that price and quantity will meet where marginal costs equals marginal revenue (X in Figure 6.3). It requires a regulatory body to force the monopolist to lower the price. Welfare maximization demands marginal costs to equal marginal utility, and since the demand curve represents marginal utility, this holds true for point A. However, the corresponding price p^w is lower

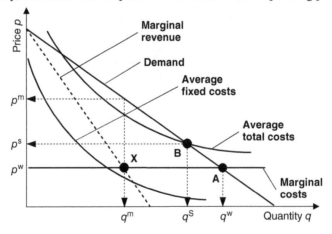

Figure 6.3 Price mechanism in a natural monopoly.

than average total costs and the monopolist would refuse to deliver the utility at that price. The second-best option is a quantity q^s at a price p^s (point B). This would allow the monopolist to avoid losses and the regulatory body to optimize welfare. Thus, the regulatory body should aim to hand the monopoly right to the producer at the price level of p^s. Since the producer is not interested in providing information about the true average total costs to the regulatory body, it is difficult to determine the location of point B.

Competition is not advisable in natural monopolies, but regulatory bodies who have the property rights over utility provision in an area can replace the competition in the market by competition for the market. Auctions are used to this purpose; well-known examples are 5G spectrum auctions.

6.2 Monopolistic Competition

Monopolistic competition is the most prevalent market configuration besides heterogenous oligopolies. Under these conditions, producers offer differentiated products – and not homogenous goods – that serve the same purpose such as T-shirts, and consumers have preferences for certain products. This allows some space for producers to set prices. They are no longer price-takers but they also do not have the freedom of normal monopolists. Their market power is more limited. Producers can differentiate products mainly by style, location, or quality. The more a producer is able to differentiate the products, the stronger the market power. Each producer provides only a relatively small quantity of goods to the market so that its actions have no impact on other producers. Thus, if a producer raises the price of a product, it will lose some customers but not all, as under the conditions of perfect competition.

However, the producer is threatened by new firms who can enter the market and attack the differentiated product. The possibility of free market entry reduces profits to zero in the long run. This result is the same as for perfect competition, but other results are different, as we will see.

The excess capacity theorem assumes a classical production function and corresponding cost curves as well as the same costs for all producers. Figure 6.4 illustrates the situation. The demand curve is again downward sloping and the slope of marginal revenue curve is twice as much as the demand curve. The marginal cost curve (supply) is upward sloping. As there are some fixed costs, total average costs are U-shaped. Profit maximizing is the quantity q^{mc} where marginal costs equal marginal revenue and the corresponding price is p^{mc}. If the average total cost curve lies below the demand curve, the monopolistic competitor generates a profit.

The strength of consumer preferences for a differentiated product is represented in the elasticity of the demand curve. The stronger the preference, the more inelastic will be the demand curve (Figure 6.4, right).

In the short run, the monopolistic competitor will act as monopolist dealing with cost curves from the classical production function (Figure 6.5, left). With market entry, the demand and marginal revenue curves will swivel around point A and become more elastic. The y-intercept moves from B to B' (Figure 6.5, right). The price is lowered and the offered quantity is also reduced. This reduces the monopolistic profit.

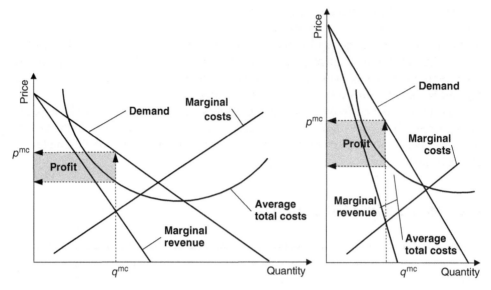

Figure 6.4 Profit maximization in monopolistic competition.

This process of market entry ends when the economic profit becomes zero. At this point, the demand curve is tangent to the average total cost curve (Figure 6.6). Production at a lower price than *p* does not make sense to the producer in the long run because it cannot cover the average total costs; such a move would generate a loss.

If we compare the results of perfectly competitive markets with monopolistic competition, we find first of all that the long-run supply curve in the perfect market is horizontal. The producer decides to produce at the level where marginal revenue equals marginal

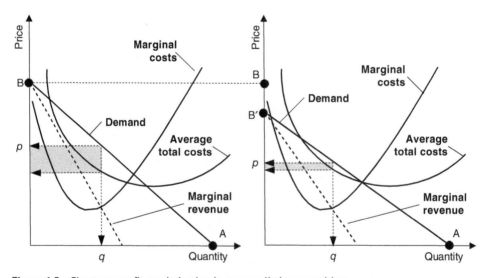

Figure 6.5 Short-run profit maximization in monopolistic competition.

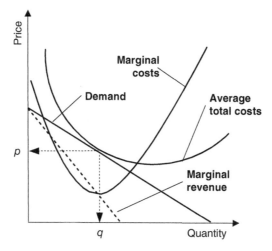

Figure 6.6 Long-run equilibrium in monopolistic competition.

costs ($MR = MC$). Here, marginal revenue also equals market price and supply is perfectly elastic (Figure 6.7, left).

In the long run, the monopolistic competitor faces a downward-sloping demand curve and decides to produce at the level where $MR = MC$. The market price for this condition equals average total costs, hence it does not realize an economic profit (Figure 6.7, right).

The difference between the two is that for quantities exceeding q_{mc}, the monopolistic competitor makes a profit, since the price is above average total costs. The price-taking producer in a perfectly competitive market faces, by contrast, higher average total costs when extending quantity beyond q_{pc}. In sum, a monopolistic competitor wants to sell more at the market price and the perfect competitor has no incentive to do so.

It is also clear that the monopolistic competitor produces a smaller quantity and sells it at a higher price than the perfect competitor.

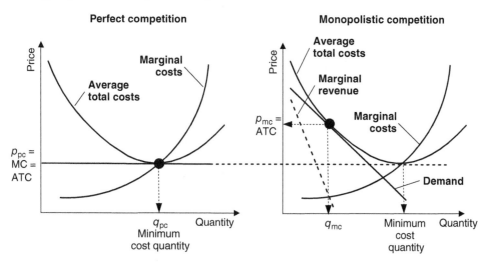

Figure 6.7 Price mechanism in perfect and monopolistic competition.

6.3 Monopsony

Monopsonies play a role in the construction sector. The government or government agencies act sometimes as the sole buyer in a market. In many countries, it is the exclusive right and duty of the government to provide public roads. As these are by far the majority of all roads, the government acts as monopsonist. The same often holds true for railroads. Thus, monopsonies are of interest. The owner in construction will not be a consumer (one-family home) but an investor and will thus act on factor markets (Chapter 7).

A typical case in the microeconomic literature to explain monopsonies is one employer looking for labor (Nicholson and Snyder 2014). This could be a large firm in a small town. If one producer is the asking for the by-far-largest supply of labor from a relatively large number of workers, then that producer holds a monopsony position. This is the inverse of a monopoly.

If we call $p \cdot \partial q / \partial L$ the marginal value product of labor (MVP_L), then we know that this must equal the wage if the producer is maximizing the profit in perfectly competitive markets. If we assume diminishing marginal labor productivity, then the MVP_L curve is downward sloping (Figure 6.8).

The labor supply curve is sloping upward because the monopsonist is not facing a negligible quantity of labor but the majority of the labor supply in a regional market. If the monpsonist hires one more worker, this worker is only willing to accept above market price. If all workers receive the same wage, this means that adding one more worker increases total cost by the pay for the worker and the increase for all other workers. The cost of hiring one more worker is always higher than the market price. Consequently, the marginal expense curve will always be above the labor supply curve.

A producer in a perfectly competitive market will hire more labor until the additional marginal costs ME_L are equal to the additional marginal revenue MR_L. If the costs were

Figure 6.8 Marginal value product curve for labor under profit maximization.

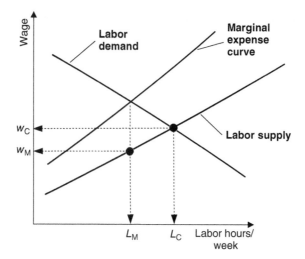

Figure 6.9 Monopsonistic pricing.

higher, then the producer would lose money; if the costs were lower, it would forgo the opportunity to increase its profit:

$$ME_L = MR_L \tag{6.7}$$

The monopsonist will add labor until the marginal revenue of labor equals the MVP:

$$MR_L = MVP_L \tag{6.8}$$

If the producer were a price taker, then the intersection of the downward-sloping demand curve and the upward-sloping supply curve would provide the market equilibrium. This is marked as L_C and w_C in Figure 6.9. Since a monopsonist is not a price taker, the marginal revenue curve (expense curve) lies above the supply curve and this determines actual labor demand by $MR_L = MVP_L$. Thus, the monopsonist demands less labor (L_M) at a lower wage (w_M) by using its market power.

The more inelastic the labor supply curve is, the lower will the wage be. To see this effect, you just have to rotate the supply curve more upward around the point L_C/w_C.

6.4 Oligopoly

In an oligopoly a few producers satisfy the demand of many consumers. Different to perfectly competitive markets or monopolies, these producers pay attention to the behavior of the other oligopolists. As such, strategies play an important role in oligopolies and game theory is the foremost tool to study the behavior. Quantity and price in an oligopoly will be somewhere between monopoly and perfect competition.

In many cases, economists study oligopolies with the help of game theory. Game theory analyzes strategic interactions between rational decision-makers with the help of mathematical models. Founding fathers of game theory were von Neumann and Morgenstern (1944). Nash (1951) defined an equilibrium solution to a noncooperative game which later became known as Nash equilibrium. This is a situation where no player can change his strategy without becoming worse off if all other players keep their strategy.

John Forbes Nash (1928–2015)

Nash was a mathematician who received the Nobel Memorial Prize in Economic Sciences in 1994 for his contributions to game theory. The recommendable movie *A Beautiful Mind* portrays his life.

Source: Elke Wetzig/Wikimedia Commons/CC BY-SA 3.0"

A simple form of an oligopoly is a duopoly, the competition between two producers. Cournot (1838) developed a solution to the problem implying that producer A chooses his output by assuming a fixed output from producer B (and vice versa) without further adjustments. This provides a stable equilibrium at a level below monopoly price and above monopoly quantity. The behavior of the two producers can be described as a simultaneous competition for quantities.

In the Stackelberg model (1934), producer A would anticipate the move of B and factor this outcome into his strategy in order to increase his own profit (and decrease the profit of B). The total profit of both producers will be smaller than the profits in the Cournot model and the monopoly but higher than in perfectly competitive markets. We are looking in difference to the Cournot model at a sequential competition for quantities.

The market outcomes are often suboptimal from the point of view of the oligopolists; the solutions do not reach an equilibrium. They might in such cases be tempted to collude by fixing prices. While this happens in real life and also in construction, it is illegal.

Oligopolies make use of advertising. These activities increase the costs of the producers and they are called transaction costs. Oligopolists and others facing such transaction costs will have to consider these in addition to the rule $MR = MC$. They create a gap between buying and selling price reducing the market quantity (Hirshleifer and Hirshleifer 1998, p. 410).

Specific market conditions need to be looked at to find a solution for an oligopoly since there are otherwise infinite possibilities of strategic behaviors. It seems to be advisable to look at the construction market in detail before advancing the theory of oligopolies (Chapter 14). We will find that oligopolies play a negligible role in construction.

References

Cournot, A. (1838). *Researches on the Mathematical Principles of the Theory of Wealth*. London: Macmillan (1897).

Hirshleifer, J. and Hirshleifer, D. (1998). *Price Theory and Applications*. Upper Saddle River, NJ: Prentice-Hall.

Nash, J. (1951). Non-cooperative games. *Annals of Mathematics* 54 (2): 286–295.

Nicholson, W. and Snyder, C. (2014). *Intermediate Microeconomics and Its Application*. Mason, OH: Cengage Learning.

Tirole, J. (2000). *The Theory of Industrial Organization*. Cambridge: MIT Press.

Von Neumann, J. and Morgenstern, O. (1944). *Theory of Games and Economic Behavior*. Princeton: Princeton University Press (2004).

7

Factor Markets

Case Study   Demand or Supply?

CASE STUDY

Out on the porch again, you have a classmate over for a visit. She works for an international contractor in Dubai on a skyscraper. The discussion turns around the migrant workers that are absolutely indispensable for the construction industry in the UAE. She explains to you that labor supply is falling short of demand, especially if you are looking for a certain level of skills.

In your own firm, you are rather suffering from a shortage in material supply. The COVID-19 crisis in 2020–2021 wreaked havoc in the construction supply chain. Luckily, you were able to hold onto your workforce during the crisis.

Your classmate will return to the UAE to finish her assignment. Then she will be looking for a new job. She knows about your interest in microeconomics and explains to you that she will be on the demand side of the labor market. This confuses you because you think that you as a firm have a labor demand and that employees are supplying labor. After some discussion, you can clarify the confusion: There seems to be a difference between consumer markets and factor markets (i.e. markets where producers buy inputs for their production). In consumer markets, individuals express a market demand and firms provide the supply of goods and services. However, in factor markets it is the firm that demands inputs such as labor and capital, and at least labor is supplied by individuals. Your classmate points out that she invested some of her overseas income in stocks of pharmaceutical companies, so, she is not only supplying her labor but also financial capital to firms.

After the goodbye, you are happy to have clarified the topic, but you determine that you want to learn more about factor markets. It is not very late and you go to your study and pick up your book on microeconomics...

Construction Microeconomics, First Edition. Christian Brockmann.
© 2023 John Wiley & Sons Ltd. Published 2023 by John Wiley & Sons Ltd.

The markets we have discussed so far, whether perfectly competitive or imperfect, were consumer markets to which producers supply consumption goods. Now, we turn to the question, where do producers acquire their inputs? We call these inputs factors of production, and accordingly, it is no big surprise that they are traded on factor markets. All producers require factors of production as input. They use a certain technology to transform these inputs into an output of higher value. Factors of production in the classical economic models are land, labor, and capital, and often I have limited the models to two inputs, labor L and capital K.

On consumer markets, producers supply goods; on factor markets producers demand goods. Households provide the labor supply and producers demand labor as factor of production. As most owners in construction demand building and infrastructure for their end purposes, they are contracting on factor markets. An exception is the one-family home demanded for consumptive purposes. Office buildings, factories, rental apartments, transportation and utility infrastructure, places of worship, sports stadia, museums and concert halls, dams and ports are all fixed assets for the achievement of other goals: These are investments traded on factor markets. For this reason, factor markets are of special importance to construction. I will investigate this in more detail in Chapter 14 on construction markets.

Land is a very peculiar factor of production in construction. It belongs to the owner and the contractor is granted access to it only at the discretion of the owner. However, as every structure is connected to the soil prevalent at a specific site, land becomes a factor of production for the contractor. The contractor does not usually have to pay a price for land (unless some additional land is being rented for the site installations) but the quality of the land impacts the output.

Labor is modeled on the form of a typical worker with a typical productivity and a typical wage. The wage, w, is the price of labor.

Capital consists of the construction and support material such as formwork or sheet piles. Construction materials become part of the structure while support material can be used again on other sites. Furthermore, we can distinguish between renewable and exhaustible materials. This is an important consideration when we take sustainability into account. Wood is a renewable material; sand in the quality required for concrete is exhaustible. Energy also belongs to the category of capital, and again we can differentiate between renewable (sun, wind) and exhaustible energy (fossil fuels). A last form of capital is equipment, and this is reusable in the same way as support material. The price of construction material and energy is the factor market price. The price of the support material and equipment is depreciation or the rental price; based on the concept of opportunity costs, they are similar.

Another factor of production is knowledge. This can take the form of advanced knowledge in management or technical processes; both impact the technology level. The knowledge or skill level of the work force affects labor productivity, i.e. the quality of labor.

Institutions influence the output on a local or national level. Culture, regulations, and norms are constitutional for institutions and these vary (Scott 2001). Some authorities might provide a building permit speedily; others take their time and still others a bribe. Democratic countries know public involvement, other countries rely on authoritarian decision-making. Institutions influence transaction costs and construction time.

The following chapter will look at the supply side of factor markets and the supply originates from households. Then, we will look at the demand side on factor markets as initiated by firms. The final chapter joins supply and demand on factor markets. In all chapters, I will restrict myself to the discussion of labor and capital as inputs.

7.1 Factor Supply of Households

The simplified flow diagram (Figure 2.3) shows the interaction between households and firms and Figure 7.1 illustrates this relationship again. In consumer markets, firms supply goods and services and consumers pay for them. In factor markets, firms demand labor and capital, and they pay for them.

7.1.1 Labor Supply

Households need money in order to be able to purchase necessary and luxury goods. The only chance for the majority is to offer their labor in the labor market and to earn a wage. To this end, individuals and households make decisions how they use their time. Time is a limited resource as the day has invariably 24 hours. An individual can choose how much time to spend working (W) and economists call the remaining part of the day leisure (L) whether it is used for cleaning, cooking, learning, or relaxing. The less an individual works, the less money he has for consumption and the more time for other activities. Instead of a budget constraint, the individual faces now a time constraint. Instead of choosing between a number of different goods, we are now looking at the time preferences of the individual. As before, constraints and preferences determine the decision of the individual in the neo-classical model (Figure 7.2).

A person can theoretically work 24 hours, and this provides the y-intercept for the time constraint; it also can work 0 hours and spend all day on leisure (x-intercept). This way we can construct the time constraint in Figure 7.2. The explanations for the construction of indifference lines and utility maximization in Section 3.3 on consumer behavior are also valid for time preferences and maximizing decisions on the use of time.

It should be clear that most people are not free to decide how many hours to work, but looking at a total work life, there certainly is some flexibility for many. Changes in preferences lead to different choices. In some Western countries, young engineers today value leisure more than the generations before them. In other words, they prefer spending time with family and friends. Figure 7.3 shows the different preferences and outcomes.

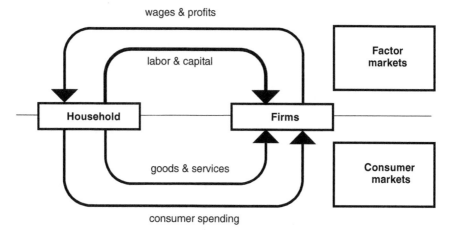

Figure 7.1 Consumer and factor markets.

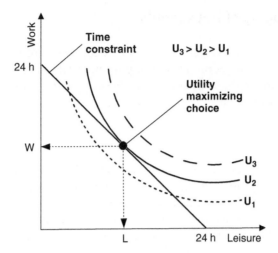

Figure 7.2 Utility maximizing choice of work hours.

The daily income (y) is the product of work hours (h_W) and wage, and if we consider theoretically 24 hours as the maximum available time, then the choice of leisure hours (h_L) influences this income:

$$y = w \cdot h_W = w(24 - h_L) \tag{7.1}$$

Differentiating Eq. (7.1) for h_L gives:

$$\frac{dy}{dh_L} = -w \tag{7.2}$$

Thus, the income-leisure combination has a slope of $-w$. An increase in income due to a higher wage can shift the choice toward more leisure time (Figure 7.4). This depends, of course, on the shape of the preference curves, but it is one possible outcome.

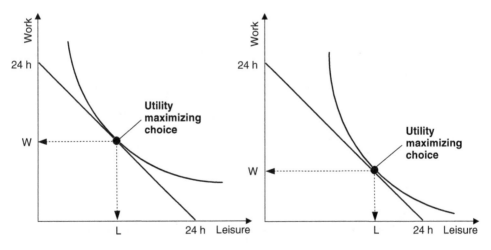

Figure 7.3 Shift in work attitudes.

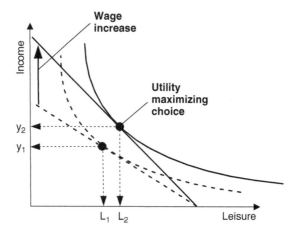

Figure 7.4 Influence of wage raise on work time.

The wage increase turns the budget constraint upwards and leads to an increase of income from y_1 to y_2. The leisure hours increase and working hours decrease in the process: Higher pay decreases work supply. We can derive a work supply function from this observation. However, the shape of labor supply is not clear. Another argument would postulate that the higher wage induces people to work longer.

A plausible shape of the labor supply curve L_s assumes rising work hours at low wages and decreasing work hours at high wages (Figure 7.5). Decreasing marginal utility of income and rising marginal utility of leisure starting at a certain level of income form the basis of the assumption.

Wages are a most important topic in economics. While many neoclassical economists (e.g. Sowell 2015) see inflexible wage rates as the root of many problems, especially in macroeconomics. Marx described wages as a means of exploiting labor. By this, he expresses the idea that the amount of labor required to produce a good determines the economic value of the good. This is known as the labor theory of value, and Marx shared the idea with Adam Smith and David Ricardo. This is in contradiction to the idea that supply and demand determine the price of a good.

Karl Marx (1818–1883) Marx is, together with Friedrich Engels, one of the foremost theoretical thinkers of communism.

"*The directing motive, the end and aim of capitalist production, is to extract the greatest possible amount of surplus value, and consequently to exploit labor-power to the greatest possible extent.*"

Marx (1867, 1996, p. 178).

Source: E. Dutertre/Wikimedia Commons/Public domain

7.1.2 Capital Supply

The capital that households supply to firms is financial capital, since by definition households are not producers; they cannot provide capital in form of goods. The basic assumption

Figure 7.5 Plausible labour demand curve.

is that the supply depends on the interest rate offered by firms: The higher the interest rate, the more are households are willing to supply financial capital.

The amount of financial capital that households can supply depends on their savings. The neoclassical savings function postulates a dependency of savings S on the interest rate i (Figure 7.6). Households have a choice how to save; it can be (broadly speaking) in savings accounts at banks or as investment in stocks. Investment in stocks is riskier so the return on investment must compensate for this. Typically, a decrease of the interest rate for savings in bank accounts will increase the investment in stocks if the return on investment is at least stable. In this case, more financial capital becomes available to firms from households. However, one should not forget that banks also invest their savings to a large extent in firms. In a closed economy with no exports or imports, investment in capital goods I is equal to savings S:

$$I = S \tag{7.3}$$

After having generated savings, households can decide whether to spend the savings in period 1 or period 2 (and keep investing in period 1). Based on an additive utility function,

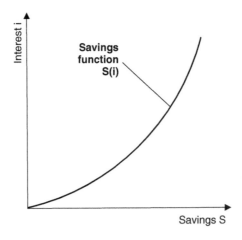

Figure 7.6 Neoclassical savings function of households.

one can show that deferred spending again depends on the interest rate i; consumption y_2 in period 2 equals in this case:

$$y_2 = (1+i)^2 \cdot y_1 \tag{7.4}$$

Thus, high interest rates influence whether we would rather spend now or tomorrow.

7.2 Factor Demand of Firms

The demand for factors of production is derived. Only if a contractor has a contract will he purchase the material and – in the longer run – employ labor. Producers supply goods in the case of direct demand from consumers but they demand inputs such as labor and capital as derived demand.

Expenditures or the rental rate for capital (v) depend on depreciation (d) and alternative uses of the capital invested, i.e. the opportunity costs of the capital or the market rate of investment (r). If the price for a machine or a building is P, then:

$$d \cdot P + r \cdot P = P(d+r) = v \tag{7.5}$$

The rate v is the same whether the producer rents or owns the equipment and whether he uses his own or borrowed money because of the principle of opportunity costs. The profit function of a producer is the product of quantity q times price p minus costs of labor (w_L) and capital (v_K):

$$\pi(q) = p \cdot q(L,K) - w \cdot L - v \cdot K \tag{7.6}$$

If we consider labor as a constant in the short run, then we get for the profit maximizing demand of capital:

$$\partial G / \partial K = p \cdot \partial q / \partial K - v = 0 \rightarrow p \cdot \partial q / \partial K = v \tag{7.7}$$

The profit maximizing demand for a factor of production is given when the price of the factor (v) equals the monetarily evaluated (p) marginal product of the factor $\partial q / \partial K$. This is nothing else but marginal expense equals marginal revenue. Demand of a factor of production takes the form of Figure 7.7.

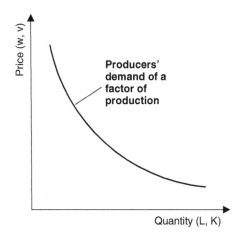

Figure 7.7 Demand of a factor of production

Profit maximizing demands that marginal expense ME_K and ME_L equals marginal revenue MR_K and MR_L (Nicholson and Snyder 2014):

$$ME_K = MR_K \text{ and } ME_L = MR_L \tag{7.8}$$

If the producer is a price taker, then the following equations must hold:

$$v = ME_K = MR_K \text{ and } w = ME_L = MR_L \tag{7.9}$$

The marginal physical productivity of a worker MP_L is the additional output that the worker produces per hour. A similar observation is possible for the marginal physical productivity of capital MP_K. The marginal revenue is defined as the extra revenue from selling one more unit. $MP_L \cdot MR$ is called the marginal revenue product of labor and $MP_K \cdot MR$ the marginal revenue product of capital. These definitions provide the following equations:

$$v = ME_K = MR_K = MP_K \cdot MR \tag{7.10}$$

$$w = ME_L = MR_L = MP_L \cdot MR \tag{7.11}$$

If the firm not only buys in a perfectly competitive market but also sells its good in one, then it is a double price-taker and the equations become simpler:

$$v = MP_K \cdot P \tag{7.12}$$

$$w = MP_L \cdot P \tag{7.13}$$

Finally, we can define the marginal value product (MVP) of labor and capital:

$$v = MP_K \cdot P = MVP_K \tag{7.14}$$

$$w = MP_L \cdot P = MVP_L \tag{7.15}$$

Figure 7.8 illustrates the considerations. If labor exhibits decreasing marginal physical productivity, then the demand curve is downward sloping. A profit-maximizing firm will demand L^* at a wage rate of w^*.

We can construct a similar curve for capital demand.

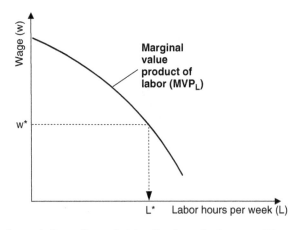

Figure 7.8 Labor demand of a profit maximizing firm in perfectly competitive markets.

7.3   Demand and Supply on Factor Markets

Supply and demand for the labor market depend on the wage, w, and the demand for capital depends on the capital costs, v. We have seen before that an increase in w or v leads to an increase in supply and a decrease in demand. The difference between factor markets and consumer markets is that firms have a demand and others provide the supply; households (consumers) supply labor as well as financial capital and firms supply goods and services (Figure 7.9).

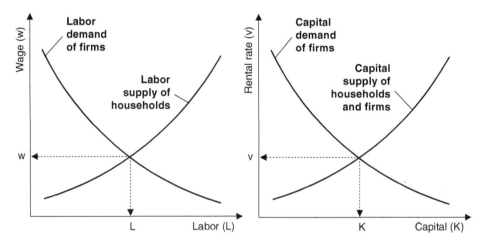

Figure 7.9   Supply and demand in factor markets.

References

Marx, K. (1867). *Das Kapital: A Critique of Political Economy*. Washington: Regnery. (1996).

Nicholson, W. and Snyder, C. (2014). *Intermediate Microeconomics and its Application*. Mason: Cengage Learning.

Scott, R. (2001). *Institutions and Organizations*. Thousand Oaks: Sage.

Sowell, T. (2015). *Basic Economics: A Common Sense Guide to the Economy*. New York: Basic Books.

8

Uncertainty, Risk, and Information

Case Study   Uncertainty?

Yesterday, you signed a new contract with a price of $2,554,982.21; so far, so good. Unfortunately, the second-lowest bidder submitted an offer roughly $120,000 higher. When you tell this to your spouse one evening on the porch, you both start dreaming about what you could have bought for that amount of money. Maybe a nice Porsche convertible? Your spouse is asking you how this price difference came about. You answer something about the uncertainty of construction markets and estimating. For yourself, you wonder whether this will turn out to be a project where you have attracted the winner's curse: Happiness today and misery tomorrow because of a project losing money. What you would give to get rid of the uncertainty! You make a mental check of the assumptions, and everything seems to be fine. Grumbling a bit, you go to bed...

During the night, you wake up and worries creep into you mind. You really hate to lose money and you find out that you are not as happy when you make a profit as you are unhappy when you make a loss of the same size. The second-lowest bidder has a good connection to the owner; maybe he knows something that you do not know. Maybe he has an information advantage? You determine to call him tomorrow; maybe then you can find out more.

While you toss around in your bed, you also promise yourself to talk to Janet, the chief information manager in your company. She keeps bugging you about implementing a risk management system for estimating, and you always felt you were too busy to listen.

The model of the perfectly competitive market rests on the assumption that producers and consumers have all necessary information at all times to make utility or profit-maximizing decisions. This assumption was always stretching reality, and it still does so today with easier access to information via the internet; however, the situation has improved.

Construction Microeconomics, First Edition. Christian Brockmann.
© 2023 John Wiley & Sons Ltd. Published 2023 by John Wiley & Sons Ltd.

The model also assumes that finding information comes without costs. We have already seen that searching for information contributes to transaction costs.

Lack of information leads to uncertainty or risks in decision-making. The question arises as to how to deal with such situations, as they are typical for the construction sector. Managers have different risk attitudes, and their decision-making will depend on these. They can employ some diverse strategies to deal with risks, such as accepting and managing risks, trading risks (insurance), or spreading risks (diversification).

Information economics analyzes situations of incomplete information. Such information can be incomplete and symmetric or incomplete and asymmetric, giving one party market power through an information advantage. This can lead to market breakdowns, wrong choices (adverse selection), and cheating (moral hazard).

Uncertainty and risk lead to the theory of transaction costs and information asymmetry to the principal–agent theory, both belonging to new institutional economics (NIE). In the discussion of these theories, we need to differentiate between exchange and contract goods. The property-rights theory provides the theoretical foundation for both.

8.1 Uncertainty and Risk

NIE shifts the focus from efficient resource allocation through perfectly competitive markets to the question of who benefits from a trade, and this depends on contracts. Contracts must deal with uncertainty, risk, and information. Neither information nor contracts are important for the model of perfectly competitive markets because of the underlying assumptions. In the construction sector, they are of paramount importance. Many consumer goods are exchange goods – we trade the good for money in one swift exchange. Construction goods are defined by contracts and contracts precede production. Economists call such goods contract goods. Analyzing construction without this difference in mind will bring wrong results.

We can differentiate between four different situations depending on the completeness of information (Figure 8.1).

If the information is complete, then it can be easily understandable and we find certainty – but it might also be that the information leads to confusion and the situation will be ambiguous. Complete information is an assumption for perfectly competitive markets. Ambiguity arises when the amount of information is large and complex, this leads to an information overflow and problems of information assessment. Ambiguity prevails in construction megaprojects (Brockmann 2021).

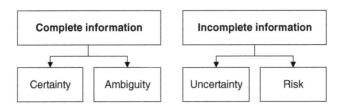

Figure 8.1 Information situations.

Incomplete information leads to uncertainty or risk. Uncertainty describes situations when a lack of information does not allow rational decision-making. Risky situations differ from uncertain ones by the possibility of assigning a probability value to the situation. We can estimate probability values by analyzing previously collected data. These data either allow to establish a trend or they are part of a simulation. In the case of uncertainty, no data are available.

8.1.1 Risk Attitudes

Individuals can welcome risks, be neutral to them, or be risk averse. Kahneman and Tversky (1979) found that most individuals suffer from a bias and are more motivated to avoid losses than to welcome gains. Accordingly, risk aversion is more widespread than risk seeking or neutrality.

Daniel Kahneman (1934)

Kahneman received the Nobel Prize in 2002 for integrating insights from psychological research and economic science, especially concerning human judgment and decision-making under uncertainty.

   "A reliable way to make people believe in falsehoods is frequent repetition, because familiarity is not easily distinguished from truth. Authoritarian institutions and marketers have always known this fact."

Kahneman (2011, p. 62).

Source: Andreas Rentz/Burda Media/Getty Images

Regardless of risk attitudes, it is always possible to calculate the expected monetary value (EMV) of a risk. It is defined as the product of the risk probability p times monetary impact I_M:

$$EMV\,[\text{currency}] = p[\%] \cdot I_M\,[\text{currency}] \tag{8.1}$$

Another explanation for risk aversion is diminishing marginal utility (Chapter 3.2.2). Depending on the endowment level, utility diminishes with higher levels of income. Utility (U) is always increasing but to a lesser degree. If we take income (Y) as endowment, we will find:

$$\frac{\partial U}{\partial Y} > 0 \quad \text{and} \quad \frac{\partial^2 U}{\partial^2 Y} < 0 \tag{8.2}$$

Figure 8.2 illustrates the results of diminishing marginal utility based on the level of income. Please remember that marginal utility is the first derivative of the utility function – or in geometrical terms, the tangent to that function.

Since we are giving up the assumption of complete information, we are now dealing with expected utility (instead of utility). With two different situations, S_1 and S_2, and the corresponding probabilities p(S), we get the following utility function based on expected value:

$$U\,(S_1, S_2, p_1, p_2) = p_1 \cdot S_1 + p_2 \cdot S_2 \tag{8.3}$$

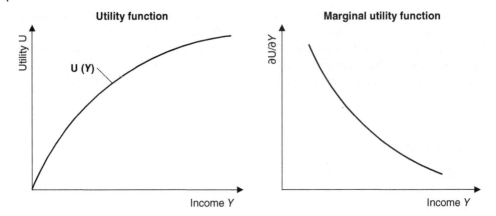

Figure 8.2 Utility and marginal utility functions.

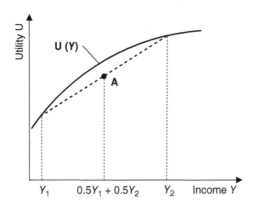

Figure 8.3 Risk aversion.

The combination of a concave utility function (diminishing returns) and expected utility gives the result of Figure 8.3. If an individual considers two income situations Y_1 and Y_2 likely to the same degree ($p = 0.5$), then he will be risk averse because he prefers the expected value of his wealth in the future (A) over a gamble ($U(Y)$).

To explain risk seeking, we would have to assume a convex utility function, and that is a rather unlikely assumption. In such a case, an individual would prefer a pay increase from 300,000 (Y_2) to 320,000 USD rather than one from 10,000 (Y_1) to 30,000 USD. As you can check in Figure 8.4, the increase in utility starting from Y_1 is smaller than that starting from Y_2. However, intuition tells us that a pay increase of 20,000 USD has more value for a person with an endowment of just 10,000 USD.

8.1.2 Risk Strategies

Risk can be accepted and managed (Kendrick 2009), it can also be insured, diluted by diversification, or shifted by contract.

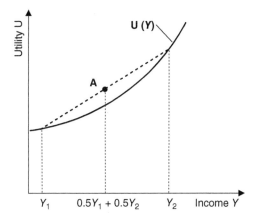

Figure 8.4 Risk seeking.

To understand the incentive to buy an insurance based on concave utility functions, we can look at the example of a person with an annual net income of 60,000 USD facing medical bills of 10,000 USD and a 50% chance of becoming ill. With no insurance, the expected utility U_{exp} would be $0.5 \cdot 50,000 + 0.5 \cdot 60,000 = 55,000$ USD (point A). With a premium of 2,500 USD, the insurance would make sense because for this point $U_{Ins} > U_{exp}$. At $55,000 + 2,500 = 57,500$ USD the insured person could jump from the straight line up to the higher concave utility function. In Figure 8.5, you can also see that there is a maximum premium where $U_{Ins} = U_{exp}$. A higher premium would leave a person with this utility function worse off than with no insurance.

Insurance is prone to two forms of abuse. One is adverse selection, where in the case above only seriously ill people look for insurance; then, the chance of becoming ill would be larger than the assumed 50%. The other problem is moral hazard when an insured person behaves more recklessly. I will discuss both points in the following chapter in more detail.

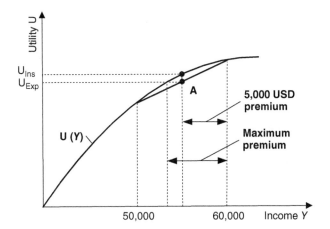

Figure 8.5 Attractiveness of insurance.

Table 8.1 Example of risk diversification.

	Both Ships are sent to same location		
	Probability	Payoff	Expected payoff
Both ships arrive	90%	£ 2000	$0.9 \cdot £\,2000 = £\,1800$
Both ships are lost	10%	0	
	Ships are sent to different locations		
Both ships arrive	$0.9 \cdot 0.9 = 81\%$	£ 2000	$(0.81 \cdot £\,2000) +$
Both ships are lost	$0.1 = 1\%$	0	$(0.18 \cdot £\,1000) = £\,1800$
One ship arrives	$(0.9 \cdot 0.1) + (0.1 \cdot 0.9) = 18\%$	£ 1000	

Diversification is possible for independent events (i.e., events that do not influence each other). Krugman and Wells (2018) provide a historical example for diversification. An English investor equips two ships in the seventeenth century and can decide to send them to the Caribbean, where pirates are lurking, or to India where typhoons are threatening. The chance of loss is the same for both destinations (10%) and the gains are also the same (£ 1000). The results are presented in Table 8.1. While the expected payoffs are exactly the same, the chance of losing all investments is in the case of diversification only 1% (instead of 10% with no diversification).

While the example is not up to date, the principle still holds true for diversification in stocks or diversification of a construction company into different markets.

8.1.3 Transaction Cost Theory

Transaction costs can be considerable. Wallis and North (1986) measured the magnitude of transaction costs as percentage of the gross domestic product (GDP) in the USA. In the hundred years from 1870 to 1970 transaction costs increased from 20% to 40%.

Thus, the use of markets is not without costs as the neoclassical model assumes. It costs money to get information, to go to the market and to transport the goods to and from the market. This applies to exchange goods. The situation becomes more difficult for contract goods such as construction goods. Table 8.2 summarizes transaction costs for construction goods.

Reasons for the transaction costs are ambiguity, uncertainty, and risk. The higher the risk aversion, the higher the transaction costs. If we define production costs in such a way that they include transportation costs, then much of the transaction costs in construction consists of management (planning, organizing, directing, controlling). A contractor's management fee typically amounts to 15%, his head office costs (mostly governance costs) to 5%, and the costs for the owner's organization 10%. This gives us a rough estimate for the transaction costs for a construction project: approximately 30% of the contract costs are transaction costs. There seems to be a trend on the owners' side to increase the layers of control, e.g. by adding a project management consultant, a quantity surveyor, a safety manager, etc. with the result of increasing transaction costs on the one hand. If everything goes well in

Table 8.2 Transaction costs for contract goods in construction.

Type	Description	Example owner	Example contractor
Search costs	Searching for transaction partners	Search for designers, contractors	Search for contract opportunities
Information costs	Getting all required information	Information about designers, contractors, materials	Information about prices, innovation
Contract costs	Setting up a contract	Design and construction contract	Subcontracts, materials, labor, equipment
Bargaining costs	Negotiating a contract	Negotiations for all types of goods	Negotiations for all types of goods
Decision costs	Concluding a contract	Agreeing to a design or construction technology	Defining a construction technology
Agency costs	Monitoring a contract	Quality control and assurance	Material control, claims
Coordination costs	Coordinating the daily transactions	Meetings, team building	Management
Measuring costs	Quantitative assessment	Progress, invoices and payments	Earned value analysis
Governance costs	Running the organization	Owner's project office	Project supervision from head office
Transfer costs	Handover of the project	Handover, commissioning	Handover, maintenance costs

the project, this money is spent unnecessarily. On the other, if the project has many problems, the additional spending might save the owner money. Unfortunately, nobody knows how a project will develop at the beginning. Evidently, the decision-making depends on the risk appetite.

The same holds true for the contractor. A contractor can add layers of internal controls and pay for it. In good times this will cost money; in bad times it will save money.

Oliver Williamson (1931–2020)

Williamson received the Nobel Memorial Prize in Economic Sciences in 2009 for his analysis of economic governance, especially the boundaries of the firm. He contributed much to the theory of transaction costs.

"The lens of contract is less a substitute for than a complement to the orthodox lens of choice. . ."

Williamson (2002, p. 4).

Source: © The Nobel Foundation, photo: U Montan

Transaction costs have the same influence as taxes in the neoclassical model of supply and demand, they increase the price and reduce the quantity (Figure 8.6). The transaction costs must be added to the production costs. Instead of the equilibrium price p^*, the owner in construction has now to pay p_T and the quantity falls from q^* to q_T. The gray area below

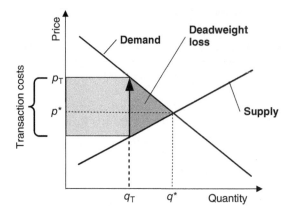

Figure 8.6 Impact of transaction costs for supply and demand.

the equilibrium price is the loss in producer surplus and the part above the loss in consumer surplus. The dark gray area defines the deadweight loss to society as a whole. Thus, lowering transaction costs does not allow only an increase in consumer and producer surplus but also social welfare.

8.2 Information

Information is the basis for communication in business, and communication is a most difficult problem to solve, especially in complex environments (Brockmann 2021). To get exactly the right information to people for decision-making is a tremendous challenge. However, this is more a management problem than an economic one. Economic questions are how we make decisions if we have incomplete information and what happens when information is asymmetric (i.e., one party has an information advantage). In the case of the neoclassical model of market interaction, both sides are assumed to have complete and clear information, thus acting in an environment of certainty. Unfortunately, this is a special case; very often, information is neither complete nor symmetric. Asymmetric information allows exercising power. Power is the possibility to enforce one's own will even against opposition in social situations. Dorman (2014) differentiates between five different types of economic power: (i) power to withhold, (ii) bargaining power, (iii) coercive power, (iv) legal power, and (v) cultural power. It is easy to imagine the use of power by withholding information in bargaining situations but also for coercion, in legal fights or cultural disputes information advantages can be used to bend the will of other people. The use of power can be constructive or destructive.

8.2.1 Satisficing Model of Decision-Making

Starting with criticism of the homo economicus, different economists developed the model of satisficing. It was Simon (1955) who introduced the idea of bounded rationality as being characteristic. The homo economicus is aware of his budget constraint, has a consistent preference order, and acts on the marginal utility equals price relation $[(\partial U/\partial x_1)/(\partial U/\partial x_2) = p_1/p_2]$

for a large number of goods. These are indeed daunting assumptions! Simon proposed instead that humans are not able to optimize their choices and instead accept solutions that are satisfactory – that is, a solution is accepted once it passes an individual aspiration level. Simon called such behavior satisficing. Instead of homo economicus, we then have to deal with a satisficing man. Bounded rationality and satisficing find their cause in unstructured problems (ambiguity), incomplete information, and transaction costs for acquiring information.

Herbert Simon (1916–2001)

Simon received the Nobel Memorial Prize in Economic Sciences in 1978 for his research into the decision-making process within economic organizations.

"Hence, a wealth of information creates a poverty of attention and a need to allocate that attention efficiently among the overabundance of information sources that might consume it. . ."

Simon (1971, p. 40).

Source: Rochester Institute of Technology/Wikimedia Commons

Satisficing man still follows the path laid out by Smith (1776) in pursuing self-interest. He does this by being opportunistic. Williamson (1985, p. 47) describes this rather uncomfortting characteristic as follows: ". . .*opportunism refers to the incomplete or distorted disclosure of information, especially to calculated efforts to mislead, distort, disguise, obfuscate, or otherwise confuse.*" This is the use of power based on asymmetric information.

Figure 8.7 demonstrates the difference in decision-making between the homo economicus and satisficing man. The homoeconomicus maximizes utility while the satisficing man optimizes utility within a set of considered solution (not all possible solutions) and doing so, he exercises bounded rationality. The precisely defined utility levels of the homo economicus are in this theory replaced by the humbler assumption of an aspiration level. On the left side of Figure 8.7, we find a decision-making process based on satisficing using the idea of a budget

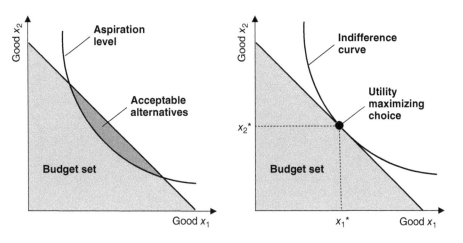

Figure 8.7 Satisficing versus utility maximizing choice.

constraint and an aspiration level. This is compared to the neoclassical model with a budget constraint and a precise preference order. It is clear that the satisficing process provides less clarity than the one in the neoclassical model. By adjusting the assumptions to be more realistic, the model becomes indeterminate. Instead of having a point for the utility maximizing choice with defined quantities (x_1^*, x_2^*) we find an area. In consequence, the demand function becomes also a space. This is the price we have to pay, and it does not allow us any longer to construct a demand curve from the model because of the indeterminate nature of choice.

Figure 8.8 provides more details of how decision-making is carried out. Satisficing will adjust an aspiration level until only one alternative remains from the set of acceptable alternatives. By doing this, the decision maker limits the effort for the search process.

Due to the singularity and complexity of most construction projects, utility maximizing by owners becomes impossible. What we can observe in construction is satisficing. The decision-making process of Figure 8.8 generally applies in construction. The information is (almost) always incomplete and often ambiguous.

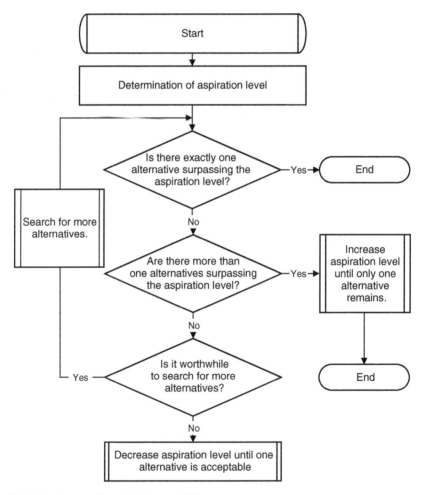

Figure 8.8 Decision-making based on satisficing.

8.2.2 Asymmetric Information

Asymmetric information can lead to market breakdowns and a number of opportunistic manifestations (hidden characteristics, hidden intentions, and hidden action); these manifestations have, in turn, certain effects (adverse selection, holdup, and moral hazard). Contract goods offer a wider range for opportunism than exchange goods. Table 8.3 gives an overview of the constellations based on the principal–agent theory. A principal undertakes a task but is not able to solve it without the help of an agent. In the construction sector, the owner is the principal in the design and construction phases and the designers and contractors are agents. The important point is that the principal and the agents do not always have the same interests and that they can use opportunism to further their goals.

After expanding the principal–agent theory and discussing an example of a market breakdown due to asymmetric information, I will explain the effects of opportunism (Table 8.3) before describing the property-rights theory.

8.2.2.1 Principal–Agent Theory

If someone (the principal) needs someone else (an agent) to reach desired goals, then we have a principal–agent relationship. The principal depends on the agent to some degree. Problems in the principal–agent relationship arise because of asymmetric information, risks, and incentives. Variants of the agency problem are adverse selection, holdup, and moral hazard.

An example is a construction owner who asks a contractor to build an office. The contractor has more information during the construction process and can use this asymmetric information to his advantage. The owner has the advantage of an information asymmetry until contract signature. The owner becomes the agent during this time; the contractor depends on the owner to sign a contract. It becomes clear that the roles in a principal–agent relationship can change with time. This is especially true since in a construction project we have to deal with team production. In a team, the result depends on the input of all team members.

To become a bit more explicit, Table 8.4 contains two examples of a low effort in principal–agent relationships of construction.

8.2.2.2 Market Breakdown Due to Asymmetric Information

Akerlof (1970) introduced a model for a market breakdown due to asymmetric information, which became known as market for lemons. He received the Nobel Memorial Prize in Economic Sciences in 2001 for his work on asymmetric information. For the illustration of his ideas, he uses the market for used cars. Asymmetric information leads to adverse

Table 8.3 Opportunism in principal–agent relationships due to asymmetric information.

Form of opportunism	Effect	Contract goods	Exchange goods
Hidden characteristics	Adverse selection	Possible	Possible
Hidden intentions	Holdup	Possible	Not possible
Hidden action	Moral hazard	Possible	Not possible

Table 8.4 Example of moral hazard and hidden action.

	Example 1: Tendering	**Example 2: Construction**
Information asymmetry	Owner knows his budget and the prices of all bidders. Contractor only knows his own estimate.	Contractor knows his efforts toward process quality. Owner cannot observe process quality continuously.
Action	Owner withholds price information.	Contractor withholds process quality information.
Hidden action	Owner distorts information during contract negotiations	Contractor hides information on process quality

(wrong) selection and then to market breakdown. In a used car market, the seller almost certainly knows more about the quality of a car than a potential buyer, creating an asymmetry of information. Then Akerlof describes the situation (1970, p. 488): "*There are many markets in which buyers use some market statistic to judge the quality of prospective purchases.*" In car markets, this is the average price for a model, the year of production, the mileage, and the general condition. There will be some cars of better and some of poorer quality in such a group. The owners of the better ones demand a higher price, which the buyers compare to the mean price as an aspiration level based on statistics. Consequently, the buyers are not willing to pay the higher asking price and the sellers take the better-quality cars from the market. What remains are lemons, for which the sellers can ask a higher price than justified based on the poor quality close to the mean value leading to adverse selection. The market will break down in the end.

George Akerlof (1940)

Akerlof received the Nobel Memorial Prize in Economic Sciences in 2001 for his analysis of markets with asymmetric information.

"*Prior to the early 1960s, economic theorists rarely constructed models customized to capture unique institutions or specific market characteristics.*"

Akerlof (2001, p. 413).

Source: MONICA M. DAVEY/AFP/Getty Images

The most prevailing procurement strategy in construction is rewarding the low bidder in a sealed-bid auction with a contract. So, the question arises, whether owners are constantly buying lemons. I will discuss this problem later in Chapter 14 (construction markets). The answer is not so easy, partly because in used car markets we trade exchange goods and in the construction sector contract goods.

8.2.2.3 Hidden Characteristics and Adverse Selection

The agent does not provide information about hidden characteristics in a transaction before signing the contract when acting opportunistically. After signature, the hidden characteristics cannot be changed.

In the market for lemons, the action of the seller is fixed (i.e., the car has defects or not) and the buyer can often find out about the defects only after signing the contract. When signing a contract, the owner of a construction project does not buy the building but the capability of the contractor to erect the building according to the contract. The incentive for contractors to admit a lack of experience in one-time transactions is zero; thus, hidden characteristics (lack of experience) are possible. Typically, owners look at reference projects, but these are deceptive because performance rests both on the ability and willingness to perform. The people in charge of the reference project might have left the company or might not be willing to provide the same effort again.

Sometimes, as in the example of the market for lemons, low-quality offers displace high-quality offers because of the cost of acquiring information or the impossibility to obtain it. This leads to adverse selection. Adverse selection is an important problem for insurance companies. If an insurance company offers a group an average premium for health insurance and only the people most likely to fall ill buy the insurance, then the insurer will most likely go broke quickly.

Figure 8.9 explains adverse selection in construction. Let the contractor as agent decide for a low-price strategy based on low efforts to build quality. The owner as principal decides to accept the lowest offer and rewards this contractor with a contract. This is a case of adverse selection that the owner can only observe after signature during construction. However, it is not necessarily so that lowest prices lead to adverse selection. An innovative contractor might offer a low price and be the best option for the owner.

8.2.2.4 Hidden Intentions and Holdup

The agent might have hidden intentions and the principal has no chance to observe them. Incomplete contracts provide an opening for hidden intentions. In the construction sector, we typically deal with incomplete contracts because the transaction costs in formulating complete contracts are prohibitively high. Construction contracts therefore often have clauses to deal with the unforeseen and important examples are change orders and claims. A contractor might find some loopholes in the contract without informing the owner beforehand. The contractor develops the hidden intention to make use of such loopholes at a later stage when the owner depends on him. This is the case when negotiating prices for

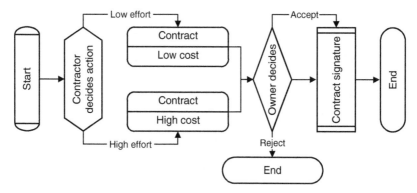

Figure 8.9 Hidden characteristics and adverse selection.

change orders or claims. It is difficult for the owner to find another contractor who can offer a lower price because the initial contractor has already set up the site installation and can use it without extra costs. By paying for the site installation, the owner has made a specific, binding investment. This is the holdup: Dear owner, you pay a higher price to me or you have additional costs by switching to another contractor. A strategy against this type of holdup is binding the contractor to the original estimate and demanding open access to this original estimate.

 Tunneling = (obs.) The owner wanted the four required tunnel-boring machines (TBM) to be fabricated by a single supplier. The parties agreed with the TBM producer on a production time with a maximum penalty of 5%. The supplier knew from the beginning that it would be impossible to finish the four TBM on time and built the expected penalty fee into the bid calculations (hidden intentions). The supplier then started fabricating the machines with his normal workforce and no overtime. Once the owner understood the delay, it was too late to switch to another supplier. The initial investment in money and time could not be regained. The supplier delivered the machines more than one year late. All the owner could do was to deduct 5% from the contract price.

8.2.2.5 Hidden Action and Moral Hazard

Hidden action sooner or later become visible to the principal. Moral hazard is a lack of incentive to take care of the interests of the principal. A simple example is an insurance with the insurer as principal and the insured as agent. Homeowners with property insurance might build too close to the water or the forest; an owner of a car with and insurance against car theft might not lock the car; a house owner with fire insurance might be negligent with a firepit, etc. Hidden characteristics and adverse selection originate before signing a contract; hidden action and moral hazard arise after signature.

Figure 8.10 shows the process of moral hazard in combination with hidden action. The principal offers the agent a contract. The agent accepts the contract. Next, the agent decides on an effort and the result corresponds to the effort.

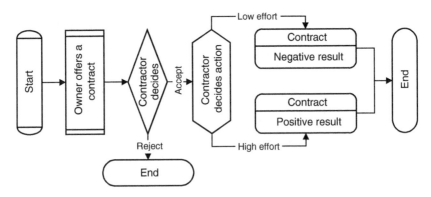

Figure 8.10 Hidden action and moral hazard.

A typical situation in construction for hidden action and moral hazard are one-off contracts. Here, both the contractor and the owner can take recourse to a hid-and-run strategy. A construction project requires inputs not only from the contractor but also from the owner (land, financing, permits, design, information, etc.) This is called a team production, and the roles of principal and agent switch. Sometimes the owner is the agent and the contractor the principal. This allows both sides to milk the other.

8.2.3 Property Rights Theory

Property rights are instrumental for the functioning of any economy. In one extreme, all property is private; in the other, all is public. Public means of production define communism; private ownership characterizes capitalism. The assignment of specific property rights in real economies is in both cases less rigid. In the 1980s, many Western countries saw a wave of deregulation and privatization of public assets (railways, telephone services. . .). Later roads and bridges have been privatized through public–private partnerships (PPPs). More dramatically, communist countries have embraced private property to a much larger degree than before (China, Russia, India, Vietnam).

The beginning of construction is marked by a contract specifying the transfer of property rights at the end of the construction period by means of a formal handover. Once the owner signs the handover, the property rights of the structure shift from the contractor to the owner. Complex contract goods such as construction projects cannot be defined unambiguously by a contract. This is the theory of incomplete contracts. There are numerous reasons: complexity, unknown soil conditions, unknown influences during construction. While the contractor keeps the property rights of the structure during construction, the owner always has the property rights to the land. Structure and land are interconnected and in most countries the land takes legal precedence in case of problems. If the general contractor for example goes bankrupt, then the owner takes possession of unpaid work and materials provided to the general contractor by a subcontractor. This subcontractor cannot remove unpaid doors from the land.

Another often-cited problem of property rights is that of externalities: the use of resources that are not part of the transaction. Many times, there are no property rights attached to the use of the air. A contractor is free to use it for emissions, to the detriment of people and the environment. Economists propose to assign property rights in order to reduce emissions. The Coase theorem states that welfare will improve whether the property rights are assigned to the contractor or the public.

References

Akerlof, G. (1970). The market for "lemons": quality uncertainty and the market mechanism. *Quarterly Journal of Economics* 84 (3): 488–500.

Akerlof, G. (2001). Behavioral macroeconomics and macroeconomic behavior. *The American Economic Review* 90 (2): 411–433.

Brockmann, C. (2021). *Advanced Construction Project Management: The Complexity of Megaprojects*. Hoboken: Wiley Blackwell.

Dorman, P. (2014). *Microeconomics: A Fresh Start*. Heidelberg: Springer.

Kahneman, D. (2011). *Thinking, Fast and Slow*. London: Penguin Books.

Kahneman, D. and Tversky, A. (1979). Prospect theory: an analysis of decision under risk. *Econometrica* 47 (2): 263–291.

Kendrick, T. (2009). *Identifying and Managing Project Risk*. New York: Amacom.

Krugman, P. and Wells, R. (2018). *Microeconomics*. New York: Worth.

Simon, H. (1955). A behavioral model of rational choice. *Quarterly Journal of Economics* 69 (1): 99–118.

Simon, H. (1971). Designing organizations for an information-rich world. In: *Computers, Communication, and the Public Interest* (ed. M. Greenberger), 37–72. Baltimore: Johns Hopkins Press.

Smith, A. (1776). *An Inquiry into the Nature and Causes of the Wealth of Nations*. Ware: Wordsworth Editions (2012).

Wallis, J. and North, D. (1986). Measuring the transaction sector in the American economy, 1870–1970. In: *Long-Term Factors in American Economic Growth* (ed. S. Engerman and R. Gallman), 95–162. Chicago: University of Chicago Press.

Williamson, O. (1985). *The Economic Institutions of Capitalism: Firms, Markets, Relational Contracting*. New York: Free Press.

Williamson, O. (2002). The lens of contract: private ordering. *American Economic Review* 92 (2): 438–443.

9

Game Theory and Auctions

Case Study   Bidding

Tonight, a classmate from high school is over for a visit. Ever since you were together in the chess club, you have played against each other. The last few years this was a fixed event, every second Tuesday of the month at 8 p.m. Today you are the host, the weather is mild, and you sit on the porch waiting for your classmate with a caipirinha in your hand. The crushed ice is softly clanking in your glass as you admire the subtleties of chess and the mastery of your classmate. You think about an opening strategy. As host, you have the right to make the first move. Should you start with the Queen's Indian Defense or the more popular Ruy Lopez Opening? There are so many strategies open to you and you still try to lay down your strategy when the bell rings and player two enters the game.

After a warm welcome and another caipirinha, you sit down and start the first game immediately. Not much talking but a lot of joyful concentration accompanies the consecutive moves. The game develops dynamically, and you both enjoy the full information about the past moves on the board. You appreciate the thinking and planning of your classmate. With the given information, you try to outguess each other. Your classmate wins more often than you and sometimes when you win, you still admire her game. Too bad that there must be a winner!

After three games, you push the chess board aside and start talking about the events of the past month. Your classmate tells you that she went to an auction for the first time in her life and actually won bidding for an antique sideboard. She was careful, just bidding a little more than the last interested person, and won at a price of 1480 USD, only 10 USD more than the second-high bidder. You tell her that last week you won a contract, but you were 120,000 USD lower than the second-low bidder. While you compare the two cases, you wish

(continued)

> *you would have had the complete information that your classmate had in her auction. She always knew the price of the last bidder as the bidding enfolded consecutively. You, on the other hand, had to send in your bid in a sealed envelope at submission date without knowing the prices of your competitors. You remember a story about the bidding for the Hoover Dam. The owner received three bids in 1931 and one bid just read "one dollar less the low bidder". Briefly, you discuss whether the story is true but then you tell your classmate that this would not be acceptable today. If it were acceptable, that strategy would have saved you 119,999 USD!*
>
> *Before you go to bed, you hope that you still will make the anticipated profit on that project. The price difference keeps you unsettled but you are also sure to have made your estimate with due care. It would at least be nice to be able to estimate your costs without doubt!*

I made already a brief reference to game theory in Section 6.4 on oligopolies. Game theory is a prominent tool to analyze strategic decision-making, i.e., decisions where the reaction of another party is of importance, just like in a chess game. Strategic games have a set of known players, strategies, and outcomes. Players in perfectly competitive markets (buyers and sellers) have no influence on market outcome. They are price takers and quantity-adjusters; for this reason, game theory is not used in their analysis. This is different in oligopolies where – as we have seen – actions of one player influence the outcome for the other(s).

Players in auctions develop their own strategies to win. Thus, game theory is an important help in understanding auctions. Auctions, on the other hand, are widely used in construction procurement. Tendering with a submission date and the award to the low bidder is pervasive in the sector and a special type of auction (i.e. a sealed-bid, first-price auction).

9.1   Game Theory

Many economists (and sociologists) introduce game theory with an example, the prisoners' dilemma: There are two burglars in this game, Bonnie and Clyde. The police are catching them on the spot. This is enough evidence for a 2-year sentence but the police know they have committed far more serious crimes, which would carry a 10-year sentence if there were proof. First, the police separate the two in different cells to prevent communication. Then they explain that if only one confesses to all jointly committed crimes by ratting the other one out, that person would earn a reduced sentence of 1 year in prison as a key witness, while the other would get the maximum sentence (10 years). However, if both confess, the sentence would be 8 years. Figure 9.1 illustrates the options to Bonnie and Clyde. Of course, if both would not confess, the sentence could only be for the last burglary (2 years).

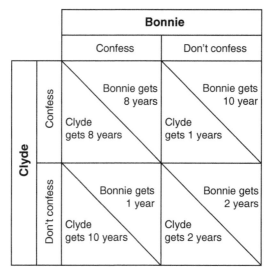

Figure 9.1 Prisoners' dilemma.

It is evident that Bonnie and Clyde can reach the best mutual outcome by not confessing; this requires trust. However, there is an incentive for each to confess by the possibility to reduce the sentence to 1 year. If Bonnie assumes that Clyde will confess, then her best option is also to confess (8 years instead of 10); if she assumes that he will not confess, then her best option remains confessing (1 instead of 2 years). The same holds true for Clyde: Confessing is for both the dominant strategy.

9.1.1 Basics of Game Theory

Many times, we can make decisions without bothering about the action of others but also very often, we have to take someone else's decisions into account. If the parties act rationally in such a situation, then we are looking at strategic decision-making. Bonnie and Clyde were looking at a strategic decision, and game theory is the study of strategic decision-making.

A game is thus any situation where individuals must choose strategically. While all games have players, strategies and outcomes do not all have an equilibrium outcome. Games differ vastly in their complexity. The prisoners' dilemma is a simple game with an equilibrium outcome (both confess).

Decision-makers are called players. These can be individuals, firms, or other entities. All players are assumed to act rationally, i.e. they are optimizing their utility or maximizing their profits. Bonnie and Clyde are the players in the prisoners' dilemma – not the police who are just setting the outcomes of the game.

Strategies are the possible actions that players can choose; they are well-defined. Sometimes one strategy is dominant: Whatever the other player is doing, such a strategy provides the best outcome for the other. Confessing is a dominant strategy for Bonnie and Clyde.

The possible outcomes of games are called payoffs. Players have a preference order for the available payoffs. The payoff for the dominant strategy in the prisoners' dilemma is an 8-year prison sentence for both players.

Jehle and Reny (2011, p. 305) provide a formal definition with x denoting the Cartesian product of all strategies. As game theory uses mathematics to describe the problems and to develop solutions to games, formal definitions become indispensable.

 Strategic games = (def.) A strategic form game is a tuple $G = \left(S_i, u_i \right)_{i=1}^{N}$ where for each player $i = 1, \ldots, N$, S_i is the set of strategies available to player i, and $u_i\colon x_{j=1}^{N} S_j \to R$ describes the player i's payoff as a function of the set of strategies chosen by all players. A strategic form game is finite if each player's strategy set contains finitely many elements.

Table 9.1 shows the payoff matrices for two players with four strategies for each player and $2 \cdot 16 = 32$ payoffs. This defines a finite strategic form game with the following parameters:

$$i = 2;\ S_i = S_A = S_B = 4;\ u_i = f\left(S_A, S_B \right) = \left(u_{A11}, \ldots, u_{A44}, u_{B11}, \ldots, u_{B44} \right)$$

There are a number of possible ways to classify the different types of games. A first distinction is between cooperative and noncooperative games. In noncooperative games, each action is an individual action, and in cooperative games, the action is jointly negotiated. The submission of tenders by different bidders is an example of a noncooperative game in construction while collusion is a cooperative game.

Next, there are zero-sum and non-zero-sum games. The players in zero-sum games are direct competitors. Whatever player A wins, player B must lose. In non-zero-sum games, a variety of solutions is possible and the combined payoffs for the players are different. Negotiating a contract price between owner and contractor after submission and before signature is a zero-sum game if everything else remains unchanged. It can become a non-zero-sum game if some parameters in the contract are allowed to change in order to find win–win

Table 9.1 Payoff matrices for two players.

		Payoff matrix player A						Payoff matrix player B			
		Strategies player B						Strategies player B			
		S_{B1}	S_{B2}	S_{B3}	S_{B4}			S_{B1}	S_{B2}	S_{B3}	S_{B4}
Strategies player A	S_{A1}	u_{A11}	u_{A12}	u_{A13}	u_{A14}	Strategies player A	S_{A1}	u_{B11}	u_{B12}	u_{B13}	u_{B14}
	S_{A2}	u_{A21}	u_{A22}	u_{A23}	u_{A24}		S_{A2}	u_{B21}	u_{B22}	u_{B23}	u_{B24}
	S_{A3}	u_{A31}	u_{A32}	u_{A33}	u_{A34}		S_{A3}	u_{B31}	u_{B32}	u_{B33}	u_{B34}
	S_{A4}	u_{A41}	u_{A42}	u_{A43}	u_{A44}		S_{A4}	u_{B41}	u_{B42}	u_{B43}	u_{B44}

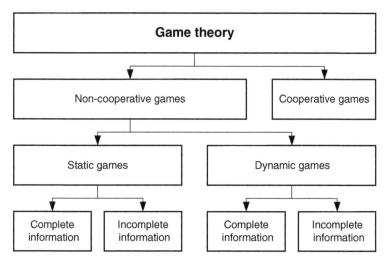

Figure 9.2 Classification of games in game theory.

solutions. In economic terms, the first contract draft is in such a case not Pareto efficient because it is possible to make at least one player better off without make the other worse off.

We can also differentiate between static and dynamic games. In static games, all players choose their strategies simultaneously without knowing the choices of the other players. In dynamic games, player A can observe the action of player B and can then adjust his own choice accordingly.

Finally, we have games with complete and incomplete information. Figure 9.2 provides a classification of games according to different characteristics (without zero-sum and non-zero-sum games). Chess is a noncooperative, dynamic game with complete information because you play against each other, make your draws consecutively and know all the past draws of the other player.

I will provide a glimpse at the basics of games with complete information, these are sufficient for the treatment of applied construction economics. The problems of incomplete information, however, are beyond the scope of this introduction. Interesting monographies by Gibbons (1992), Binmore (1994, 1998) and Tadelis (2013) offer deeper insights.

9.1.2 Static Games with Complete Information

In static games, players choose their actions simultaneously and if the payoffs are common knowledge, then the players have complete information. The prisoners' dilemma is a static game with complete information. Static games are also called simultaneous-move games.

Real-life situations are in game theory transferred into a normal-form game. The payoff matrix of the prisoners' dilemma represents such a normal form. We have seen that there is one dominant strategy for each player in this game. Not all games have strictly dominated strategies but if they have, all rational players will use them.

The following formal definition of a strictly dominated strategy is from Gibbons (1992, p. 5):

 Strictly dominated strategy = (def.) In the normal-form game $G = \{S_1, \ldots, S_n; u_1, \ldots, u_n\}$, let s_i' and s_i'' be feasible strategies for player i (i.e. s_i' and s_i'' are members of S_i). Strategy s_i' is strictly dominated by strategy s_i'', if for each combination of the other player's strategies i's payoff from playing s_i' is strictly less than i's payoff from playing s_i''...

I introduced the idea of a Nash equilibrium in Section 6.4 and now provide a formal definition. You should remember that not all games have an equilibrium solution (Gibbons 1992, p. 8):

 Nash equilibrium = (def.) In the n-player normal-form game $G = \{S_1, \ldots, S_n; u_1, \ldots, u_n\}$, the strategies $(s_1^*, \ldots, s_n^*)$ are a Nash equilibrium if, for each player, s_i^* is (at least tied for) player i's best response to the strategies specified for the $n - 1$ other players.

A Nash equilibrium provides for an unequivocal and stable solution to a game. An example is Cournot's model of a duopoly.

9.1.3 Dynamic Games with Complete Information

Dynamic games differ from static games in the characteristic that the moves are not simultaneous but consecutive. In the business case, you were playing a static game when submitting your tender, as all players had to submit their bids at the same time. Your classmate had to deal with a dynamic game when winning the auction for the antique sideboard. Her winning bid was the last of a sequence.

The basic example of a dynamic game is (Gibbons 1992, p. 57):

- Player 1 chooses an action a_1 from the feasible set A_1.
- Player 2 observes a_1 and then chooses an action a_2 from the feasible set A_2.
- Payoffs are $u_1(a_1, a_2)$ and $u_2(a_1, a_2)$.

The grenade game provides an example. Player 1 has two possible actions, to give player 2 1000 USD or nothing. Player 2 observes player 1's action and then has the possibility to detonate a grenade that will kill both or not. To kill both is a noncredible action (player 2 would be better off alive and without 1000 USD). The solution follows from backwards induction. Player 1 knows that player 2 has a noncredible threat and will decide to offer nothing to player 2.

In backwards induction, player 2 as rational decision-maker must maximize his utility:

$$\max_{a_2 \in A_2} u_2\left(a_1, a_2\right) \tag{9.1}$$

If the maximization problem has a unique solution, the reaction $R_2(a_1)$, then player 1 can anticipate the reaction and must solve:

$$\max_{a_1 \in A_1} u_1\left(a_1, R_2\left(a_1\right)\right) \tag{9.2}$$

Stackelberg's model of a duopoly belongs to this type of games.

In repeated dynamic games with complete information we often encounter a tit-for-tat strategy: If player 1 decides to cooperate in round 1, then player 2 will also cooperate in round 2. However, if player 1 defects in round 1, then player 2 will do the same in round 2.

9.2 Auctions

We trade in markets and how we organize the trading is up to norms and rules. The easiest way is the trading between two individuals one offering a good and the other a price followed by some haggling about the price and finally the conclusion of the transaction. This type of procedure is called a double-auction market. It becomes cumbersome in larger markets and over longer distances, and under such circumstances posted offer markets develop. Here, a seller offers a good at a fixed price (Dorman 2014). As we see, auctions have a long tradition although they have been replaced in most cases by anonymous exchange markets. This is most obvious in internet markets but also when shopping in a supermarket. Buyer and seller never meet face-to-face.

Auctions belong to the group of matching markets where sellers of special goods need a matching partner as buyer with similar special interest. Matching markets need a market design (Roth 2015). Therefore, it cannot come as a surprise that we encounter different types of auctions. The market design can follow different goals among which are efficient allocation, maximizing sellers' income, minimizing transaction costs, guarding against corruption, or mitigating effects of collusion (Milgrom 1989).

9.2.1 Basics of Auctions

The first distinction regarding auctions is that between private-value auctions and common-value auctions. Bidders in private-value auctions appreciate the good in question according to their taste differently. The worth of the good is different to each individual, and each one is willing to offer a different maximum price; you might think about an auction with an original from Rembrandt offered. All bidders in common-value auctions essentially assign the same price to the good. Drilling rights in an oil field provide an example. The amount of oil does not depend on the preferences of the bidders. There is a certain amount available, and the market price of oil is independent of the production from this specific oil field. Sometimes bidders know their private values, sometimes they have doubts about them. The same holds true for common values. Before exploitation of the oil field, all bidders can only rely on estimates of the yield. Figure 9.3 provides a classification of auctions.

Besides the discussed private and common value auctions with or without knowledge of the values, there are also common-value auctions with known or unknown private values.

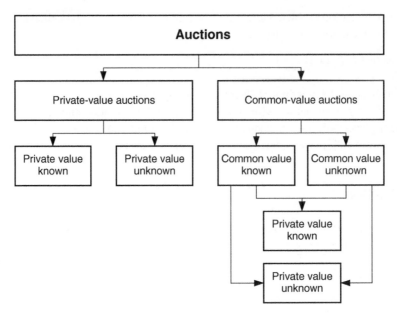

Figure 9.3 Classification of auctions regarding value.

Bidding for construction project most often takes place as common value auction with unknown private values. This assumes that the construction costs of the project are very similar to all possible contractors but that the individual estimates are imprecise. The differences in the estimates are far larger than the differences in costs. Under such circumstances, the winning bidder might suffer from the winner's curse: The construction costs for the contractor exceed the contract price and the contractor makes a loss. Such a contractor has underestimated his private value.

In ascending price auctions, the auctioneer and all bidders have reservation prices, i.e. private values. The maximum theoretical yield of an auction is the difference between the auctioneer's reservation price and the one from the bidder with the highest private value (Figure 9.4).

Depending on the bidding rules (the market design), we can distinguish four major types of auctions:

- English auction. The auctioneer starts with a low reservation price and the bidders successively raise their offers. Typically, there is a minimum bid increment. The highest bidder will win the auction.
- Dutch auction. The auctioneer starts with a high price and gradually lowers it. The first bidder to accept a price offered by the auctioneer will win the auction. It is the highest bid.
- Sealed-bid auctions. All bidders submit a price to the seller, and the highest offer will win the auction. This type is common in construction, but the award will be to the lowest offer. It is also called a first-price, sealed-bid auction.

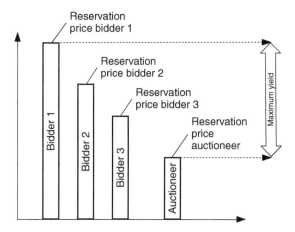

Figure 9.4 Maximum yield in an auction.

- Vickrey (or second-price, sealed-bid) auctions. All bidders submit a price to the seller and the highest offer will win the auction, just like in the sealed-bid auction. However, the high bidder must pay second-highest price. In construction, the low bidder will win the auction but at the second lowest price.

Each bidder must consider the actions of the other bidders regardless of the type of auction: This is the definition of strategic decision-making, and thus game theory is helpful in analyzing auctions. While contractors pay attention to the actions of competitors, they cannot observe them during the ongoing auction (sealed-bid).

9.2.2 English and Vickrey Auctions

Bidders in English auctions keep bidding until the asking price reaches their private value v_i. If they win the auction at this price level, their payoff is zero; thus, it makes sense to bid up to this level. Doing so is the dominant strategy for all bidders. If all bidders increase their bids in small steps, the bidder winning the auction will pay approximately the price of the second-highest value. At this point the second highest bidder drops out and it does not make sense for the remaining bidder to increase his offer. The payoff to the winner is the difference between the second-highest bid and his own private value. In an English auction, the actions of other bidders are visible (Figure 9.5).

Vickrey auctions have the same result as English auctions. The high bidder wins and pays the price of the second-highest value. This follows from the auction design. The winner in a Vickrey auction is a price-taker; the price depends only on the actions of the other bidders. For a bidder in a Vickrey auction, the offers by other bidders can be seen as offers to the focal bidder, this is called the industry perspective; the focal bidder will accept any offer up to his reservation price. Therefore, it is a dominant strategy for each bidder to put their reservation price into the sealed-bid envelope. Since it is a dominant strategy, the actions of the other bidders have no impact.

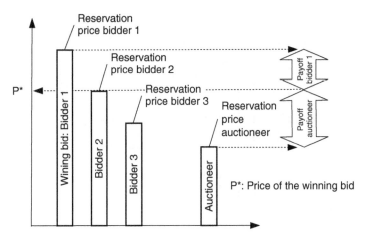

Figure 9.5 Results of an English auction.

9.2.3 Dutch Auctions and Sealed-Bid Auctions

Bidders in Dutch auctions must make a single decision – at what price to raise their hand. The first bidder to do this will be the high bidder and win the auction. The same holds true for sealed-bid auctions. There is no way to observe the actions of other bidders. In a Dutch auction, only the winning bidder acts, and in sealed-bid auctions, the action is sealed. There is no optimal bidding strategy in both types of auctions.

Dutch and sealed-bid auctions can be inefficient. Milgrom (1989) provides an example. Bidder 1 has a reservation price of 101 USD. Bidder 2 has a reservation price of 50 USD 80% of the time and of 75 USD 20% of the time. Bidder 1 does not know the reservation price of bidder 2 but he does know the distribution. If bidder 1 bids 51 USD he will win 80% of the time 50 USD (101 – 51 USD), thus the expected profit is $0.8 \cdot 50 = 40$ USD. If he bids 62 USD or more, the expected profit will be less than the 40 USD (101 – 62 = 39 USD), to do this is not rational. In the end, bidder 2 will win sometimes with the reservation price of 75 USD. The difference between the two reservation prices of 26 USD (101 – 75 USD) provides a chance to make someone better off without making someone worse off. Thus, the allocation is not Pareto efficient.

The revenue equivalence theorem postulates that the Dutch, the English, and sealed-bid auctions all yield identical revenues to sellers and expected profits to bidders.

9.2.4 Competitive Bidding

Competitive bidding is the typical procurement form in construction. The seller (owner) offers a good with a fixed quantity (design) in a sealed-bid auction to a few buyers (bidders, contractors). The winning bid is the one with the lowest price. The contractor submits a bid which represents the contractor's private value. If the costs to finish the project were c, then the contractor's profits would be $\pi = (b - c)$. Figure 9.6 illustrates the results of a sealed-bid auction in competitive bidding. The owner has a payoff in this case and contractor 1 must hope that the construction costs c are lower than the bid price b once the project is finished.

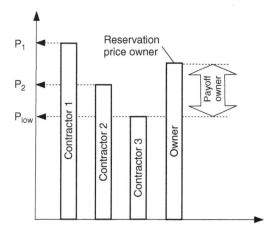

Figure 9.6 Results of a sealed-bid auction in competitive bidding.

Contract goods (such as construction goods) are produced after signing a contract, they are most often single units of considerable complexity (Brockmann 2011). This implies three major problems for estimating such projects (i.e. determining the reservation price): (i) There is no repetitive production of the same good and thus no experience about the costs. When someone produces a million pencils, it is of little importance whether the initial price is correct, it can be adjusted within short time. In single-unit production the initial price cannot be changed because the contract is signed and binding before production starts; (ii) the inherent complexity of many construction projects makes it hard to consider and judge all relevant facts; (iii) there is no complete control over the production conditions; productivity is influenced by the environment as well as by the process evidence of the owner. As construction is a highly integrative process with the owner being an important external factor of production, he must know what is required of him (process evidence). As this is a case of team production, productivity also depends on the relationship between owner and contractor.

Milgrom (1989) discusses two premises in conjunction with pricing of complex contract goods: the private and the common values assumption. The private values assumption under these circumstances states that contractors can determine their cost correctly (labor, materials, equipment, subcontractors, indirect cost) and Milgrom believes that this assumption does not hold. He assumes estimating errors by all bidders (ε_i) with a normal distribution about the mean (i.e. no bias). All detailed analyses of single estimates and the bid-spread of submissions support the statement with the exception of bias; there are relatively often biased submissions. The estimating approach takes errors into consideration and deals with the problem by detailing the project into an extensive work breakdown schedule. Judgment mistakes occur for most items; however, they should not be systematic. If there is no bias, then errors will cancel each other out over a large number of items and there is a tendency towards a mean value. In an example of a post-construction analysis of a construction project, the differences in single items reached almost 300% (planned vs. actual) while the overall difference was only 3%. The contractor was lucky; he had overestimated the total cost by this amount.

The second assumption is accepted by Milgrom for complex construction goods: All companies face approximately the same cost (C), the common values assumption holds. In different segments of the market companies of equal size tend to compete against each other, therefore the purchasing power of the companies is similar. Short-term advantages of one competitor (e.g. use of cheap foreign labor) must be imitated by the others due to the competitiveness of the market. Since all contractors use the same subcontractors and buy from the same suppliers yielding the same amount of market power, prices can vary only due to a superior technology (same output with less input). Again, companies falling behind in the technology race must catch up or face bankruptcy.

With these considerations Milgrom can formulate the components of a bid X_i from bidder i:

$$X_i = C + \varepsilon_i. \tag{9.3}$$

While the estimating error (ε_i) is unbiased overall, this does not hold true for the successful bid. The lowest bid lies below the mean value and therefore below the market price p^*. What we see in Figure 9.7 is exactly the result of a monopsony: Too little demand and too low a price in comparison with the result from a perfectly competitive market. This is no surprise – the sealed-bid auction creates a market for a single good, and in this market, the buyer acts like a monopsonist. The bids are not independent because the common value is shared by all bidders.

With Milgrom's formula $X_i = C + \varepsilon_i$ and the underlying assumptions, we can calculate the expectancy values of winning a bid depending on the number of bidders. The expectancy value of a bid E(b) for a number of contractors (i) depends on this number i and is in all cases except for $i = 1$ below the price level of the equilibrium price p^* in competitive markets (Leitzinger 1988). Winners are faced with a price below equilibrium in competitive markets (Table 9.2).

A negative expectancy value in competitive bidding means that the owner always realizes a positive payoff if he chooses to organize a sealed-bid auction with more than

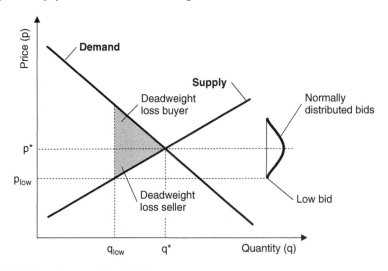

Figure 9.7 Pricing in competitive bidding.

Table 9.2 Expectancy values for bids in sealed-bid auctions.

Number of bidders	1	2	3	4	5	6	7	8	9	10
Exp. value E(b)	±0.00	−0.56	−0.85	−1.03	−1.16	−1.27	−1.35	−1.42	−1.48	−1.54

one bidder. The opposite holds true for contractors – they will have to forgo some of the expected profits due to the market design.

References

Binmore, K. (1994). *Game Theory and the Social Contract I: Playing Fair*. Cambridge: MIT Press.

Binmore, K. (1998). *Game Theory and the Social Contract II: Just Playing*. Cambridge: MIT Press.

Brockmann, C. (2011). Collusion and corruption in the construction sector. In: *Modern Construction Economics* (ed. G. de Valence), 29–62. London: Taylor & Francis.

Dorman, P. (2014). *Microeconomics: A Fresh Start*. Heidelberg: Springer.

Gibbons, R. (1992). *Game Theory for Applied Economics*. Princeton: Princeton University Press.

Jehle, G. and Reny, P. (2011). *Advanced Microeconomic Theory*. Harlow: Pearson.

Leitzinger, H. (1988). *Submission und Preisbildung: Mechanik und ökonomische Effekte der Preisbildung bei Bieterverfahren*. Köln: Carl Heymanns.

Milgrom, P. (1989). Auctions and bidding: a primer. *Journal of Economic Perspectives* 3 (3): 3–22.

Roth, A. (2015). *Who Gets What – and Why*. Boston: Houghton Mifflin Harcourt.

Tadelis, S. (2013). *Game Theory: An Introduction*. Princeton: Princeton University Press.

Part II

Applied Construction Microeconomics

10

Construction Sector

Before we can talk about the construction sector, we must define it. To this purpose, we focus on nominal definitions which cannot be true or false but rather, useful or not. This becomes highly important when comparing data between countries, as they sometimes use different definitions for the same term. Comparison makes sense only if the data collected refer to the same statistical population. Figure 10.1 shows a general classification for economic activities, from global level on down to individual firms.

The global economy encompasses economic activities within and between countries. A national economy limits these activities to one country, often measured by gross domestic product (GDP) or gross national product (GNP). GDP measures the value of goods and services produced within a country's borders; the concept is based on national borders. GNP measures the value of all goods and services produced by a country's citizens; the concept is based on citizenship. Today, economists typically use GDP, and I will follow this convention.

National economies are structured into sectors for statistical purposes. The UN uses the International Standard Industrial Classification (ISIC) of all economic activities, and the EU has an adjusted sectorial classification system with the Statistical Classification of Economic Activities in the European Community (NACE, from the French Nomenclature générale des Activités économiques dans les Communautés Européennes).

However, the breakdown of sectors into industries is not clearly defined. All firms producing similar products form an industry. This issue is, the definition of *similar* is open to interpretation. While an industry comprises only competitors, a sector will also include nonrivals. Firms, finally, are legal entities.

It is important to understand that the construction sector is a statistical construct. Many available government data are based on this construct. To show the importance of this observation, one example might suffice: Repair and maintenance of structures is part of the construction sector; repair and maintenance of cars is not part of the automobile sector. In many developed economies, investments for repair and maintenance approach 50% of construction spending. It should be clear that productivity of new build structures is much higher than for repair and maintenance. Thus, by attributing repair and maintenance to the construction sector but not to the automobile sector, productivity in construction is lower by definition. It is also lower for other reasons, which we will discuss later. Chapter 10.2 will

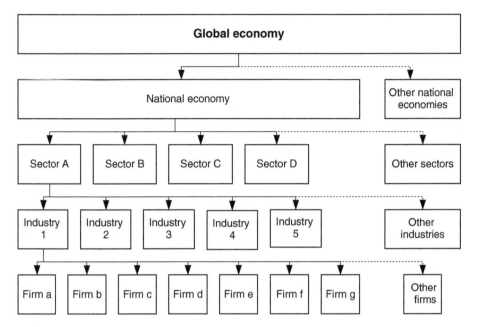

Figure 10.1 Classification of economic activities.

look at two other concepts related to the construction sector and highlight the consequences of defining the construction sector for statistical purposes.

10.1 Definition

NACE (Eurostat 2008) is, like ISIC, a statistical classification system that aims to clarify the attribution of an economic activity to a sector. To this purpose, it is organized into sections, divisions, groups, and classes. If an activity could be attributed to more than one category, then it will be assigned to the category that provides the highest percentage of value added (Figure 10.2).

A rough classification divides the economy into three sectors (primary, secondary, and tertiary). Table 10.1 shows how the sections of NACE correspond to these three sectors.

If you look at a construction project, you will find the materials in section A (wood) and B, equipment and other prefabricated elements in section C, energy in section D, water and waste in section E, real estate in section L, professional services of architects and engineers in section M, as well as many other services in other sections. This highlights that the NACE definition of construction is a rather narrow one.

Figure 10.2 is based on the fact that NACE (Eurostat 2008, p. 53) is using the concept of value added and it defines this as: "*The gross value added at basic prices is defined as the difference between output at basic prices and intermediate consumption at purchasers' prices*". Thus, the concept of value added measures the economic contribution of a sector as its output minus the inputs from other sectors.

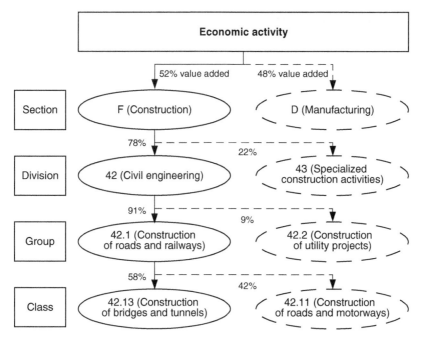

Figure 10.2 Classification by NACE.

As microeconomic concepts were especially developed for manufacturing, it is important to see that this section C of NACE (manufacturing) combines a large variety of industries in its divisions: Manufacture of food products, beverages, tobacco products, textiles, wearing apparel, leather and related products, wood products, paper products, printing, petroleum products, chemicals, pharmaceuticals, rubber and plastic, nonmetallic mineral products, basic metals, fabricated metals, computers electronic and optical products, electrical equipment, machinery, vehicles and trailers, other transport

Table 10.1 Economic sectors and NACE sections.

Primary sector	A: Agriculture, forestry, and fishing
Secondary sector	B: Mining and quarrying; C: Manufacturing; D: Electricity, gas, steam, and air-conditioning supply; E: Water supply, sewerage, waste management, and remediation; F: Construction
Tertiary	G: Wholesale and retail trade; I: Transportation and storage; I: Accommodation and food service activities; J: Information and communication; K: Financial and insurance activities; L: Real estate activities; M and N: Professional, scientific, technical, administration and support service activities; O, P, and Q: Public administration, defence, education, human health and social work activities; R, S, T, and U: Other services

equipment, furniture, other manufacturing, repair and installation of machinery and equipment.

The divisions of construction are much more limited in numbers: Construction of buildings, civil engineering, and specialized construction activities.

10.2 Economic Contribution

We can estimate the economic contribution of the construction sector based on three concepts. The first one is the concept of value added as used for NACE and ISIC. The second one asks how much money is spent on construction. This is the investment concept, and because of the narrow definition of construction in NACE and ISIC, it is more inclusive. The third is the multiplier concept which determines how much GDP increases if a dollar is invested in construction. It is based on investments but it still encompasses more than the investment concept.

10.2.1 Value-Added Concept

The economic contribution of construction based on the value-added concept of NACE or ISIC measures the value creation by firms engaging in the process of the construction of buildings, civil engineering projects, and specialized construction activities. Table 10.2 provides data from the Organization for Economic Co-operation and Development (OECD) on

Table 10.2 Value added by sectors in different countries (2020).

Country	Agriculture (A) (%)	Mining, manufacturing, energy (B, C, D, E) (%)	Construction (F) (%)	Services (G–U) (%)
Australia (2019)	2.0	19.8	7.7	70.5
Brazil (2018)	5.2	17.8	4.1	72.9
China (2019)	7.4	32.0	7.2	53.4
EU 27	1.8	19.4	5.7	73.1
EU low: Greece	4.	14.1	1.7	79.5
EU high: Finland	2.8	19.3	8.0	69.9
Japan (2019)	1.0	23.6	5.4	70.0
Russia	4.1	27.7	5.7	62.5
South Africa	2.7	25.1	3.2	69.0
UK	0.6	12.7	6.0	80.7
USA (2019)	1.0	14.5	4.3	80.2

the economic contribution of different sectors to the economy in the most recent year for which data are available (Organization for Economic Co-operation and Development 2021a). As Table 10.2 clearly demonstrates, construction is an important part of the secondary sector in most countries.

Another indicator of the importance of a sector is the number of people employed. Table 10.3 provides OECD data for selected sectors (Organization for Economic Co-operation and Development 2021b).

Together, Tables 10.2 and 10.3 give ample proof that in most OECD countries the service sector is most important. It is interesting to note that construction contributes a higher percentage to employment than to value added.

To highlight the importance of construction, it might be advisable to compare different industries. I have picked three industries in Germany and assembled the data from different sources: The automobile industry – which is a global leader; the chemical industry – which is the strongest in Europe; and construction (Table 10.4).

While the automobile industry is a key industry in Germany for its impact on exports, construction contributes more to GDP and employment.

Table 10.3 Employment by sectors in different countries (2021 or latest available data).

Country	Agriculture (A) thousand persons	Manufacturing (C) thousand persons	Construction (F) thousand persons	Services (G–U) thousand persons
Australia	303	908	1157	10,300
Finland	100	328	189	1898
France	683	3079	1809	21,410
Germany	528	7994	2866	30,653
Italy	949	4150	1391	15,545
Japan	1940	10,391	4871	49,047
Russia	4272	—	4909	47,989
South Africa	792	1497	1096	11,117
UK	324	2692	2140	24,426
USA	2305	14,410	11,187	122,923

Table 10.4 Comparison of the economic contributions of three industries/sectors in Germany 2020.

Industry/sector	Automobile industry	Chemical industry	Construction
Value added	4.7%	1.8%	6.1%
Employment	833,000	304,000	2,866,000

10.2.2 Investment Concept

We have seen that the NACE definition of the construction sector does not include all activities for which managers, architects, and engineers are trained and for which construction is required. Missing are:

- Wood from forestry (A)
- Raw materials from mining (B)
- Prefabricated components from manufacturing (C)
- Energy (D)
- Water (E)
- Architectural and engineering (M)
- Renting and leasing of equipment, landscaping (N)
- Public services (O)

All these activities contribute to the planning and construction of buildings and infrastructure. We are well advised to consider these activities when determining the economic impact of construction. The concept used for this purpose is construction investment, valued at the amount of money that owners spend until taking possession of a building or infrastructure. It should come as no surprise that the corresponding figures are considerably higher than those from the value-added concept.

Table 10.5 shows construction investment for several countries, ranked highest to lowest, together with the corresponding contributions to GDP in percent (European Construction Industry Federation 2021). The contributions of construction to GDP almost double when we change the framework for data collection from value added to the investment concept (EU 27: 5.7 vs. 9.5%, shown in Table 10.2). It seems to me that the picture emerging from the investment concept better portrays the contribution of the construction sector.

Still missing from these data are all real estate activities, and this also makes sense to me because construction within the investment concept includes all activities until handover of the project to the owner (with the exception of the owner's expenses for developing the real estate) and then other real estate activities (buying, selling, renting, operating) follow. Note that real estate together with the corresponding land value make up the largest amount of wealth in most countries.

Table 10.5 Construction investment in the EU (2019).

Country	Construction investment (billion Euros)	Contribution to GDP (%)
Germany	373	10.9
France	194	8.0
Italy	130	7.3
Spain	125	10.1
Netherlands	79	9.7
Finland (high %)	34	14.2
Greece (low %)	8	4.1
EU 27	1324	9.5

10.2.3 Multiplier Concept

The multiplier concept asks for the effect of a unit investment in an industry or sector and the corresponding increase in GDP (GDP multiplier) or in employment (employment multiplier).

To understand the concept, we must look at the (macroeconomic) Keynesian consumption function (C) with an autonomous consumption part C_a and an income-dependent consumption ($c \times Y$) where c denotes the propensity to consume and Y the income.

$$C = C_a + cY \tag{10.1}$$

In a closed economy without import and export, the national income is equivalent to GDP and can be spent on consumption (C) and autonomous investments (I_a).

$$Y = C + I_a \tag{10.2}$$

Replacing C in Eq. (10.2) by Eq. (10.1) provides:

$$Y = C_a + cY + I_a \tag{10.3}$$

Rearranging the equation for Y yields:

$$Y = (C_a + I_a)/(1 - c) \tag{10.4}$$

The derivative of Eq. (10.4) for the autonomous investment I_a provides the change in GDP (Y):

$$dY/dI_a = 1/1 - c = 1/s \tag{10.5}$$

This tells us that the multiplier (dY/dI_a) is larger when the marginal propensity to consume is large or the propensity to save is small. Both values are time-dependent; in Germany, the propensity to save was 12.6% in 1991 and 9.7% in 2016.

An extension of the model can include imports and exports as well as taxes; both influence the multiplicator. Taxes (t) reduce income and consequently also consumption. The multiplicator thus becomes:

$$dY/dIa = 1/(1 - c(1 - t)) \tag{10.6}$$

The propensity to import (m) influences the multiplicator because spending for imports reduces spending at home. Considering imports, the multiplier becomes:

$$dY/dIa = 1/(1 - (1 - t)(c - m)) \tag{10.7}$$

In an open economy, savings, taxes, and imports reduce the multiplier. With a propensity to save of 10% ($s = 0.1$, $c = 0.9$), a tax of 25% ($t = 0.25$) and a propensity to import of 10% ($m = 0.1$), we find the multiplicator to be:

$$dY/dIa = 1/(1 - (1 - 0.25)(0.9 - 0.1)) = 2.5 \tag{10.8}$$

This means that for every monetary unit invested, GDP will grow by 2.5 units.

Table 10.6 Example of the multiplier effect.

Rounds	Increase in GDP	Total increase in GDP	Multiplier
1	100	100	1.00
2	60	160	1.60
3	36	196	1.96
4	21.6	217.6	2.18
...	...	...	...
n	0	250	2.50

Table 10.7 Multiplier for construction investments in Germany, 2007.

Project type	Direct effect (million Euros)	Indirect effects (million Euros)	Total effects (million Euros)	Multiplier
One- and two-family homes	1000	1359.1	2359.1	2.36
Multiple-family homes	1000	1361.7	2361.7	2.36
Industrial buildings	1000	1592.0	2.592.0	2.59
Public buildings	1000	1390.0	2390.0	2.39
Infrastructure	1000	1549.6	2549.6	2.55

The initial autonomous investment leads to an increase in spending, and the money spent goes to someone who again spends it, and so on. There are several cycles of additional spending but they are always reduced by savings, taxes, and imports. If $dY/dI_a = 0.6$ and we look at an autonomous investment of $I_a = 100$, then we can see the effect in Table 10.6.

The determination of real-life multipliers requires a detailed input–output analysis. Results from such an analysis in Germany are shown in Table 10.7. The economic Institute RWI used data from 2007. The basis for the analysis is an additional investment of 1000 million Euros as direct effect (Rheinisch-Westfälisches Institut für Wirtschaftsforschung 2011, p. 47).

If politicians were interested in increasing GDP (i.e., furthering growth), they would invest in infrastructure directly or further investments in industrial buildings indirectly. The effects in these areas are most profound. This is an example of normative macroeconomic policy advice. The multipliers in Germany (2013) for services (1.54) and manufacturing (1.99) were considerably lower than for construction (Bundesverband der Deutschen Industrie 2013, p. 25).

Policy makers could thus focus on construction as a vehicle for growth because of the high values of the multiplier. In addition, it creates mostly local jobs on the construction site in high-wage countries such as most Western countries. In these countries, labor costs on site make up almost 50% of the total construction investment.

10.3 Actors in the Construction Sector

The number of actors in the construction sector depends on the underlying definition of the sector. ISIC and NACE provide a narrow definition and the investment concept a wider one. Actors in the construction market are not only the owners as buyers and contractors as sellers as it applies to the value-added concept. Looking at the investment concept, we have to add architects, engineers, suppliers, and equipment producers. However, there are still others with important roles: administrative, legal, and professional institutions, as well as bankers and insurers. They pursue different activities during the life cycle of a built structure. Figure 10.3 is a modification of a proposal by Carassus (2004). It contains three constructs, i.e. the life cycle, the construction sector according to ISIC/NACE, and the construction sector according to the investment concept:

- *Construct 1 – life cycle.* The life cycle consists of three main categories with more subcategories. Typically, a structure is erected and then operated or used before final demolition. New construction is usually built in months, but it can be a rather long phase for some infrastructure megaprojects; the development and design phases alone might take decades. Operation will often be by far the longest phase – one only has to think of castles or bridges that have been in use for centuries. There will be several rounds of renovation and modernization, with continuous maintenance during the operation phase. Only construction belongs to the construction sector, not facility management (operation). Demolition does not take much time.

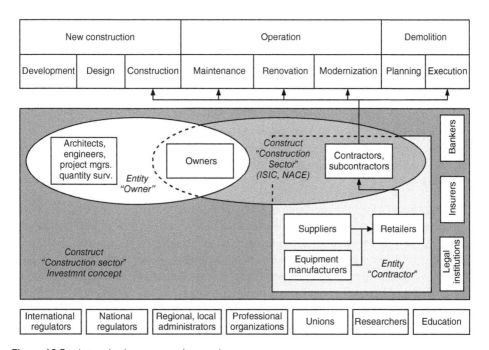

Figure 10.3 Actors in the construction market.

- *Construct 2 – construction sector according to ISIC/NACE.* This is the construction sector as defined by the value-added concept and discussed in Section 10.2.1. It includes owners and contractors as well as subcontractors. Of course, ISIC and NACE do not refer to the owner explicitly, but there would be no contract without an owner. A firm can only become a contractor by signing a construction contract with an owner.
- *Construct 3 – construction sector according to the investment concept.* In this view, the owner receives support of architects, engineers, project managers, quantity surveyors, and other consultants. Owners might also need help from lawyers to draw up a contract, from bankers to provide financing, and from insurers to limit risks. Contractors rely on the resources of suppliers and equipment manufacturers – for which they often need the help of retailers. Like owners, contractors often require help from lawyers, insurers, and bankers. As noted in construct 3, we find in Figure 10.3 two entities called *owner* and *contractor*.
- *Entity owner.* In very rare cases, a single layperson will want to contract a construction firm. While this is the normal case for manufactured products, it is an exception in construction. To plan and complete a construction project is far too complex for a layperson, and in many countries, it is not even possible because of legal constraints; only professionals have the right to submit a building application to the authorities granting a permit. Thus, the owner in construction is an entity. The owner in the narrow sense must recruit help by architects (form and function), civil engineers (stability and usability), specialized engineers (building services, geo-technology), project managers (organization, quality, quantity, schedule, and cost), quantity surveyors (price forecasts, tender documents, plan and construction changes), and other consultants (ecology, acoustics, etc.). In addition, most owners require external financing by banks, legal advice by lawyers and maybe help by courts in case of litigation as well as risk minimization through insurance. In the end, the owner with all the help from professionals has access to a formidable amount of specific knowledge.
- *Entity contractor.* Contractors also form a network to fulfil the tasks of the contract. Subcontracts can make up 100% of all construction work and must at least cover all the tasks that the contractor cannot complete by himself. Contractors must buy all materials and equipment required for the works. Most will be purchased through retailers but some direct purchases are also common; ready-made concrete is a prime example. Some equipment might be rented. Contractors also need a bank to finance expenditures, lawyers for legal advice, and insurers. The most common insurance is a contractors-all-risks (CARs) insurance.

Construction is a highly regulated field. The built environment created by owners and contractors encroaches on the natural environment on which we depend for a sustainable life. It also is the environment in which people pass most of their time, and as such, it influences society. This requires urban and rural planning and good architecture. An unfriendly built environment will not cause direct damages, but collapsing buildings will lead to deaths – another reason for regulation. Regulatory agencies can be national or international institutions such as the EU commission department for competition. As all construction is in the end local, centred around the construction site, local or regional authorities (zoning, utilities) will also be involved.

Professional organizations set standards such as methods and rules on ethical behavior. They support continuing education, licensing, and networking. Unions allow the organization of labour to counter the market power of employers. The degree of unionization varies from country to country. Researchers push the boundaries of our subject knowledge and advance theory. The link between academics and practitioners in construction is unfortunately not very strong. However, university education for bachelor's or master's degrees provides the fundamental knowledge for all practitioners in construction.

In the following Chapters 11 (Theory of the Owner) and 12 (Theory of the Contractor), I will develop the ideas on actors in construction to a higher level.

10.3.1 Market Demand

There are no statistics on the number of owners per year. What we do find in some countries are the number of building permits per year, and this number is substantial. At the same time, it only reveals owners who undertake new construction or renovation requiring a permit. For many small projects, this is not necessary. More unfortunately, data are available only on buildings. There were 232,069 permits issued in Germany in 2020 for buildings (Statistisches Bundesamt 2021, p. 89). This number indicates the transactions in the market; it does not calculate the number of owners – as one owner can request more than one building permit per year. If we assume somewhat arbitrarily that for residential construction each owner requests three permits per year and for nonresidential construction five permits per year, the result is shown in Table 10.8.

What does this mean? With all the uncertainty about the exact number of owners, and considering all the buildings and infrastructure constructed without permits, both currently and in the past, it seems clear that the number of owners is large on the national level. However, it is small when compared to the manufacturing sector, where almost certainly every single person out of a population of 83,000,000 in Germany will make at least one purchase per year (i.e., act as buyer).

Figure 10.4 shows the distribution of construction investment according to categories of construction demand for Germany in 2020. Figure 10.5 illustrates the same data for 2016 (Hauptverband der Deutschen Bauindustrie 2017, 2021). As we can see, there are noticeable shifts over a rather short period. The business cycle for building and infrastructure are not always coupled.

The European Construction Industry Federation (2021) uses slightly different categories, but the general picture that buildings dominate the market is also visible in this case. Figure 10.6 provides data for the 27 EU countries.

Table 10.8 Number of building permits in Germany, 2020.

	Permits	Permits/owner (estimated)	Owners (estimated)
Residential buildings	186 990	3	62 330
Nonresidential buildings	45 079	5	9 015
Sum:	232 069		71 345

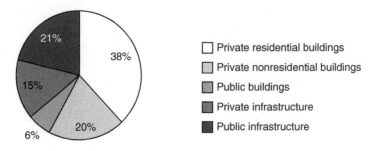

Figure 10.4 Market demand in Germany 2020.

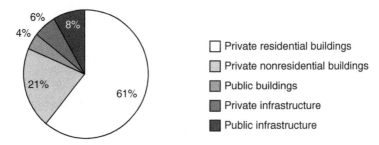

Figure 10.5 Market demand in Germany 2016.

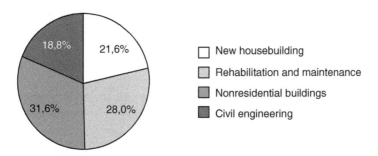

Figure 10.6 Market demand in the 27 EU countries 2020.

The development status of a country has also an influence on market demand. The UN distinguishes between developed economies, economies in transition, and developing economies based on per capita GNP. Ruddock (2008) provides some data for the value-added concept and based on data from the UN (Figure 10.7).

The largest demand for construction is in developing economies, especially for housing and infrastructure, as the industrial sector still has to develop. Unfortunately, these countries have also the biggest problems to finance construction. All available funds often go into new construction and very few resources are left for maintenance. Economies in transition have larger funds available and an industrial sector to support. Therefore, construction value added reaches a maximum. Developed countries have a large stock of structures,

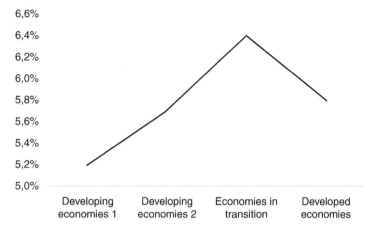

Figure 10.7 Market demand according to economic development status.

both buildings and infrastructure. The need for construction diminishes and funds are diverted from new construction to maintenance and rehabilitation.

10.3.2 Market Supply

The structure of construction is disaggregated. Most countries have many construction firms. Table 10.9 illustrates this statement with data from the larger European countries (European Construction Industry Federation 2020).

The data (Table 10.10) show that in all European countries, there are many construction firms with a rather small number of employees, with low average turnover per firm (European Construction Industry Federation 2020).

Table 10.9 Number of construction firms in selected EU countries (2019).

Country	Construction firms	Population (million)	Firms/ thousand people	Investment (billion €)	Average output per firm (€)
Italy	501,000	60,000,000	8.4	130	259,000
Spain	422,000	47,000,000	9.0	125	296,000
France	419,000	65,000,000	6.5	194	463,000
Germany	340,000	83,000,000	4.1	373	1,097,000
Poland	325,000	38,000,000	8.6	49	151,000
Czech Republic	190,000	11,000,000	17.3	21	111,000
Netherlands	190,000	17,000,000	11.2	79	416,000
Belgium	114,000	12,000,000	9.5	48	421,000
Hungary	109,000	10,000,000	10.9	21	192,666
Sweden	107,000	11,000,000	9.7	53	495,000

Table 10.10 Average number of employees in construction of the EU. 2019.

Country	Construction firms	Employees	Average number of employees per firm
Italy	501,000	1,343,000	2.7
Spain	422,000	1,278,000	3.0
France	419,000	1,790,000	4.3
Germany	340,000	2,551,000	7.5
Poland	325,000	1,212,000	3.7
Czech Republic	190,000	375,000	2.0
Netherlands	190,000	483,000	2.5
Belgium	114,000	286,000	2.5
Hungary	109,000	400,000	3.7
Sweden	107,000	356,000	3.3

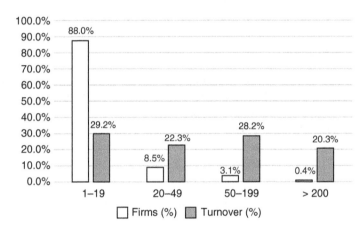

Figure 10.8 Construction firm structure and turnover in Germany 2020.

A closer analysis of the firm structure is provided in Figure 10.8 for the percentage of firms and the corresponding percentage of turnover for structural firms in Germany in 2020 (Hauptverband der Deutschen Bauindustrie 2021). It is obvious that small firms dominate the market in number but not in turnover. A miniscule percentage of larger firms (0.4%) creates a sizable amount of turnover (20.3%). I will analyse this disparity in Chapter 14 on construction markets in more depth.

Table 10.11 summarizes the number of employees in the different types of firms in construction by percentage in Germany in 2015 (Baulinks 2016). Unfortunately, there are no data available for 2019, so the total number of employees in 2015 (2,243,000) slightly differs from that in 2019 (2,551,000, Table 10.10). However, this affects the overall percentage only little.

The largest number of employees finds work in the finishing trades and in building service trades (electrical, mechanical, plumbing).

Table 10.11 Number of employees in the construction industry in Germany, 2015.

	Employees	Percent
Finishing works, building services	1,167,000	52
Structural works	804,000	36
Civil engineers	144,000	6
Architects	128,000	6
Sum:	2,243,000	100

Table 10.12 Herfindahl–Hirschman Index for different sectors in Germany 2008.

Industry	HHI
Finishing works	2
Structural works	3
Manufacturing, metal products	11
Products for gas, water, heating	139
Coal mining	2942

The Herfindahl–Hirschman-Index (HHI) describes the concentration in a sector (Stiglitz and Walsh 2002). The US Federal Trade Commission uses the index to control competition and determine whether to allow mergers, for example. An HHI <1.000 describes a sector which is not concentrated. A value between 1.000 and 1.800 signals a moderate and any higher value a strong concentration. The value is calculated as the squared sum of the market shares α_i in percentage of all firms in a market. A monopolist ($n = 1$) has a value of 10.000 – (100% × 100%).

$$R = \sum_{i=1}^{n} \alpha_i^2$$

(10.9)

The values for some sectors in Germany for 2005 are shown in Table 10.12 (Monopolkommission 2008). They are very low for the construction sector and much higher for manufacturing. Again, data are not available for all years.

10.4 Summary of the Construction Sector

There are several preliminary takeaways from this chapter. They will be expanded in the coming ones on the theory of the owner (Chapter 11), the theory of the contractor (Chapter 12), construction goods (Chapter 13), construction markets (Chapter 14), and contracting (Chapter 15).

- The economic contributions of construction to GDP are strong for the value-added concept, averaging somewhere around 6% in developed countries. Using the investment concept, the contribution is approximately 10%.
- Statistical definitions such as those used by ISIC and NACE determine the data. Comparing manufacturing and construction does not completely make sense since construction is a homogenous sector producing buildings and infrastructure (albeit of a great variety!) and the manufacturing sector produces a large variety of products from food to chemicals, from automobiles to furniture. Attributing repair and maintenance to construction and not to the automobile industry makes productivity comparisons useless.
- The multiplier concept proves that the construction sector can be a primary tool to foster economic growth. For an investment of a dollar in the construction industry, the total economy will profit more than for the same investments in services or manufacturing.
- The construction sector contributes even stronger to employment than to GDP.
- The number of annual transactions in construction is small compared with manufacturing. The number of owners is comparatively small; however, the number is still considerable on a national level. Owners have a much stronger position than buyers in manufacturing because they sign up a number of professionals as support.
- The number of firms in construction is large; the concentration very low. Thus, many construction firms have little market power. However, despite the fact that large firms are limited in number, they can yield more market power than small firms.
- In almost all countries, many owners have access to many contractors on the national level. Here, we have the situation of a perfectly competitive market. As we will see later, this is not the determining factor in contracting, when market transactions are concluded. The overall ratio of contractors to owners will be between four firms/owner and eight firms/owner.

References

Baulinks (2016). Baukonjunktur / Baumarkt aktuell. http://www.baulinks.de/baubranche/baukonjunktur-bauwirtschaft.php; accessed January 2017.

Bundesverband der Deutschen Industrie (2013). *Industrielle Wertschöpfungsketten – Wie wichtig ist die Industrie?* Berlin: Bundesverband der Deutschen Industrie.

Carassus, J. (ed.) (2004). *The Construction Sector System Approach: An International Framework. CIB W055-W065 Construction Industry Comparative Analysis Project Group.* Rotterdam: CIB.

European Construction Industry Federation (2020). https://www.fiec.eu/application/files9016/0190/8790/FIEC_Key_Figures_Edition_2020.pdf; accessed December 2021.

European Construction Industry Federation (2021). https://fiec-statistical-report.eu/eu-en; accessed December 2021.

Eurostat (2008). *NACE Rev. 2: Statistical Classification of Economic Activities in the European Community.* Luxemburg: EU.

Hauptverband der Deutschen Bauindustrie (2017). *Bauwirtschaft im Zahlenbild.* Berlin: Hauptverband der Deutschen Bauindustrie.

Hauptverband der Deutschen Bauindustrie (2021). *Bauwirtschaft im Zahlenbild*. Berlin: Hauptverband der Deutschen Bauindustrie.

Monopolkommission (2008). *Siebzehntes Hauptgutachten der Monopolkommission 2006/ 2007*. Baden-Baden: Nomos.

Organization for Economic Cooperation and Development (2021a). https://data.oecd.org/ natincome/value-added-by-activity.htm; accessed December 2021.

Organization for Economic Cooperation and Development (2021b). https://data.oecd.orgemp/ employment-by-activity.htm; accessed December 2021.

Rheinisch-Westfälisches Institut für Wirtschaftsforschung (2011). *Zukunft Bau – Multiplikator- und Beschäftigungseffekte von Bauinvestitionen*. Essen: Rheinisch-Westfälisches Institut für Wirtschaftsforschung.

Ruddock, L. (2008). Assessing the true value of construction and the built environment. In: *Revaluing Construction* (ed. P. Barrett), 67–82. Oxford: Blackwell.

Statistisches Bundesamt (2021). *Ausgewählte Zahlen für die Bauwirtschaft – Dezember und Jahr 2020*. Wiesbaden: Statistisches Bundesamt.

Stiglitz, J. and Walsh, C. (2002). *Economics*. New York: Norton.

11

Theory of the Owner

We can develop a theory of the owner from consumer theory in Chapter 4 (consumers in perfectly competitive markets). Consumers maximize their utility, and while doing so, they act as price-takers whether they face perfectly competitive markets, monopolies, monopolistic competition, or oligopolies.

This conjures up the image of a consumer very much at the mercy of powerful producers. Hollywood successfully adapts such stories of David (consumer) against Goliath (producer). Doubtlessly, there is often a mismatch between consumer and producer, and for this reason, we have extensive legislation for consumer protection. An example of such legislation are rules for the general terms and conditions that producers often add to their products. The question arises whether an owner in construction is also often in a weak position; can we interchangeably use the terms consumer and owner?

In Chapter 4 consumer theory was constructed with five building blocks:

1) Consumers have stable preferences.
2) Consumers face a budget constraint.
3) Consumers maximize utility by choosing the consumption bundle where the marginal rate of substitution equals relative price.
4) We can construct the individual demand by varying the prices and thus the budget constraint.
5) The market demand curve is the aggregate of the individual demand curves.

In the following chapters, I will analyze whether these assumptions also apply to the owner in construction.

After extending the definition of the entity owner more precisely than in Chapter 10, I will describe the tasks of an owner. From these two building blocks, we can deduct the owner's behavior based on available information. After this, we will follow how an owner develops a construction contract, engages in and concludes the procurement of a contractor, and finally supervises the construction process. At the end, I will summarize the main points.

Construction Microeconomics, First Edition. Christian Brockmann.
© 2023 John Wiley & Sons Ltd. Published 2023 by John Wiley & Sons Ltd.

11.1 The Owner as an Entity

I have started with a discussion of the person described by consumer theory and raised the question of whether a consumer and an owner in construction are similar to each other. They are not, as we will see. Therefore, we need a suitable word to describe a person who engages others to fulfil specific construction needs. Words manifest prejudices and conjure images. Thus, the question arises as to what images are aligned with a term like customer. Finally, we need a clear definition of the owner, i.e. the customer in construction.

11.1.1 Terminology

Throughout the previous chapters, I have used the term *owner* to designate someone who has a structure planned, designed, and built. Now I will explain why. Words are not arbitrary; they have meaning, and they transport this meaning in communication and sensemaking.

For this reason, it is impossible to call such a person a buyer. This term presupposes a transaction in the form of an exchange, as in the case of shopping: We go into a store, pick a good, and pay for it. Such a good is called an exchange good because we exchange the good for money in the same instance. Manufacturing the good before the sale makes this exchange possible. Construction inverses this process and requires first a definition of the good, then a contract, and finally production. Such a sequence describes a contract good. Since it is the owner who defines the good, specifies the contract, and supervises construction, the owner cannot simply be called a buyer. If someone buys an existing building, then she acts by definition in the real estate sector and not in construction.

Similar in meaning is the term customer, as it defines someone who purchases a good or service customarily. The term assumes previous transactions with exchange goods and is for this reason no better suited than buyer.

The difference between consumption of and investment in a good depends on the use. People who have a one-family home built under their own supervision – with their own money and for their own private use – are consumers. Other owners who engage in planning, designing, and construction for business purposes are investors. Thus, the use of the term consumer for all people acquiring a structure is confusing. Most of them are investors, and the term investor obscures most of the responsibilities of an owner.

There are two further terms to designate the owner: client and employer. Merriam-Webster defines a client first of all as one that is under the protection of another and secondly as someone who engages the advice of a professional. The first definition is always in the background and accordingly a client is seen as deficient with regard to power. This is not even true for relationships between architects, engineers, quantity surveyors, and project managers on the one side and an inexperienced person who wants a structure on the other. When we determine the relationship between the same person and contractors, we must understand that the full professional advice of architects, engineers, and project managers supports the owner so that it becomes even more misleading to address him as client, both parties depend on each other during construction. It is absurd to address the contractor as client of the owner and the same is true in the other direction. The term client can neither describe the relationship between owners and professionals nor the one between owner and contractor.

The Fédération Internationale des Ingénieurs Conseils (FIDIC) uses in its standard contracts the term employer. This describes a situation where the owner is in charge of design and construction. The owner hires the employer's personnel (i.e., professionals and a contractor). To me, this overemphasizes the role of the owner. An employer has the right to give direct instructions to an employee. This is not directly possible for the relationship between owner and contractor. However, an owner in construction certainly has the right to hire and fire.

The term owner is not as strong as the term employer and it signifies a person who owns the construction process. This is, to my understanding, the term that best describes the corresponding relationships and responsibilities.

11.1.2 Images and Prejudices

A well-known comic can illustrate the problem of words and images as well as their suggestive strength (Figure 11.1). The cartoon first of all shows that there are many people involved in construction: architects, engineers, the Occupational Health and Safety Administration (OSHA), the sales force, contractors (field teams), and finally, the customer. The cartoon is nice for a joke but catastrophic for understanding construction. It tells a story of different groups engaging in construction who completely disregard what the others want, plan, or do. No reasonable owner will pay a single cent for the services and works delivered.

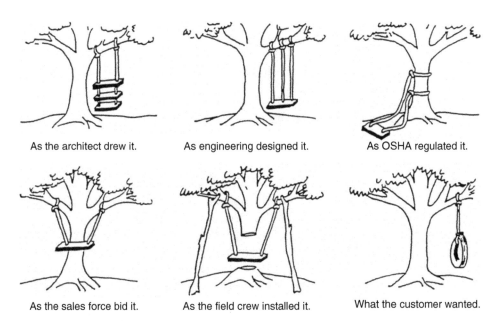

As the architect drew it.	As engineering designed it.	As OSHA regulated it.
As the sales force bid it.	As the field crew installed it.	What the customer wanted.

Figure 11.1 The construction process as wrongly but funnily depicted in a cartoon.

Table 11.1 Comparison between a buyer of an automobile and an owner in construction.

	Buyer of an automobile	Construction owner
Idea and design	Seller (car producer)	Owner (buyer)
Ownership of place of production	Seller	Owner
Contract definition	Seller	Owner
Type of market exchange	Seller	Owner
Unilateral right to changes	Nobody	Owner

The cartoon is wrong in almost every detail, as it describes a customer who is at the mercy of ruthless foes.

The reality is more like the following: The owner initiates a building process and communicates his wishes to the architect. The architect elaborates different drawings and models requesting approval by the owner. Sensible owners (they usually act rationally) will make sure that the design meets their wishes. Architects define form and function of the building and engineers ensure safety and usability. Any change required by engineering must again meet the owner's approval. A competent architect knows the norms and laws of OSHA and will incorporate them into the design from the beginning. An incompetent one must do it later it to get approval. Again, the owner's approval is required. The contract with the contractor is based on the design, as agreed by the owner. Fulfilment without defects is the main contractual obligation of the contractor. If a contractor's estimator (sales force) got the offer price wrong, this remains the contractor's problem; contractors must abide by the contract. If the construction crews (field crews) make mistakes, that is also on the contractors – they must remedy the errors at their own expense. In the end, owners get what they want.

The cartoon transports the idea of a weak customer. Comparing the responsibilities between a buyer in the car industry and an owner in construction brings up many differences (Table 11.1).

The owner is in control of the production process in construction, while this is the responsibility of the seller in manufacturing (automobile industry).

11.1.3 Organization

The organization of construction developed over many centuries. When the cathedrals of the Middle Ages were erected by master builders who combined the role of architect and engineers, they could already rely on a specialized workforce organized in guilds (Gruneberg 1997). Today, there are 45 formalized different apprenticeships in construction in Germany. In the middle of the nineteenth century, science had provided theory for structural analysis, deepening the difference between architect and structural engineer. Approximately at the same time, innovations such as the electric bulb by Thomas Edison and the elevator by Elisha Otis made the advice of specialized engineers necessary. Architects took the lead on behalf of the owner for buildings, and civil engineers did the same for infrastructure.

In the 1970s, quantity surveying became more widely used, especially in combination with the introduction of alternative procurement methods. This was most true for commonwealth countries (Langford and Male 2001). In other countries, owners became frustrated with the architects' focus on aesthetics and function and their increasing disregard for money and time. This provided a chance for project management firms to close the gap. In many projects, these project management firms replaced architects as closest advisers to the owner. The growing complexity of structures, the stronger interrelation between structure and the environment, and new technologies possible by digitalization such as smart buildings keep adding specialists to construction. Geotechnical engineers, hydrologists, environmental engineers, computer specialists, specialists for acoustics or earthquakes and tsunamis – the list is long and keeps growing. Today, construction is a very specialized field with numerous participants and far more more stakeholders (i.e., the public). While it is the task of the contractor to organize the trades on the construction site, the owner remains responsible to sign up all the professional help required for a specific project. Thus, we are facing two entities in a construction project: A group around the owner and another around the contractor.

How do owners organize their part of the project? Many (but not all) owners are laypersons. Some of them have a construction background – this is especially true for public works or industrial works by large companies with their own construction department. However, in many Western countries, the trend is to outsource most of the construction planning. There are very few organizations that still command their own construction workforce.

Figure 11.2 summarizes the type of owners that are prevalent, not including owners with their own workforce. Construction experience might come from formal training or learning by doing; therefore, the frequency of construction transactions plays an important role.

Figure 11.2 clarifies that the need for professional help depends on the competence of the owner – and on the complexity of the project.

When the owner contracts with professional help, the professionals are hired to serve the owner's goals. That should be their own goal as well, but of course there are sometimes

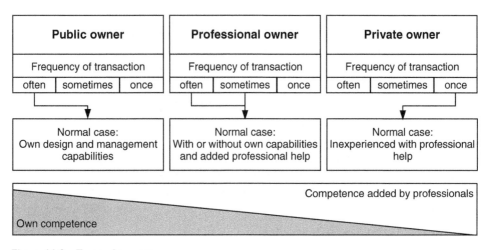

Figure 11.2 Types of owners.

The entity "owner"

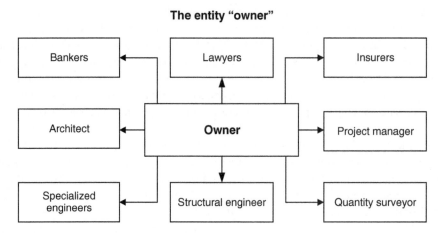

Figure 11.3 The entity owner.

conflicts of interests. Sometimes famous architects have their own agenda when designing a unique building. However, this is not the typical case. The old adage states that *"he who pays the fiddler calls the tune,"* and as the owner pays all professionals, the owner is in a position to get his will. This applies strongly with regard to architects, engineers, project managers, quantity surveyors, and lawyers, but less so toward bankers and insurers. Banks and insurance companies have their own rules and terms, which they will not bend to the wish of a single customer, but there is a wide range of service from which the owner can pick. Figure 11.3 shows the composition of the entity *"owner"*.

11.2   Tasks of the Owner

The owner is the entity who takes responsibility for planning and supervising construction. Owners can do this by themselves or with the help of others. They must follow all regulations and ensure safety during construction. Owners must organize the enterprise of construction and determine the market conditions by drafting a contract. For small projects, architects support the owners to this end; for larger projects, project managers perform this role.

The minimum involvement of an owner includes the following tasks:

- Develop the initial business idea.
- Provide a construction site.
- Secure project financing with the help of banks.
- Define the structure with the help of architects and engineers.
- Obtain a building permit with the help of architects and project managers.
- Formulate the contract and its conditions with the help of lawyers, quantity surveyors, or project managers.
- Determine the rules of the market exchange by the contract.
- Use the owner's one-sided right to change the design during construction if necessary.

Table 11.2 describes the typical construction process for a traditional procurement approach, which has three phases: design, procurement, construction (design/bid/build).

Other procurement approaches notably include early contractor involvement, concurrent design overlapping construction, design/build contracts, and design/build/operate contracts. Greenhalgh and Squires (2011) provide much more information on this topic.

Private owners have absolute discretion to choose the procurement that best suits their purposes, which means in economic terms that they can maximize their own utility. Public owners might have limited options due to government regulations.

The tendering approach also organizes the construction market, and again the owner has absolute freedom of choice within the legal framework. If the owner wants the strongest competition possible, then open tendering is the way to go. In this case, any willing contractor can submit a bid. If the owner wants to control the quality of the contractors more closely, then the owner will opt for selective tendering with a hand-chosen number of bidders (typically, 5–10). In some instances, the owner might deem it preferable to negotiate the contract with just one bidder.

There are more ways to shape the market. Two important considerations are the specificity and the size of the contract. It is possible to bundle the project in one contract, including structure, finishing works, and building services by awarding it to a general contractor. Another choice is the letting of separate contracts to special trades. The size of the project puts financial demands on the contractors by requiring bonds, and it increases risks. Therefore, smaller contractors will be excluded from contract lots judged too large and risky.

Private owners enjoy in many countries the right of contractual choice. They can formulate any condition to their advantage as long as they do not propose immoral contract conditions. They are able to shift rights, obligations, and risks in a way to suit their own ends. Of course, there are standard contract forms as those prepared by FIDIC, which look for an equitable distribution of rights and responsibilities, but no private owner is bound to them. In many cases, this is different for public owners.

11.3 Behavior of the Owner

The first difference between consumer and owner theory is the fact that owners more often buy investment goods and not consumer goods. Consumer theory can only apply to those buying construction goods for consumptive purposes. The archetypical consumer good in construction is the one-family home. Most other construction goods are used as investment by private or public owners. They use these goods as part of their own productive purposes.

11.3.1 Consumers Buying Construction Goods

It seems unproblematic to assume utility maximizing behavior for owners and consistent preference orders when buying an existing home (real estate sector) because they can compare the features of such homes. More difficult is the case when an owner is designing a one-family home with the help of professionals. In many cases, the preferences will develop in conjunction with the advice of the professionals, especially the architect. The complexity

Table 11.2 Typical design and construction process.

1) Conceptual design	Owner and architect (engineer) develop the first design concept for buildings (infrastructure)
2) Preliminary design	Owner, architect, and engineer develop the design to the next maturity level
3) 35% detailed design	Owner, architect and engineer finalize form, function, and constructability of the structure
4) 65% detailed design	Architect and engineer advance the design to guarantee stability and usability
5) 100% detailed design	Final design for construction
6) Tender preparation	Owner prepares tender with professional help (architect, quantity surveyor, project manager, lawyer); tender documents include contract, drawings, specifications, and at times a bill of quantities
7) Bid preparation	Contractors prepare bid, and owner provides information if required
8) Submission	Owner opens the contractors' bids
9) Bid evaluation	Owner evaluates bids with professional help (project manager, quantity surveyor)
10) Contract negotiation	Owner negotiates the contract supported by professionals with one or more contractors simultaneously
11) Signature	Owner and contractor sign a binding contract
12) Construction period	The contractor erects the structure supervised by the owner and professionals
13) Handover	The owner takes possession of the structure
14) Maintenance period	The owner with the help of professionals checks the structure for defects; the contractors correct defects
15) End of contract	The contractual relationship ends with the maintenance period

and intangibility of the one-family home at the design stage are the reason for preference uncertainty.

Owners of one-family homes display satisficing behavior. While preference uncertainty is in contradiction to consumer theory, the budget constraint poses a strong restriction on all owners. Budget constraints, together with clear preferences, allow us to develop the market demand curve. In the case of new-built homes, the demand curve can only be constructed once the design is finished and preferences are firmly established. At this point, the one-family home is unique. Consequently, there is no aggregate demand curve for such one-family homes. Product homogeneity was one of the assumptions used to derive a market demand curve and utility maximizing (not satisficing) another one; thus, it is impossible to construct the demand curve in a similar way.

11.3.2 Producers Buying Construction Goods

To assume that an owner who procures an investment good maximizes profits also seems highly plausible. Present value analysis is an excellent tool for comparing investments.

Buildings and infrastructure require an initial investment and provide continuous services over their lifetime. Using the concept of opportunity costs, investors will procure a building only if it provides the best possible return among all alternatives. Construction goods are for investors assets, which they need to pursue their business goals – be it the production of consumer goods in factories, the delivery of services in office buildings, or the provision of transportation facilities as a public good.

We have come across the strange fact that most buyers of construction goods are not consumers but producers. In 2019, just 18% of construction investment was spent on one-family homes in Germany. In such a case, the owner – together with at least an architect – face a small general contractor or even smaller specialty contractors on the market. This is definitely not a situation of David against Goliath, which often applies to the case of consumer goods.

Market power of the owner is even stronger in case of professional investors. When Airbus tenders for a new production facility, who is Goliath and who is David? This abnormality of market power set aside, consumers and producers who acquire construction goods do not differ from other consumers or producers in their behavior; they maximize utility or profit and must observe their budget constraint.

Another difference between most construction transactions and many consumer decisions is the degree of involvement. The consumer of a one-family home expresses a direct need and is highly, sometimes even extremely, involved. It's an emotional purchase, investing not just money but hopes for a happy future – they don't call it *"building my dream house"* for nothing. Investors satisfy a derived demand to fulfill their primary goals (production, service). There is little emotional involvement in the majority of procurement decisions, which makes establishing relations difficult (relationship contracting).

11.4 Information of the Owner

Complete information is one of the core assumptions for perfectly competitive markets and an information asymmetry one for monopolies. How does this square with construction markets?

Much information about construction is publicly available; no detailed information can possibly exist for singular projects before publishing the tender documents. Contractors deal with product, process, and price information on a daily basis, and so do the professionals in the construction industry (i.e., architects, engineers, quantity surveyors, or project managers). All owners rely on such services, be it inhouse or from the outside. The one-time owner for the one-family home will certainly require help from professionals. The professional owner might have a building department as part of the company and also has access to professionals. In general, market information is complete and equally accessible.

This is quite different in particular with regard to the one singular project at hand. During the design stage, the options are limitless. There are at least 200 faucets available in most countries and 200 sinks. This provides for almost 20,000 different combinations. Alas, faucets and sinks must be combined with thousand of other items in a one-family home. A good architect will favor maybe 10 of these faucets/sink combinations and propose them to the owner. The owner will choose one of these combinations without being able to keep in mind all the other items related to installing this faucet, let alone the dozens of other faucet

Table 11.3 Use of information asymmetries.

Stage	Owner	Contractor	Means
Pre-project stage	Symmetric information		Hidden qualities
Design stage	Information advantage		Adverse selection
Tendering stage	Information advantage		Hidden information
Negotiation stage	Information advantage		Hold-up
Construction stage		Information advantage	Hidden action
Warranty stage		Information advantage	Moral hazard

options not even seen. This example shows that nobody has a full knowledge of the opportunities because of the number of parts and the complexity of construction goods. Even with the best help, owners cannot maximize the design; they must use satisficing behavior.

While the owner faces an infinite set of product possibilities, the contractor faces an infinite set of process possibilities. Neither of them has complete information. This is a problem of complex contract goods. It is no problem in commodity markets; the buyers of Coca-Cola do not need to know the secret ingredients. All they must know is whether they prefer Coca-Cola over all other available soft drinks – and this is easy to test.

Once the owner opens all tenders on submission date, the owner gains a clear and decisive information advance over the contractors. Even if the submission is public and all offers are announced, the contractors only know their own price details while the owner knows detailed price information of all bidders. Typically, owners use the price information to pressure contractors during negotiations. This is a strong information asymmetry favoring the owner.

Once the contract is signed, the information advantage shifts toward the contractor. Now the contractor has detailed process information. Typically, contractors will try to use that leverage for their benefit in hopes of greater profit.

We can often observe opportunism on both sides, albeit at different times. Table 11.3 summarizes the information status along a project timeline and the means each party can use to gain an advantage. The information in the table is based on a design/bid/build contract. It will be slightly different for other contracts, but the main statements remain unchanged.

Again, we can see that the owner is not always in a position of David. In the very important front-end of the project, the owner is Goliath. This is not what we typically observe in commodity or service markets.

11.5 Developing a Contract

In most countries, laws restrict the producers when formulating the general conditions of purchase. These are preformulated clauses, the fine print. It pays for companies who provide commodities or repetitive services to have lawyers formulate such conditions, thus safeguarding their own interests. The consumer has neither the time nor the knowledge to read and understand the fine print. Therefore, it makes sense that legislators protect the weaker party (consumer).

It might not come as a surprise at this point of the discussion that things are different in construction. Here, it is the owner who formulates the contract conditions, and the contractor can accept them or abstain from bidding. Public owners are often bound by procurement laws, but this is not the same for private owners. For them, the freedom of contract with its customary limitations (immoral contracts) set the boundaries.

To level the playing field, there exist numerous standard forms of contract. One group of them are FIDIC contracts. These consist of general and particular conditions. While the general conditions should remain unchanged, the owner is free to set the particular conditions.

Overall, we find that the rights of owners to formulate a construction contract with the help of architects, quantity surveyors, or lawyers allow them to wield considerable market power.

11.6 Procurement of a Contractor

It is the owner who creates the market and determines the rules by which a contractor is procured. Contractors can agree to these conditions or abstain; very seldom they can adjust the conditions in their favor. Procurement is embedded in the overall project phases. Typically, it is proceeded by design and followed by construction (Figure 11.4).

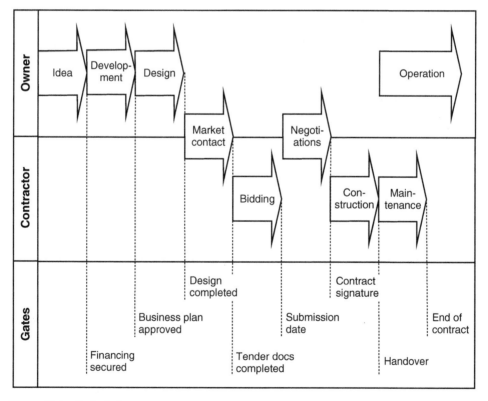

Figure 11.4 Project phases.

Owners choose the market for their singular and self-designed goods by choosing a tendering approach. If an owner resolves to go for public tendering, then the market will be in normal times thick. The only exceptions are times when the sector is already working at capacity, which happens in boom times. In a recession, this market might become congested. Fifty or more offers are not unheard off. If the choice is a selective tendering route, then the owner can determine the thickness of the market. Most owners decide to choose 5–10 selected bidders. This provides enough competition to get results that are at least equivalent to the price level of supply and demand in perfectly competitive markets; most often they are lower. Almost all owners will use submissions (sealed-bid auctions) as an organizational tool, thus creating a small commodity market. The last choice is a negotiated contract, where the owner discusses with just one contractor. This is, of course, a matching market, and the match is concluded before the negotiations.

Last but not least, owners must choose between a confrontational or a partnering approach. In partnering, the use of power in contraproductive. The assumption of profit-maximizing behavior by the owner and the contractor does not fit well into arguments for partnering as long as self-interest is the guideline at any one moment. There exists a host of literature on construction procurement (e.g., Masterman 2003 or Greenhalgh and Squires 2011). Three important procurement choices are shown in Figure 11.5.

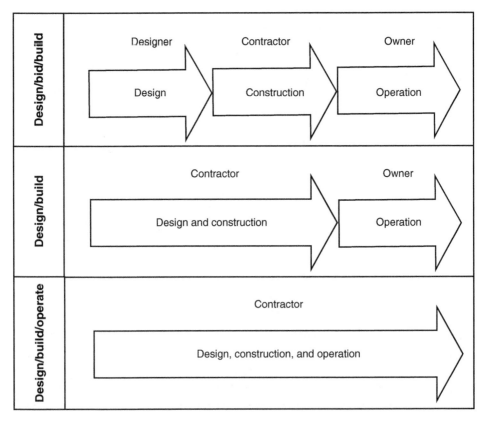

Figure 11.5   Basic contractual options for the owner.

Pros and cons are plentiful in the discussion without a clear tendency toward a master approach. Missing in the discussion is mostly an economic analysis of the different choices, and this is what is important in our context. Concentrating on production costs alone, there is much to say in favor of the traditional design-bid-build approach. The owner can choose conditions under these circumstances that will provide in most cases the lowest contract prices. However, the costs for the owner are the sum of production and transaction costs, and the latter ones are considerable when drafting, negotiating, and enforcing the contract. While production costs are clear on submission day, the transaction costs develop over time until the end of the project. This is especially true if we consider changes and claims as part of the transaction costs.

The choice of payment methods in the construction contract has little to do with the market design. It influences the organization and sets incentives. Possible contracts can be based on unit prices, lump-sums, guaranteed maximum prices, target cost prices, and reimbursable costs.

Table 11.4 summarizes the impact of the owner's procurement decisions on the market design. It also shows very clearly that the owner is designing his own market without the influence of contractors. Owners must, however, consider the business cycle; the following observation hold true for normal times. In case of extreme booms or busts, the evaluations will be different.

Table 11.4 makes it overwhelmingly clear that the possibility to design the market favors the position of the owner. The argument hinges on the fact that construction projects are singular and that each project creates its own market. This claim applies only to microeconomic analysis, not to macroeconomic perspectives. As we are considering prices and quantities in this microeconomic analysis, macroeconomic views are not relevant. There are plenty of data on the macroeconomic level for the housing market, the office market, the warehouse market, and so on. These are expressed in some unit, most often square meters or cubic meters and based on average qualities; the data are aggregated from different projects. Such data are helpful in the initial design stages before a contract is signed but no owner buys an average building nor does the owner pay an average price. To provide an analogy: Once a car buyer has configurated a BMW M5 limousine and made the decision to buy one, all prices of the other cars lose direct influence. The direct market is limited to the BMW dealers in the vicinity of the buyer.

Table 11.4 Procurement decisions and market designs.

Procurement	Buyer	Contractors	Possible problems
Open tendering	One	Many	Market congestion
Selective tendering	One	5–10	Collusion among contractors
Negotiated contract	One	One	Adverse selection
Design-bid-build	One	Many	Process quality suboptimal
Design-build	One	Many	Design/process quality suboptimal
Design-build-operate	One	Many	Operation suboptimal

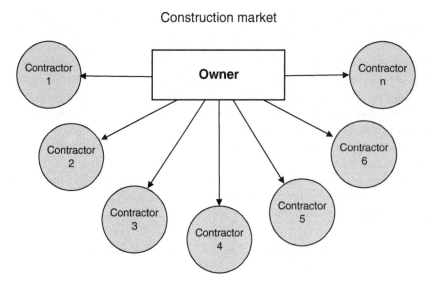

Figure 11.6 Construction market for a singular project.

The same happens once an owner has decided on a design. This does not deny that prices from other projects procured by other owners do not have an indirect influence on the price determination of the contractors. Whatever the arguments, the owner creates in most cases a single-product market with a competitive number of contractors.

By the procurement method, the owner creates a market for a singular good, and by the contract, the owner can determine the rules of the game. This is a very powerful position. It is easy to say that contractors have all the right to abstain from signing a contract, no owner can force them. However, except in boom times, contractors must chase contracts in order to keep the level of production and to cover their fixed costs. Most of the time they must accept contracts that would be very different if the contractors were in a position to formulate them.

Figure 11.6 shows the typical market for a singular construction good.

11.7 Supervision of the Construction Process

Different from all manufacturing happening behind the closed doors of a factory by exclud-ing the buyer, construction sites belong to the owner, who will be present most of the time, day in and day out. Construction supervision is a professional task and is delegated to architects, engineers, or project managers. In case of defects, action typically happens immediately unless the defects are hidden.

During construction, the owner not only has the right to grant or deny access to the site where the contractor has set up production and to supervise all work but also is involved in the construction process. The owner must provide permits, control safety, and provide information. Typical organizational tools are weekly progress and coordination meetings.

Table 11.5 Consumer in commodity markets and owners in construction markets.

Category	Buyer of commodities	Owner
Entity	Single person	Multi-personal Multi-organizational Professional
Behavior	Utility maximizing	Satisficing as consumer Profit optimization based on satisficing behavior as investor (producers)
Procurement	Buying of defined goods and services in existing markets	Designing the construction good Designing markets
Demand	Direct Emotional involvement Construction of demand curve possible	Derived No emotional involvement No individual and no market demand curve
Information	Information advantage on the seller's side	Information advantage depends on phase; until signature, the owner has the advantage; later, the contractor
Contracts	Provided by seller	Provided by owner
Market	Many or few sellers	Many contractors

A buyer of a commodity has no influence during production. The owner of a construction project has far-reaching rights and responsibilities. This type of production is called team production. The result in team production does not only depend on the contractor but also on the owner. The contractor is not the producer; the team is.

11.8 Summary

The discussion of the position of the owner in construction markets highlighted some rather important differences in comparison to the consumer in commodity markets from whom consumer theory was developed. It becomes clear that consumer theory cannot be applied to the owner in construction markets. Table 11.5 condenses the information developed in the previous chapters.

References

Greenhalgh, B. and Squires, G. (2011). *Introduction to Building Procurement*. London: Spon Press.

Gruneberg, S. (1997). *Construction Economics: An Introduction*. Houndsmill: Macmillan.

Langford, D. and Male, S. (2001). *Strategic Management in Construction*. Oxford: Blackwell.

Masterman, J. (2003). *An Introduction to Building Procurement Systems*. London: Spon Press.

12

Theory of the Contractor

If competition is strong in a certain sector, we would expect low profits. Profits as percentage of turnover in Germany from 2002 to 2013 varied between 4.0% (2003) and 6.7% (2013). A recession lasting until 2006 is the reason for lower profits in 2003 and a boom for the higher profits in 2013. The structure of the sector with its small average firm size leads to the legal organization of the vast majority as individual proprietorships. This means that we have to subtract compensation to the owner, interests on equity, and taxes to find the net profit. For a company with a turnover of 2 million USD, a profit of 4% amounts to 80,000 USD. Managing a company with a turnover of 2 million USD certainly should generate an income of 80,000 USD. This means that interest and tax payments cannot be paid from this margin. Equity is low, maybe 15%. Assuming investments in the firm of 500,000 USD and 7% interest (equivalent to average profit with stocks) yields another 35,000 USD. This hypothetical amount is the loss of an average individual proprietorship in our case based on opportunity costs. Let the turnover of the same company increase in a boom (2013) to 3 million USD and we will get a profit of 201,000 USD (3,000,000 × 0,067). Increasing income and interest proportionally gives an income of 120,000 USD and 52,500 USD interest, a total of 172,500 USD. The profit in 2013 of 28,500 USD before taxes is not enough to cover the loss of 2003 (35,000 USD).

In 2015 during a boom, profit on turnover for the largest contractor in Germany, a public limited company, was 1.58%. Half of this was used to pay dividends. Both the average small company and large contractors certainly are not facing windfall profits in boom times. It rather seems that construction markets show textbook behavior with zero economic profit for the contractors in the long run.

A theory of the contractor must explain how such firms act to survive. To this purpose, I will analyze the contractor as an entity, its tasks, behavior, available information, bidding, pricing, and finally production.

12.1 The Contractor as an Entity

Construction firms range from very small one-man firms to global players such as China State Construction Engineering Corporation with a revenue of 235 billion USD in 2020 as the world's largest construction firm with a profit on turnover of 1.5% (Fortune 2021). The

Construction Microeconomics, First Edition. Christian Brockmann.

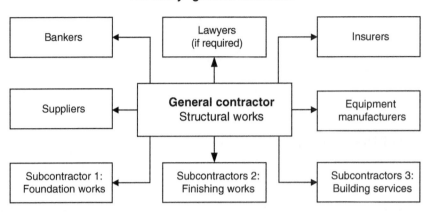

The entity "general contractor"

Figure 12.1 shows the diagram with the following boxes:

- Bankers
- Lawyers (if required)
- Insurers
- Suppliers
- **General contractor** Structural works
- Equipment manufacturers
- Subcontractor 1: Foundation works
- Subcontractors 2: Finishing works
- Subcontractors 3: Building services

Figure 12.1 Contractor as an entity.

wide range of the phenomenon construction firm makes it again necessary to define a typical case, and this will be neither the one-man firm nor China State. The typical case depends on the choice of the owner, whether he lets a single contract to a general contractor or many different contracts to specialty contractors. A general contractor takes responsibility for coordinating and delivering all construction activities. For this purpose, the general contractor relies on the help of many specialty contractors. Figure 12.1 shows a general contractor who completes the structural works with his own workforce and all other works with the help of subcontractors.

Owners often decide to choose a general contractor. They also can opt to employ management contracting where the management firm controls all construction activities (Greenhalgh and Squires 2011). In such an arrangement, it is also possible that the owner himself takes the role of managing all construction works (Figure 12.2). Sometimes, owners even provide important equipment and supplies. In such a case, it seems to be justified to speak of an employer instead of an owner. This setup shows how powerful the role of an owner can become by his own choice.

Each way of organizing the construction project has certain advantages and disadvantages. Of economic importance is the power distribution between owner and contractor and again, there is no way of characterizing the owner has David who faces a Goliath in form of one or more construction firms. I will take the organization with a general contractor (Figure 12.1) as exemplary for the following discussions. In such a case, the owner and the general contractor team up to complete the construction works, albeit with contradicting economic goals: Own utility versus own profit maximizing. This is a zero-sum game; one's gains will be the other's losses.

12.1.1 Cooperation

Contractors can decide to cooperate by forming a joint venture or a consortium. I reserve the term joint venture for a cooperation between firms from the same industry and the term

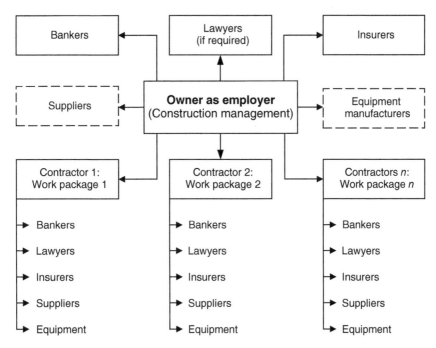

Figure 12.2 Owner as employer.

consortium for firms from different industries. Construction joint ventures (CJVs) are the typical case, and consortia are formed under special circumstances for the construction of complex projects such as airports, oil refineries, or power plants. In all these cases extensive works must be provided by plant manufacturers.

CJVs are always contractual joint ventures with an internal joint venture agreement and an external construction contract. Equity joint ventures only have an internal joint venture agreement, and they are used in other sectors. CJVs can be formed by contractors from the same or from different countries. In the latter case, we talk about international construction joint ventures (ICJVs). Most owners will demand from CJVs or ICJVs a clause of joint and several liability, i.e. the owner can address a claim against a single contractor or all contractors in a joint venture. Figure 12.3 shows an example of an ICJV.

CJVs allow contractors a choice between hierarchy (subcontractor) or cooperation (CJV/ICJV). This choice does not substantially change the relationship between owner and contractor.

12.1.2 Organization

A general contractor must perform different activities in order to complete a construction contract. The staff functions comprise a bidding team, a purchasing department, a contractor's yard with storage areas, a mechanical workshop, and administration. Project teams work in line functions (see Figure 12.4).

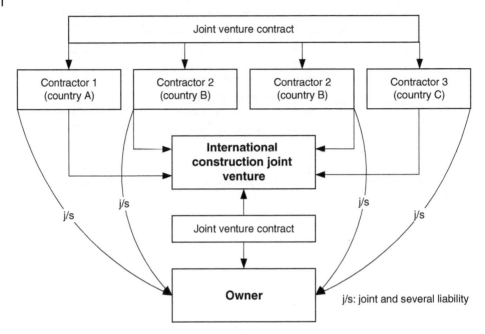

Figure 12.3 International construction joint venture.

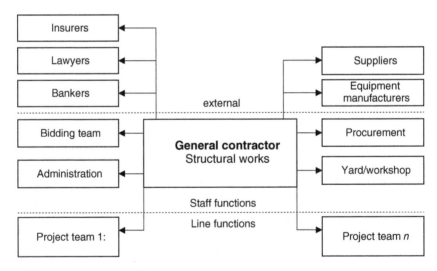

Figure 12.4 Contractor's organization.

12.2 Tasks of the Contractor

We have seen that the owner develops the product design together with architects and engineers and that he determines via the contract the construction period. For submitting a bid, the contractor can only optimize the construction process. He must find the cost minimum for a given construction period (Figure 12.5). Each contractor will find a

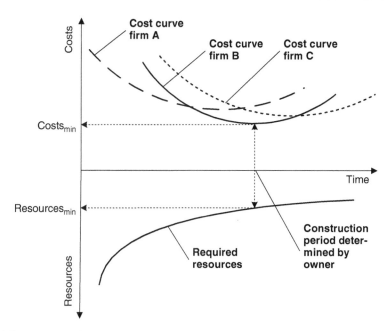

Figure 12.5 Cost minimization through technology.

somewhat different process technology and one of these will have a cost minimum at the contractually stipulated time or be close to it. In Figure 12.5, firm B has found a technology with the cost minimum for the contract period. A technology also has a fixed set of resources (labor, materials, equipment). From one singular project to the next there will be adaptations to the technology. Even in the case of consecutive projects of the same type, there might be some technological advances, these are innovations with the goal to lower the costs. With the product design provided by the owner, only process innovations are possible for the contractor and these can only have the goal of cost minimization. A contractor who delivers a higher quality than defined by the contract through advanced technology is not maximizing profits and thus acts not rational.

Figure 12.5 suggests that the most innovative contractor with the most suitable technology for the project will win the contract. This would be the ideal economic result. Unfortunately, contractors are not able to estimate the contract price correctly and they have no idea about the exact shape of the cost curve. In the worst case, the contractor with the largest estimating error will come out on top of the bidding contest. Instead of the smartest firm winning the contract, this will be the most erroneous. The whole sector including the owner is suffering from this problem.

12.3 Behavior of the Contractor

Recall the assumptions for producers' behavior from Chapter 4: (i) factors of production (inputs) are scarce; (ii) producers face investment constraints; (iii) producers maximize their profits; and (iv) producers decide rationally.

Table 12.1 Differences between accounting and economic profits.

Accounting profit	Economic profit
Revenue	Revenue
• Labor costs	• Labor costs
• Material costs	• Material costs
• Equipment costs	• Equipment costs
• Subcontractor costs	• Subcontractor costs
• Site installation costs	• Site installation costs
• Company overhead	• Company overhead
	• Interest on equity
	• Owners' salary

Contractors work in a highly competitive environment, and certainly inputs are scarce and the cost of these inputs reflect the relative scarcity. The costs of inputs differ, however, around the world. Most prominent is the range of wages, from less than 1 USD per hour to more than 40 USD per hour. Construction firms on average have a low equity which leads to strong investment constraints. The low profit margins require constant attention to profit maximization. So far, the assumptions for producer behavior hold also for contractors. The last assumption needs modification: The complexity and singularity of projects lead to incomplete and ambiguous information and in turn to mistakes. Rationality for contractors is bounded, i.e. they intend to behave rationally but fail to do so at times. Any modeling of contractors' behavior must refer to bounded rationality. A result of bounded rationality is the inaccuracy of estimates and schedules.

Further data support the conclusion that I have drawn on profits and equity. Runeson (2000) reports for Australia profits of 4% of turnover without specifying the point of reporting in the business cycle and without reference to the legal form of the firms. These small profits lead to short lifespans for the firms: Only 44% of the Australian building firms have been in business more than 10 years. The data by Runeson refer to profit according to accounting principles.

There is a difference between accounting profits and economic profits, as shown in Table 12.1. Accounting profits are calculated on the basis of generally accepted accounting principles, while economic profits are calculated according to the principles of opportunity costs. One difference is the evaluation of assets. This is based on book value in accounting and on market value in determining economic profits. In addition, opportunity costs comprise interest on equity and payment of the owner's salary in case of owner proprietorships.

The theory of perfectly competitive markets posits that economic profits are zero in the long run because the lack of entry and exit barriers attracts competition in economic upturns (booms), which in turn drives economic profits toward zero. The opposite happens in downturns (busts), when exits lower competition and take pressure from pricing, thus allowing an upward trend toward zero economic profits.

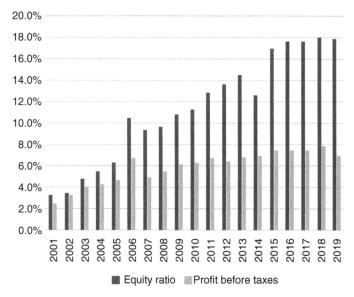

Figure 12.6 Equity ratio and profits for structural works in Germany. *Source:* Data from Hauptverband der Deutschen Bauindustrie 2020

While profits are low in construction, so is equity, as data from Germany for the period between 2001 and 2019 show (Figure 12.6, Hauptverband der Deutschen Bauindustrie 2020).

The first part of the period until 2005 was a downturn of construction activities, followed by a slow and later accelerating upswing. Low equity is one of the reasons why entry into the construction sector is easy. Profit before taxes as a percentage of turnover lingers below 4% in downturns (2001–2005) and grows to 7% in upturns (2012–2019). Equity never exceeds 18% and is clearly below 10% in a recession. The profits are determined by accounting principles. Coupled with the introductory discussion at the beginning of the chapter, the data suggest zero economic profits for the sector on average. The figure also highlights the influence of the business cycle.

12.3.1 Strategy

Porter (1985) distinguishes three main strategic options for producers:

1) The focus (cost leadership or differentiation)
2) The rules (change or acceptance)
3) The place of competition (main or niche market)

To a large degree, Porter describes the option of competition in perfect markets (cost leadership, acceptance, and main) or in imperfect markets (differentiation, change, niche). What are the options in construction? Since the product design is not in the hands of contractors, differentiation becomes difficult for them to say the least. It rests on the promise to provide more process quality in the future and the validity of this promise is hard to ascertain for the owner. An owner who decides for a low-cost award assumes either no influence of process quality on the product or equal process quality provided by all bidders.

The reputation of a contractor is similar to beauty: It lies in the eyes of the beholder and most often owners refuse to recognize it: Differentiation is typically not an option for contractors.

As the owner writes the contract, the contractor can only choose between the options to accept or to abstain. Contractors must accept the rules as stipulated by the owner or they are irrelevant for the project at hand.

Thus, the remaining strategic option for contractors is their choice of the place of competition: Main market or niche. General contractors opt for the main market, specialty contractors for a niche.

Many authors of marketing (e.g. Kotler and Armstrong 2020) describe the importance of the marketing-mix consisting of product, price, promotion, and placement (4Ps). As discussed, contractors cannot influence the product. Placement is also not an option; the owner provides the landed property. Promotion in construction suffers from the heterogeneity of construction products and the limited amount of purchases. Thus, the price is in this context the main variable for the contractor.

Since we find that contract goods predominate construction markets, contractors cannot produce in advance. This means that they have no control over their production program. The market is demand driven without the possibility to level production by temporarily producing on stock. On the other hand, contractors must be able to start construction on short notice, which either means that they must hold back resources for possible new contracts or that they must rely on markets to provide the necessary resources. Flexibility is the recipe for survival, and contractors achieve flexibility in many instances by outsourcing.

12.3.2 Legal Organization

A major influence is the legal organization. Basic options are stock corporations, limited liability corporations, and owner proprietorships. The legal organization influences ownership, management, control, reporting, as well as the ability to procure loans and profit. The last point is of special economic importance, and it allows us to return a third time to the question of contractor's profits.

I assume for the following analysis that firms with more than 200 employees are all stock corporations, firms with at least 20 and less than 199 employees all limited liability corporations, and firms with less than 20 employees all owner proprietorships. Taxes are comparable for all types of legal organization (around 30%). This allows us to calculate average profits and taxes per class by using published information.

Table 12.2 provides data for German contractors with different legal organizations for the year 2019, a boom year. Assume the sum of financial assets per class allows calculating the average equity per class. With an estimated interest rate of 5% and a dividend of 25% for stock corporations, it is possible to calculate the financial burden per firm. Owners in proprietorships must subtract their salary from profits as opportunity costs; a value of 85 000 € per owner is conservative. It is not surprising that the net profits for owner proprietorships are negative in Table 12.2. We started from the average profit for all firms. The actual profit

Table 12.2 Estimated economic profit for structural works in Germany in 2019.

Firm	>200 employees	50–199 empl.	20–49 empl.	1–19 empl.
Assumed legal structure	Corporation	Limited liability	Limited liability	Owner proprietor
Number (2018)	260	2287	6345	65,945
Turnover 2019	29,000,000,000 €	40,300,000,000 €	31,900,000,000 €	41,800,000,000 €
Average turnover	111,538,462 €	17,621,338 €	5,027,581 €	633,862 €
Average profit (7%)	7,807,692 €	1,233,494 €	351,931 €	44,370 €
Assumed taxes (30%)	2,342,308 €	370,048 €	105,579 €	13,311 €
Assumed capital/firm	5,000,000 €	3,000,000 €	1,000,000 €	100,000 €
Assumed equity	900,000 €	540,000 €	180,000 €	18,000 €
Interest (5%) on equity		27,000 €	9000 €	900 €
Dividends (25%)	1,951,923 €	—	—	—
Proprietor salary	—	—	—	85,000 €
Net profit/firm	3,513,462 €	836,446 €	237,351 €	−54,841 €
Total net profits	913,500,000 €	1,912,951,000 €	1,505,995,000 €	−3,616,475,500 €
Economic profit		715.970.500 €		

is highest for small firms, exactly because management salary is subtracted later. Different from the assumptions in Table 12.2, the profit for owner proprietorships is higher than 7% and that for stock corporations is lower. Only the overall economic profit of roughly 716 million € is meaningful.

The result of the calculation is a net profit of not even 10,000 € per year and company in boom times. This is a clear signal of a very competitive result, close to the ideal of zero economic profits for competitive markets.

Interesting is a comparison with the year ending the recession from 1995 to 2005. The comparison between 2019 (boom) and 2005 (bust) allows us to have a glimpse at the long-term economic profits in construction. Using the actual data for 2004 (number of companies) and 2005 (turnover, average profit) and the same assumptions as above allows calculating the economic profit. This amounts to a loss of €2.600 million during a recession compared with a profit of €716 million at boom times. The conclusion of the several discussions of profit must be that construction sector works close to the ideal of a perfectly competitive market with zero economic profit. There is even a small doubt that the long-term average economic profit is below zero.

This might be different for the individual contractor. Some will still make a profit in a downswing and others will go bankrupt. All are fighting to survive in a competitive industry.

It is not surprising that contractors fight for their profits; it is a question of survival. This also impacts the often-discussed adversarial climate in the construction sector. It is not easy to be generous for contractors when the alternative is insolvency.

12.3.3 Growth of the Firm

Contractors have in principal the choice between five different markets: regional, national, international, multinational, and global (Brockmann 2009). For each market, different characteristics can be identified for the firm's orientation, its behavior, the segments for which it produces goods, and for the type or organization. Table 12.3 gives a synopsis of the market characteristics.

Construction firms can grow from within or by mergers and acquisitions. In the first case, companies can start on a regional level and then expand to span the globe such as China State Construction Engineering Corporation. However, most firms remain on the regional level.

Figure 12.7 shows an expansion path for contractors. As a firm grows from one level to the next, it must acquire additional know-how and resources. At the same time, it will lose some competences for smaller projects. A global player is not the best choice if you plan your one-family home. Each type of project requires specific knowledge and an organization to put this knowledge to work. Global players have too big an overhead for one-family homes. Accordingly, they are not interested in such projects and lose the required know-how. Given the low equity and small profits of contractors, it will take a long time to grow from a regional to a global contractor. This growth is much faster if someone from the outside is willing to invest money to buy know-how and resources by mergers and acquisition.

Global contractors must have specific characteristics (Brockmann 2009). Among them are high competencies on the organizational, cultural, social, diplomatic, technological, financial, estimating, and negotiating level coupled with a global reputation. This allows them to

Table 12.3 Synopsis of market characteristics.

Market	Orientation	Behavior	Segments	Organization
Regional market	Local network	Local	Few segments, specialized	Local headquarters
National market	Many local networks within the home country	National	Many segments, diversified	National headquarter, local subsidiaries
International market	Many local and some international networks	Ethnocentric	Many national segments, few international segments, diversified at home, specialized abroad	National headquarter, local subsidiaries, overseas department
Multinational market	Many local and many international networks	Polycentric	Many national and international segments, diversified	National headquarter, local subsidiaries, overseas department, international acquisitions
Global market	Megaprojects	Polycentric	Different segment for megaprojects	Same as multinational

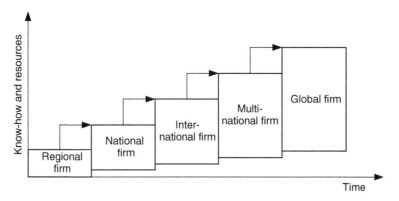

Figure 12.7 Expansion path for contractors.

tackle a megaproject in any place of their choosing. Global contractors will not sign up for small or midsize projects, here the local or national competitors have a cost advantage.

Runeson and de Valence (2009, p. 210) describe global contractors as the *"new construction industry"* and they continue: *"The new construction industry is an oligopolistic industry, with a small number of very large firms selling a differentiated product in a global market, competing through value for money"*. I disagree for several reasons, which I have explained above. The new construction is not selling a differentiated product, instead owners request singular products, especially on the level of megaprojects. Contractors offer in response a corresponding performance with the characteristics listed above. There is strong competition most of the time with prequalification, selective tendering, and a large enough number of competitors fighting for the job. Contracts are awarded mostly on price and not on value. This is especially true for heavy civil projects, as the owners are public entities.

Risks and rewards are both high. Nobody knows the average profit of the global players. I have been involved in projects where the contractor had a win of 20% and in others where the loss was 30%. Successful global players will probably have above-average profits and unsuccessful ones will close in on bankruptcy faster than usual.

12.4 Information of the Contractor

Contractors can only react to the demand from owners. Typically, they are well informed about the owners' activities in their area. This is part of the acquisition efforts. It is also in the interest of the owner to make their intentions known to create a thick market. Through participation in submissions, contractors usually know very well the general price level of their competitors. They do not know the details of the competitors' estimates, nor their underlying chosen construction technology.

During contract negotiations, a contractor faces an owner with access to detailed information from all submitted estimates, while he only knows his own estimate. In contract negotiations, an owner will typically pressure a contractor to lower the price by referring to the overall price of competitors or the comparative prices of certain items. The contractor is in no position to ascertain the truth of the owner's statements and cannot talk to the

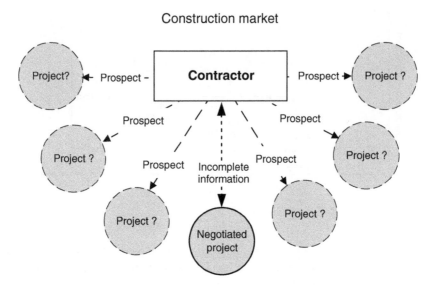

Figure 12.8 Information and prospects of the contractor negotiating a contract.

other competitors (collusion). This is an especially unbalanced situation so the estimate is just that: a price idea without certainty. The information of the contractor is much more limited and ambiguous than that of the owner. Contractors drop out of the bargaining process when they feel that the price does not cover the costs; they are not sure about it.

Figure 11.3 sums up the information situation of the owner with full and detailed information about all offers. The contractor is not in a similar situation; he has only part of the information that the owner has and all other possible projects on the market are just that – possibilities or prospects (Figure 12.8). Contractors must decide under highly ambiguous circumstances.

Once the two parties sign the contract, the information advantage shifts to the contractor. Even if the owner or a couple of his representatives supervise the site continuously, they will not have all the full information that the contractor can access. The breadth and depth of information that is available to the contractor is not obtainable to the owner. Labor productivity is one example. Many contractors use a reporting system with daily updates for the work completed by each worker, which in turn allows calculating productivity (output/hour). This information remains privy to the contractor.

12.5 Bidding

Bidding starts in most cases with the contractors acquiring the tender documents. There, they will find a product design, a contract, and tendering conditions advanced by the owner. Few will be the owners who do not draft the documents to their fullest advantage. Thus, the first task for contractors is to study of the documents and decide whether the risk distribution is acceptable.

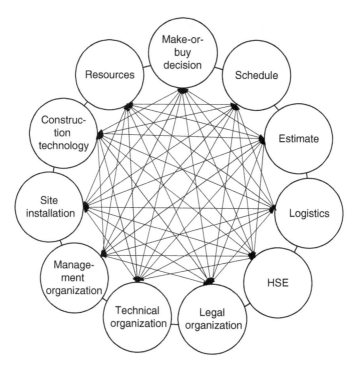

Figure 12.9 Integration of partial plans during bidding.

The next step includes the development of a production plan and an organization together with the associated costs. Doing this, the contactor relies on his past experience but needs to expand it into the unknown because of the singularity of product design, location, time, and group of stakeholders. The singularity might vary to large degrees between a garden wall and a subway project. As nobody knows the future, this involves guesswork. To a certain degree, it is not science but art.

The bid that the contractor hands over at submission date must integrate 11 partial plans (Figure 12.9). These are interdependent and must be aligned. They include the amount and type of work to subcontract (make-or-buy decision), resources (labor, material, equipment), construction technology (transformation of inputs into outputs), site installation (production factory), schedule (sequence of activities), logistics (transportation and storage), HSE (health, safety, and environmental protection), plus the management, technical, and legal organization. All is condensed by the estimate into the financial offer.

Considering time constraints during the bidding period ranging from 3 weeks for a mid-size project to 12 months for a megaproject, all partial plans cannot be developed to more than a sketchy detail. It takes a leap of confidence to submit the offer.

12.6 Contractor Pricing

The practice of estimating varies across the world and among firms in a given country. Every firm can determine the submission price in any way it deems suitable.

Having stated this, there are probably three main approaches: (i) parametric pricing, (ii) unit pricing, and (iii) resource pricing. The degree of detailing increases from (i) to (iii) and the information requirements also increase in the same direction.

Parametric pricing uses information from completed projects and is only tentatively adjusted to the project at hand. It is based on a price per unit, most often $/m^2$, sometimes $/m^3$. Minimal adjustments are made according to the required quality.

Unit pricing refers to the prices of parts of the structure such as walls, slabs, windows, or heating systems. Prices are again per unit of wall (m^2) or door (piece). Quality adjustments can now be made on the level of units. There is no need for an average quality assumption for the structure. Again, the price information is taken from the past and not much adjusted to a specific project.

Resource pricing still requires more information. A wall consists maybe of ready-made concrete, rebars, and tying wire. Formwork with ties and supports guarantees the required shape. Labor is necessary to erect the formwork, to cut, fix, and tie the rebars and to pour and compact the concrete. A crane transports the heavier materials, a pump might place the concrete and vibrators will compact it. In resource pricing, all this is considered separately, including loading, storage, and transportation costs. A single item such as labor for pouring concrete cannot be estimated with greater accuracy than parameters, but misjudgments will tend to cancel each other out. The secret of resource pricing lies in the fact that the project is broken down into as many items as possible and each item will be priced independently to the best knowledge of the estimator. The prices will be inaccurate but if there is no bias, they will also be uniformly distributed. Overall, this enforces the tendency to determine costs in such a way that the overall difference between planned (estimate) and actual (execution on site) becomes small.

Several studies at the University of Applied Sciences Bremen showed that for all single units there were differences between planned and actual costs typically by $\pm 20\%$. However, the variation of the contract price was in the range of $\pm 2\%$. The projects were typical building projects of no great complexity. Depending on complexity, these variations can be much larger. I have participated in a tunnel project where the construction costs to the contractor were more than three times higher than the contract price; another tunnel project had a cost difference of €100 million and a transportation project a loss of €400 million to the contractor. Such data are typically private, the press publishes only cost overruns for the owner (Flyvbjerg et al. 2003).

Besides a bias, the most important problem in estimating is the determination of labor productivity, wages, and indirect costs. The labor productivity must be an educated guess as nobody knows the workers who will do the job in the future. While labor productivity is estimated for each activity separately, the wage is determined as an average for the complete labor force. Indirect costs mostly comprise the costs for the site installation (i.e. the factory and headquarter overheads).

The estimate is a model of what is expected to happen at a later date on site – it is a model of the future.

The typical process in resource pricing is shown in Figure 12.10.

Point 9 in Figure 12.10 describes markup pricing. The marginal cost approach that is so important to mainstream microeconomics is not used in construction. The reason is simple: The marginal costs cannot be determined. Marginal costs depend on mass production

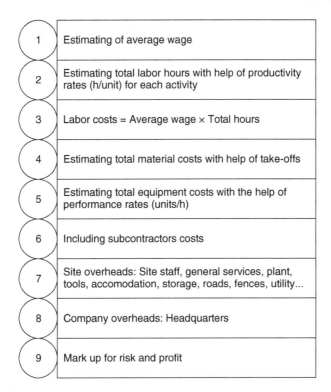

Figure 12.10 Markup pricing in construction.

and the cost data that come along with this continuous experience of repetition. Contractors do not add to turnover by increments but by leaps. Take the contractor of one-family houses who builds 5–10 houses per year. The decision to add one house increases yearly turnover by a minimum of 10%. Thus, one of the central tools of microeconomics (marginal cost) is not available in construction.

Markup pricing contains a considerable number of risks:

1) It is impossible to determine the exact average wage. The estimator envisages a team of workers with different qualifications and different wage levels. The team that will work on site will differ from the estimator's model.
2) The labor productivity rates cannot be exact. They have a commanding influence in high-wage countries and pose the largest risks, but productivity depends not only on the qualification of the workers but also on their motivation, on the production process, on weather conditions, and on changes to plans.
3) Determination of productivity rates for equipment is less risky. Equipment needs no motivation by itself. However, the qualification and motivation of the operators play a role. Further impacts that are hard to determine are the efficiency of the production process (i.e. the amount of waste) and weather conditions.
4) Material costs are less risky in general. Take-offs carry no other risk that lack of time or care. Estimators will use price information from the market while estimating. Purchases

will be made right after signature for the duration of the project. Nevertheless, there are times when material prices fluctuate wildly.

5) What holds true for materials also applies to subcontractor prices. They are rather predictable, because many of the risks are transferred to the subcontractor, e.g. labor productivity.
6) Site installation costs carry the double risk of wrong quantities and wrong prices. The estimator must determine the quantities based on the assumed construction technology.
7) Company overheads depend on yearly turnover. This is only predictable to a certain degree.

The list makes it clear that estimators face many uncertainties, and information is incomplete. The more typical the building or structure is for a contractor, the smaller are the risks.

12.7 Production

The production of contractors is determined by the character of construction goods: they are contract goods which are produced according to the owner's design.

In the case of exchange goods, producers supply the market. They consider the potential success of a product and, if they are convinced of it, they start production. Doing this, they can harmonize the determinants of the product and the production process. Contractors must develop a production process for a given product. Ideally, the contractor with the best construction technology will win every bid. This is the main point where contractors can gain a competitive advantage. Due to the risks and uncertainties of the estimate, often the contractor with the greatest estimating error outperforms the one with the brightest idea.

12.7.1 General Characteristics

Today's markets (not only in construction) are mostly buyers' markets. The buyers decide about the success of a firm. In times of product scarcity, markets can become sellers' markets. This used to be the general case in communist countries and directly after the Second World War in Europe. The idea of supply and demand simultaneously determining quantity and price focuses on the interaction between buyer and seller. This balance shifts in construction toward the owner's product design. Figure 12.11 describes how the product design determines the contractor's output; the transformation direction is not from inputs (factors of production) to outputs (structure) but the inverse.

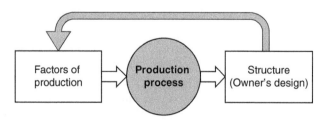

Figure 12.11 Production process in the construction sector.

Table 12.4 Production differences between construction and manufacturing.

	Construction	Manufacturing
Firm	Headquarters and production in separate locations	Headquarters and production often in joint location
Personnel	Assembly work on different sites	Factory work in fixed place
Financing	Low equity	Higher equity
Accounting	Construction accounts	Industry accounts
Purchasing	Small packages, short-term contracts, decentralized delivery	Large packages, long-term contracts, centralized delivery
Marketing	Before production	After production
Organization	Project organization	Process organization
Production	On demand, single batch, site production	On stock, mass factory production
Buyer	Integrated in production	Separate from production
Production process	Mechanized, driven by labor as production factor	Automated, driven by materials as production factor

The production characteristics for contractors differ considerably from those for manufacturing of exchange goods (Table 12.4).

Separation of headquarters and sites lead to information and controlling problems. Assembly work requires daily transport of the workforce or on-site accommodation with impacts on private lives; transportation costs additional money and time; motivational influences are possible. Low equity puts constraints on financing factors of production; consequentially not the best but the available equipment is used. Differences in accounts sometimes lead to incomparable data sets. Small-batch, short-term purchases with delivery to variable sites increase costs and make logistics more complicated. Marketing (acquisition) before production is typical for contract goods and leads to all problems connected with asymmetric information, including opportunism. Project organizations are often more complex, do not allow to spread fixed costs, and as a consequence, do not warrant detailed planning nor purchase of specific asset. On demand, single batch and site production all increase production complexity; site production also depends on the weather. The integration of the owner causes multiple problems. He has to be aware of his contributions (process evidence) and must provide them on time. From the contractor's point of view, the owner is a production factor from the owner's perspective it is the opposite. Labor intensity of construction processes leads to an increased importance of motivation, skills, and information.

Some things are out of the contractors' control. Owner behavior, soil conditions, and weather all influence the production process in unpredictable ways. The production results are stochastically distributed due to these outside influences. Motivation of the workforce also has a high and variable impact on production. Supply chain fragmentation affects production to a much larger degree than in manufacturing. The main influences on production are shown in Figure 12.12.

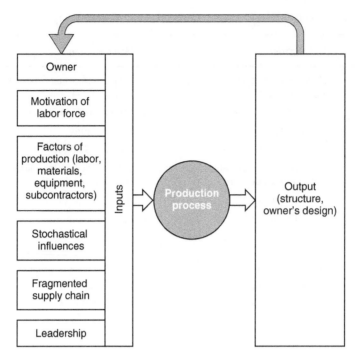

Figure 12.12 Detailed production process in the construction sector.

12.7.2 Production Determinants

Production processes differ with regard to certain determinants, and these influence the complexity of the processes. Most determinants make construction processes complex. Among the different production types are:

- Production line/work shop/site construction
- Parallel/variable production
- Manual/automated production
- Continuous/discontinuous production
- Mass/single-batch production

12.7.2.1 Production Line, Work Shop, Site Construction, Parallel, or Variable Production

These process-oriented determinants are influenced by the location of production and the possibilities to support production. Features of these determinants are provided in Table 12.5.

The production line features a product that is assembled while moving from workstation to workstation. While the product moves, labor and equipment are stationary. Production lines require detailed planning and high investments. The activities at each workstation are repetitive. We can find production lines in construction only on megaprojects such as tunnels or bridges with a large number of similar elements (tunnel or bridge segments).

Table 12.5 Locational features of production methods.

Method	Location	Output	Subprocesses	Activities
Production line	Fixed (factory)	Movable	Stationary	Homogenous
Workshop	Fixed (work shop)	Movable	Stationary	Heterogenous
Construction site	Variable	Stationary	Mobile	Heterogenous
Parallel prod.	Fixed	Mobile/immobile	Mobile	Homogenous
Variable prod.	Fixed	Mobile/immobile	Stationary	Heterogenous

Production lines are exemplary for mass production in manufacturing but unusual in construction.

Products are stationary in workshops and the required equipment and labor are brought to the product as required. The consecutive activities vary in type and intensity. A typical example is a garage where cars are repaired. A workshop in construction might be a room in a building where different activities are carried out by different trades. In a garage, the mechanics are doing all the work; in a constructed room we find first the concrete and rebar people, then electricians, plumbers, tilers, joiners, and painters.

A construction site encompasses many workshops and the activities at the different ones are interlinked; coordination thus becomes a problem. While production lines feature technical coordination, management coordination is the choice on construction sites. The output per site does not allow the magnitude of investment that is characteristic of mass production.

One management answer to the coordination problem is scheduling, another lean construction. In lean construction, the sequence of activities is fixed (same as scheduling) but the resources are also aligned in such a way that all successive activities in one area take the same amount of time. Thus, a production train with an engine and coupled trades as wagons pass through the building or structure at constant speed. There are similarities to the production line (constant speed) while in contrast, labor and equipment are mobile and the product remains stationary.

These determinants influence productivity, flexibility, and susceptibility to interference. There is a trade-off between productivity on the one side and flexibility/susceptibility to interference on the other side. Mass production in a factory is shielded against interference. On the other hand, an owner who is present on the construction site every day will frequently interfere to achieve the intended results. Other potential causes of interferences in construction are the soil conditions, the embeddedness of the construction site into the social and ecological environment, as well as weather conditions: production in construction must be flexible.

Parallel and variable production differ in the temporal attribution of resources and product. In parallel production, there are at the same time enough resources available so that the same work processes can be executed in different locations simultaneously. There is no need to change the resources. The characteristics of variable production are that successive work processes are executed in one location only; therefore, resources must be used in a

Table 12.6 Advantages and disadvantages of production methods.

Method	Productivity	Flexibility	Resilience
Production line	High	Low	Low
Workshop	Middle	Middle	Middle
Construction site	Low	High	High
Parallel production	High	Low	Low
Variable production	Low	Middle	High

sequential way. Typical examples for parallel production are two or more production lines next to each other and for variable production the garage for car repairs; even if a garage has two or more hoisting platforms, the work processes at each one will be different.

In construction we can find both, parallel and variable production. However, variable production is predominant. We most often find variable production in buildings; in infrastructure projects, parallel production takes place often; for a long bridge, work can be executed on several piers at the same time. Even in tunneling, quite often more than one tunnel boring machine is used. Industrialized construction (prefabrication) also has many characteristics of parallel production. The advantages and disadvantages of the different production methods are summarized in Table 12.6.

12.7.2.2 Automatization

Depending on the equipment used in production, we can differentiate between tool-supported, equipment-supported, mechanical, automatized, and digitalized production. Economists do not follow fads, but rather make a cost/benefit analysis to decide for the type of automatization that best suits the purpose. This requires mentioning because some authors complain about the lack of digitalization in construction as if this would be a goal by itself.

The more equipment a contractor uses, the higher are the fixed costs. High fixed costs can only be justified by a large output in one production facility (i.e., in one construction site). Even large construction projects with maybe 100 million USD production value per year and a construction time of three years have a relatively small output compared to car production. The level of output does not support the same investment in equipment. At the low end in construction are renovation jobs such as exchanging 20 windows in a building or even just one door. It is certainly possible to conceive some digitalized equipment for the job; unfortunately, the costs would be too high as the same equipment might be of little use on other projects and because of relatively high transportation and setup costs. Large investments need a steady and large flow of employment.

Depending on the work at hand, we find in construction tool-supported, equipment-supported, and mechanical production. Most activities are tool-supported. Equipment-supported functions often include moving earth or lifting activities. Examples of mechanical production are tunnel boring machines, paving machines in road construction or some bridge girders. Automatized production can be found in some prefabrication factories but they are seldom.

The right choice depends on the costs of the factors of production. In high-wage countries it makes sense to substitute labor by equipment. This is not the case in low-wage

Table 12.7 Characteristics of the degree of automatization in construction.

Method	Productivity	Flexibility	Resilience
Tool-supported manual production	Very low	Very high	Very high
Equipment-supported production	Low	High	High
Mechanical production	Middle	Middle	Middle
Automatized production	High	Low	Low
Digital production	Very high	Very low	Middle

countries. Observers who complain in low-wage countries about the lack of equipment show only that they have not understood economic principles. Table 12.7 summarizes some important points about automatization.

12.7.2.3 Mass or Single-Item Production

There exists a continuum from mass to single-item production. While the construction contract typically defines single-item production, there are many parts in that contract which are mass produced (windows, faucets, nuts and bolts, etc.). To the contractor, these are typical material parts. Some mass production can happen in prefabrication on site under the control of the contractor.

To the largest degree, construction produces tailor-made goods. The lack of repetition impacts learning and extensive use of fixed capital as well as specific tools or equipment. Mass production means always working in the same place, single-item production moves to new places all the time. Breadth of knowledge replaces depth of it. It also impacts the experience of the stakeholders; some like the comfort of the known (mass production) and others the excitement of discovering the unknown in single-item production.

In mass production, extensive design and planning precede the production of prototypes with the possibility to improve. Construction has to deliver an acceptable product without such advantages. Table 12.8 shows the effects on productivity, flexibility, and susceptibility to interference, i.e. process resilience.

12.7.2.4 Continuous and Discontinuous Production

The distinction between continuous and discontinuous production describes the flow of the product. Structural steel is an example for continuous production where the steel is transformed into the ultimate product in several steps. The difference between continuous production and the production line is the fact that transformation replaces assembly (production line).

Table 12.8 Advantages and disadvantages of the amount produced.

Method	Productivity	Flexibility	Resilience
Mass production	High	Very low	Low
Single-item production	Low	Middle	High

Table 12.9 Advantage and disadvantages of continuous/discontinuous production.

Method	Productivity	Flexibility	Resilience
Continuous production	High	Very low	Low
Discontinuous production	Low	Middle	Very high

Construction goods are clearly produced in a discontinuous way. The characteristics of both types are shown in Table 12.9

12.7.2.5 Summary of Production Types

It has become clear that flexibility and a high resilience are of primary concern in construction processes. This is a question of survival for contractors. As productivity is often a competing goal with these two complementary goals, the market linking contractors and owners must decide which one to prioritize. A contractor can mass produce prefabricated housing and this is done. However, the market does not overwhelmingly accept the technology and the resulting housing in many countries. Communist countries were leaders in industrialized construction. This lasted as long as the planned economy. In market economies, uniformity of construction is a hard sell. Contractors must follow market demand and will sacrifice some possible productivity to gain flexibility and resilience.

The following Figure 12.13 shows the impact of the typical production types used in construction (solid dots connected by black line) on complexity. It becomes clear that opting for flexibility and resilience also means accepting high complexity.

With the variety of construction goods there are always exceptions to any rule. Mechanical production is identified as typical for large engineering projects. However, under certain circumstances, such projects can be rather fully automated, i.e., the schedule is driven by equipment.

An example for an automated bridge project is the BangNa expressway in Thailand. It is a 55 km long elevated expressway built with segmental construction. The foundations and piers were constructed in a typical way with different workshops combined into a construction site featuring parallel as well as variable production, support by tools and equipment, single-item and discontinuous production.

The segments for the superstructure were prefabricated and with a volume of more than 22,000 segment with dimensions of 27.2 m × 2.5 m × 2.6 m, the precast yard approached production line quality. Transport from one station to the next was mostly by cranes, sometimes on rail. The segments were moved through different work stations with stationary (and specialized) workers, tools, and equipment. Figure 12.14 displays the schematic process. First some components (struts, deviators, transversal post-tensioning) were assembled and transported by trucks to the rebar cage preparation. The molds were surveyed and then the rebar cage lifted by tower crane into the molds. Ready-mixed concrete came by truck mixers and were placed by conveyor belts. After initial curing, the segment was shifted on rails into the match-cast position and next into the curing and repair position. Finally, shuttle lifts transported the segments into storage. All core activities were repeated within 24-hour cycles. Total expenditures for the yard (factory) were around 80 million USD.

Production Type	Construction	Production complexity		
		Low	Middle	High
Production line	*Not possible (immobile product)*			
Workshop	Used for subprocesses			
Construction site	Typical for overall project			
Parallel production	*Used for some subprocesses*			
Variable production	Typical for overall project			
Tool-supported manual p.	Typical for small projects			
Equipment-supported p.	Typical for average projects			
Mechanical production	Typical for large engineering projects			
Automated production	*Seldom*			
Digital production	*Not economical (high fixed costs)*			
Mass production	*Some components*			
Single-item production	Typical for overall project			
Continuous production	*Not possible (immobile product)*			
Discontinuous production	Typical			

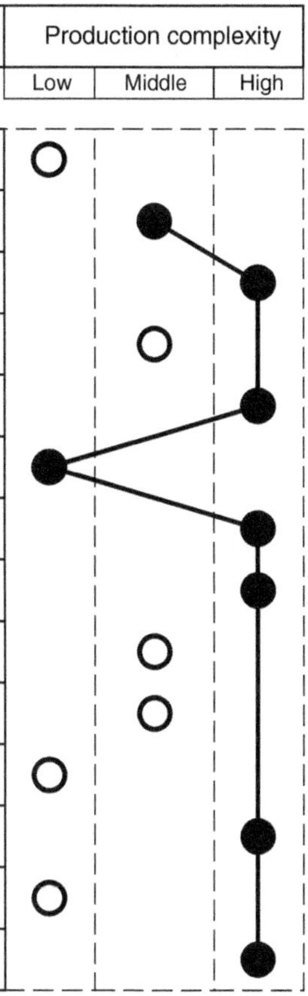

Figure 12.13 Production and complexity for construction.

A second and even older example (1989) is the segment factory of the Great Belt Tunnel. Production took place in a factory that shielded the works against inclement weather. The rebar cage was assembled using welding robots and then they were transported by an overhead belt to the basin for epoxy coating, drying, and placement in the molds. Concrete was brought to the molds by a different overhead belt system. The hardened segments were next taken to the curing and outfitting stations, again by overhead belts with vacuum grips. This production process resembles even more closely an automated production line.

Both processes were highly productive but not very flexible. In both cases, large storage areas were required as buffers to soften the impact of delays in tunnel construction or superstructure erection. Few were the disruptive impacts in the precast yard or factory; manifold they were on the construction site. Overall resilience could only be achieved by costly storage facilities.

These examples also show that contractors are not averse to process innovation. However, the innovation must bring a cost-cutting advantage. This was true 40 years ago and it still is.

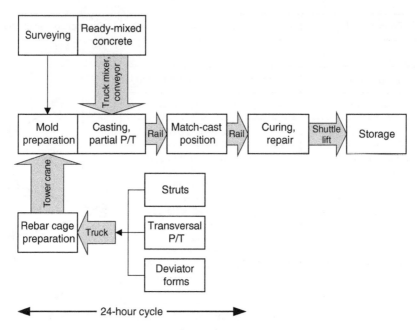

Figure 12.14 Schematic process in the precast yard in BangNa.

A modern example (2021) is the planning for the tube construction of the sunken-tube tunnel Femern Belt (https://www.youtube.com/watch?v=XmP9Ez-u9SM). The production line extends from the factory to sea floor of the construction site.

12.7.3 Production Functions and Cost Curves

We have seen that there are three major types of production functions:

1) Substitutional, Type A
2) Substitutional, Cobb–Douglas
3) Limitational, Leontief

The first question to answer is whether the typical production function in construction is substitutional or limitational.

The amount of material is fixed and determined by the owner's design. Major remaining variables are therefore labor and equipment. In many projects, labor and equipment can be substituted for each other. It is possible to excavate a construction pit by labor only or by equipment predominantly (an excavator driver is still necessary). A more refined example is the use of systems formwork instead of self-made formwork. Systems formwork is more elaborate and more expensive; it reduces the cost of labor input. Thus, the decision for systems formwork is one for substituting equipment for labor. Whether a contractor will make this decision depends on the costs of the inputs. In general, the decision for systems formwork pays off in high-wage but not in low-wage countries. What we can learn from these examples is that, typically, we can substitute labor for equipment in construction to a

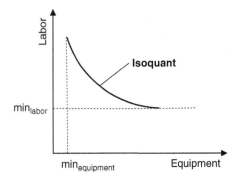

Figure 12.15 Range of substitutability.

certain degree. A minimum amount of labor and equipment remains necessary. Figure 12.15 shows the range of substitutability with a minimum input of labor and equipment.

The example of the BangNa Expressway above (Figure 12.14) shows that sometimes in megaprojects, the production is limitational; the inputs form a fixed ratio. With the inputs shown, we have the ratios as given in Table 12.10 for the production of one bridge segment and for the daily production of 50 segments.

The table shows us that the ratios of all n inputs extends into an n-dimensional space that we cannot depict graphically.

Table 12.10 Limitational production function for the BangNa Expressway.

Elements	Number	Daily production
Struts	2	100
Transversal post-tensioning	6	300
Deviator blocks	2	100
Rebar cage	1	50
Mold	1	50
Concrete	$34\,m^3$	$1700\,m^3$
Rail (from A to B)	1	50
Match-cast position	1	50
Rail (from B to C)	1	50
Curing and repair position	1	50
Storage area	1	50
Trucks		Multi-use
Tower crane		Multi-use
Surveying station		Multi-use
Truck mixer		Multi-use
Conveyor belts		Multi-use
Shuttle lift		Multi-use

The important observation in limitational production is that the addition of 10 truck mixers without an increase of the other resources does not increase production at all. To produce another segment, another set of resources as given in the middle column of Table 12.10 is necessary. Fixed relationships apply also to the erection of the segments. Productivity was very high. With the resources at hand, it was possible to build 2500 m (68,000 m²) of bridge superstructure per month. On the other hand, flexibility was low because it would have taken six months to order all components required to increase the construction speed by just one day. Speed depends in this case ultimately on the five erection girders; they were the drivers of the process. Adding one girder with the corresponding additional resources would have taken six months and increased speed by 20%, this is neither a quick nor a subtle adjustment.

In sum, most construction technologies are substitutional and only some, especially for megaprojects, are limitational. I believe that this is an important idea for cost engineers, quantity surveyors, and project managers.

The second question is what type of a substitutional production function applies to construction. Consider this hypothetical example: We must complete the structural works for an office building, and we have enough skilled workers available in our company. We need not rely on subcontractors. If we start with one worker and add a second one, it would be plausible to expect an increase in productivity. Just think of the poor lonely worker who wants to hoist a load. First, he has to hook it, then he has to climb the tower crane to lift the load, get down, etc. Overall productivity would be higher with one person on the crane, one loading at times and others using the load for construction purposes (fixing formwork, tying rebars, placing concrete). This is the reason why crews are doing the work on many construction sites and not solitary workers. Thus, we can assume at the beginning increasing marginal returns.

Now, consider the other end of this hypothetical experiment: In a limited space, workers are fixing rebars. Let the space be 50 m² (7 m × 7 m). Maybe 6 workers have the optimal productivity, but adding another 20 will definitely decrease productivity. We are facing decreasing or even negative marginal returns. This experiment describes a substitutional production function of type A (Figure 12.16).

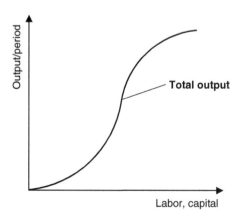

Figure 12.16   Type A production function with two variable inputs.

The cost function is a mirror image of the output function around the 45° line. In addition, we need to add the fixed costs (head office and site costs). In the upper part of Figure 12.17, the marginal costs decrease because of the increasing productivity of labor and equipment. From point A onward, they increase because of the decreasing productivity.

The average fixed costs decrease with an increasing quantity of output; the larger the output, the smaller the influence of the fixed costs. The average variable costs decrease from zero output until point B, where the tangent from the origin of variable cost curve touches the curve, then the average variable costs increase. Finally, the minimum average costs can be found at point C of the variable cost curve where the tangent from the origin touches the variable cost curve.

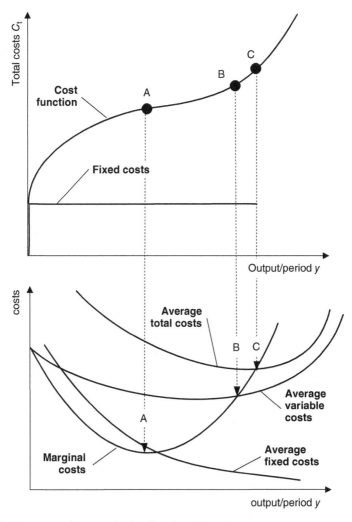

Figure 12.17 Cost curves for a production function of type A.

The minimum of the total average costs multiplied by the quantity produced per period allows us to determine the total costs per period and, accordingly, the necessary resources that make up the total costs. So, the production function could in the end solve many problems for contractors. There is only one problem in construction: the output in the type A production function is homogenous but construction output is not. In many buildings or bridges, many units (m² floor or bridge deck) are similar while other parts might be very different (the lobby in a building or the abutment for a bridge). Therefore, the results of the production function can only provide general ideas. The most important one is the U-shaped form of the cost curve, as it varies with output. It should always be clear that this type of cost curves pertain only to a single project, not the yearly production of a firm.

These thoughts on production functions and cost curves apply only to a single project, and they show us that there is a cost-minimum solution to every construction project. When working on the tender, contractors try to reach this ideal combination of inputs for the given circumstances. Nobody knows the exact shape of the production function and the cost curves, but the success in a submission can serve as a proxy: The low bidder is the one with the cost minimum solution. This, of course, does not take estimating errors into consideration.

The actual cost curve for a company with several project in a given period allows us to deduct the theoretical cost curve of a contractor. Figure 12.18 shows six contracts that start and finish at different times over a period. The contractor will put his best crew on the first contract, and this crew has the highest productivity. The second-best crew goes to the second contract, and so on. When crew 1 finishes its job, it moves to the next contract, in our

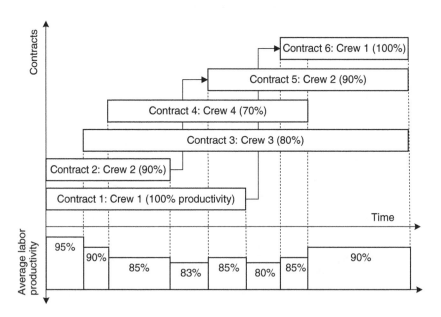

Figure 12.18 Contracts and average productivity for a contractor.

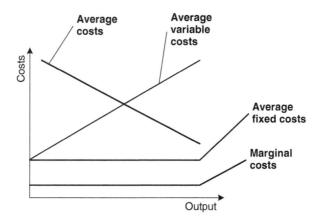

Figure 12.19 Cost curves for contractors.

case to contract 6. From this information, we can construct the average labor productivity in the form of a histogram. The average productivity goes up and down. This is also true for equipment: The best-suited will go to contract 1, and so on. Thus, labor and capital costs will vary over a period and the best approximation will be a constant line. The cost of the last unit produced, the marginal costs, will accordingly also be constant: The law of diminishing marginal returns does not work in construction.

From this thought experiment, we can derive the cost curves for a company and as these are the sum of the project cost curves also for those. When marginal costs are horizontal, then variable costs have a constant slope. Fixed cost can also be modeled as horizontal. In reality, this is not true – fixed cost at the beginning and end of a project are higher because of mobilization and demobilization.

Figure 12.19 shows the cost curves for construction projects and contractors.

12.7.4 Production Decisions

As discussed previously, demand triggers production directly. Other than exchange goods where demand and supply (production) are constantly interacting, construction contract goods follow a sequence: The owner decides to publish procurement documents and the contractor has to decide how to react. At the beginning lies the business decision whether to bid or not. If the answer is yes, the contractor must develop a plan how to fulfil the demand of the owner. For this purpose, he has to work out a solution to align the 11 partial plans (Figure 12.9). The contract price is the most important information that the contractor will provide. It determines not only the chances to win the bid but also the chances to make a profit.

Nothing is easier in construction than to increase contract volume: the contractor must only undercut the prices of all competitors. Unfortunately, this will drive the contractor into bankruptcy. He must find a way to lower the production costs by use of the most adequate technology. If there were no estimating mistakes, the production technology is the most important influence that the contractor has for reducing costs.

Table 12.11 Manufacturers and contractors in comparison.

Category	Manufacturer	Contractor
Entity	Single company Rather fixed supply chain	Sometimes the owner acts as contractor Single company or joint venture Often changing supply chain
Tasks	Product development Production	On-demand production
Behavior	Profit maximizing Growth orientation	Profit maximizing Survivorship orientation
Strategy	Cost leadership or differentiation Acceptance or change of rules Niche or main market	Cost leadership Acceptance of rules Niche or main market
Procurement	Large orders	Small orders
Sales	Selling	Bidding in submissions
Pricing	Full cost information	Limited cost information
Information	Information advantage on the seller's side	Information advantage depends on phase; until signature the owner has the advantage, later the contractor
Contracts	Provided by seller	Contractor must accept the contract provided by owner
Market	Mostly oligopolistic	Perfectly competitive for the contractor
Production	In advance Mass production Production characteristics rather simple	On demand Single batch Production characteristics complex

12.8 Summary

Table 12.11 highlights the differences between the archetypical manufacturer from mainstream economics and a contractor.

References

Brockmann, C. (2009). Global construction markets and contractors. In: *Economics for the Modern Built Environment* (ed. L. Ruddock), 168–198. London: Taylor & Francis.

Flyvbjerg, B., Bruzelius, N., and Rothengatter, W. (2003). *Megaprojects and Risks: An Anatomy of Ambition*. Cambridge: Cambridge University Press.

Fortune (2021). Global 500 China state construction engineering, rank 13. Fortune. https://fortune.com/company/china-state-construction-engineering/global 500/; accessed January 2022.

Greenhalgh, B. and Squires, G. (2011). *Introduction to Building Procurement*. London: Spon Press.

Hauptverband der Deutschen Bauindustrie (2020). Finanzkennzalhen, https://www.bauindustrie.de/zahlen-fakten/preise-ertraege/finanzkennzahlen; accessed January 2022.

Kotler, G. and Armstrong, P. (2020). *Principles of Marketing*. Harlow: Pearson.

Porter, M. (1985). *Competitive Advantage*. New York: Free Press.

Runeson, G. (2000). *Building Economics*. Melbourne: Deakin University Press.

Runeson, G. and de Valence, G. (2009). The new construction industry. In: *Economics for the Modern Built Environment* (ed. L. Ruddock), 199–211. London: Taylor & Francis.

13

Construction Goods

When economists talk about goods, they often refer to goods and services together (Hirshleifer et al. 2005). One aspect differentiating the two is their tangibility. A bus is a tangible good; bus rides are intangible services. In this chapter, I will draw a clear line between goods and services. Doing this, the question arises whether construction products are goods, services, or a mix of both. I will add a time dimension to the discussion and thus define the output of the construction sector primarily as transitional performance bundles. However, there are other perspectives that bring different results (Section 13.1).

The heterogeneity of construction goods demands some order that is of use in a way to help us understand observable phenomena. To this purpose, I will introduce in Section 13.2 a theory-based typology of construction goods. Section 13.3 will finally summarize the main points.

13.1 Goods and Services

Construction goods differ from others in many ways. They are visible and long lasting. Some define our legacy (Pyramids of Gizeh, Angkor Wat, Great Wall of China, Eiffel Tower, and Sydney Opera House). They all constitute the built environment in which we live. The aesthetics, form, and function of buildings and infrastructure impact the quality of our daily lives.

We organize the production of construction goods in projects because they are difficult to transport and because they become part of the land in a specific location. The properties of the land are part of the good and influence its behavior: We are dealing with fixed assets.

Much of microeconomic theory assumes the presence of homogenous goods such as endless bottles of Coca-Cola Classic rolling off a production line. A look out of your window will convince you quickly that construction goods are everything but homogenous. How to deal with this problem is the topic of Section 13.1.1. Economists distinguish between tangible goods and intangible services. This leads to the question, whether we are facing in construction the former or the latter (Section 13.1.2). Manufactures typically produce on stock and consumers can inspect any such good before buying it. We call such goods exchange goods, as we exchange money for the good. Contractors sign a contract before production; they exchange the promise to deliver a structure without defects against the owner's promise to pay on time. In construction, we deal with what new institutional

Construction Microeconomics, First Edition. Christian Brockmann.
© 2023 John Wiley & Sons Ltd. Published 2023 by John Wiley & Sons Ltd.

economics (NIE) define as contract goods. I will discuss the resulting problems in Section 13.1.3. The characterization of construction goods as investments provides yet another perspective (Section 13.1.4).

13.1.1 Heterogeneity

Remember the view from the observation deck of the Empire State Building? From up there, you look down at a multitude of buildings with different shapes. You will have the same impression by looking at any skyline. In Figure 13.1, you find a view of Dubai with the Burj Khalifa in the middle, the tallest building on earth in 2022 with a height of 830 m. This figure provides a visual proof of the heterogeneity of construction goods. Try to find two buildings that are the same. Of course, there are very uniform housing divisions in the USA or identical row houses in the UK or in other countries. However, with regard to the total building stock of one city, the degree of uniformity is minimal.

How come most of us can identify the producer of a car and not of a building? This is not only a question of branding – it is the heterogeneity that prohibits branding. Most people dislike the idea of living in a uniform built environment. Heterogeneity is the owners' preference, and one that comes with considerable opportunity costs – widespread standardized building could lower construction prices considerably.

Table 13.1 shows a categorization of different types of buildings and infrastructure. It is split into three categories:

1) Buildings for housing, social purposes, and services
2) Buildings for production
3) Infrastructure

Buildings are the nodal points in our built environment (vertical construction) and infrastructure provides connections between the knots (horizontal construction). The categorization is not very sophisticated, and attribution to any one category is not absolutely clear. The purpose of the exercise is to show the diversity of construction goods. It is easy to add more examples – slaughterhouses or retaining walls, for example.

For each category, we can define subcategories. Just to name a few, for detached homes: bungalow, courtyard, or atrium house, cottage, farmhouse, gable-front, housebarn, I-house,

Figure 13.1 Heterogenous forms of buildings in the skyline of Dubai. *Source:* Rastislav Sedlak SK/ Adobe Stock

Table 13.1 Different types of buildings and infrastructure.

Structures of the built environment					
Buildings	Habitation	Detached homes	Townhouses	Apartment build.	
	Business	Offices	Stores		
	Culture	Theaters	Cinemas	Concert halls	
	Education	Schools	Universities	Laboratories	
	Religion	Churches	Mosques	Synagogues	Temples
	Sports	Athletic grounds	Gymnasia	Natatoria	Arenas
	Health	Hospitals	Doctor's offices	Fire stations	
	Traffic	Stations	Airports	Ports	Parking
	Public offices	City offices	County offices	State offices	Federal offices
	Law	Courts	Prisons	Police stations	
	Military	Barracks	Training areas	Facilities	
	Tourism	Hotels	Resorts	Restaurants	Bars
Industry	Energy	Power plants	Heating plants	Refineries	Wind parks
	Production	Factories	Storage	Silos	Chimneys
	Logistics	Distrib. Centers	Warehouses		
	Traffic	Bus depots	Train depots	Plane depots	Shipyards
	Agriculture	Farms	Storage		
	Services	Shops	Offices	Banks	
Infrastructure	Roads	Highways	Inner-city roads	Rural roads	
	Rail	High-speed rail	Passenger rail	Industrial rail	Private rail
	Water struct.	Canals, locks	Dredging	Harbors	Offshore
	Bridges	Viaducts	Aqueducts	Railway bridges	Expressways
	Tunnels	Road tunnels	Rail tunnels	Utility tunnels	
	Dams	Earth/rock dam	Concrete dam		
	Water supply	Rain basins	Treatment plants	Sewers	Fresh water
	Electricity	Power lines	Substations		
	Gas supply	Gas lines	Substations		
	Information	Telephone	Internet	Television	

mansion, ranch, split-level, stilt house, and villa. How many villas or mansions look alike? This discussion proves the point of heterogeneous construction goods.

However, even manufactured goods are not all completely homogeneous. The times of the Tin Lizzy, when Henry Ford promised you any color of car as long as it was black, are long past. Cars come with many variations. Yet, the body of a car is clearly distinguishable. In the end, manufactured goods tend to become ever more diverse and construction goods become somewhat more standardized with a wide gap in between.

Treating construction goods as homogenous in microeconomic analysis would be a serious mistake. Treating most manufactured goods as homogenous can still pass as an admissible simplification. Goods that are essentially the same around the world are called global goods. Those where marketing and branding are universal but content differs are called transnational goods (e.g. Coca-Cola).

13.1.2 Construction Goods as Transitional Performance Bundles

If we distinguish between investment and consumer goods, then we are looking at three general market forms in construction:

1) Construction investment goods
2) Construction consumer goods
3) Construction services

Manufacturing markets differ mostly in the type of interaction between producer and consumer. In consumer markets, producers use a marketing mix for the interaction (product, price, place, promotion), in investment markets comparative competitive advantages, and for services costumer integration. In construction, the same one-family home might be an investment for the developer, a consumer good for the family buying it, and a service (brokering) by the real estate agent.

Original demand, a large number of consumers, and many individual decisions identify consumer goods. Derived demand, few investors, and many group decisions mark construction investment goods. Services are intangible and, accordingly, we cannot transport or store them. The customer becomes part of the value performance, and this is called customer integration – the customer is an external factor. It also means that the quality of the service depends partially on the customer (Table 13.2).

We find that contractors of one-family homes (consumption) face an original demand from some local customers when applying these categories. Decision-making is individual, distribution direct and market contacts personal. Parts of the production are intangible (planning, management), other parts tangible (bricks, concrete, steel). With the exception of prefabricated houses, one cannot store or transport one-family homes. Customer integration is of high importance for success.

When building a bridge (investment), the contractor faces quite different conditions. In a specific region the public owner might be the only one and demand is derived, i.e. the original task is facilitating traffic flow. Decision-making is formalized and made by a group of independent persons (buying center), distribution is direct, and market contacts are personal. Again, parts of production are intangible, others are tangible. Entire bridges cannot be transported, nor can they be stored. Customer integration remains of high importance.

Table 13.2 Characteristics of different types of goods.

	Consumer goods	Investment goods	Services
Demand	Original	Derived	Original/derived
Customer	Many	Some	Many/some/few
Decision-making	Individual	Group	Individual/group
Distribution	Multi-stage, indirect	Direct	Direct/indirect
Market contacts	Anonymous	Personal	Anonymous/personal
Substance	Tangible	Tangible	Intangible
Storability	Yes	Yes	No
Transportability	Yes	Yes	No
Production	Autonomous	Autonomous	Integrative

Common to all cases is that contractors offer performance bundles consisting of many different activities when signing the contract. After these initial discussions, the question arises as to how to qualify and summarize the activities of contractors. Hillebrandt (2000) defines construction as a service. This is hardly convincing when looking at the built environment. Economists often make use of an ex-ante (before the fact) and an ex-post analysis (after the fact), and this is very helpful for construction. When signing the contract, the contractor and owner are looking at a pure service (intangible, integrative) and after handover at a pure product (tangible, autonomous). In between, there are many transitional stages with a decreasing amount of services and an increasing amount of product: Contractors offer transitional performance bundles (Figure 13.2).

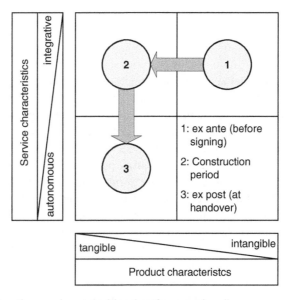

Figure 13.2 Construction goods as transitional performance bundles.

13.1.3 Construction Goods as Contract Goods

The NIE provide a very seminal categorization of goods by distinguishing between exchange and contract goods. Of interest are the different types of transactions leading to those denominations. Transactions describe a contract and the passage of property rights. In the case of exchange goods, contract signature, and the passage of property rights coincides while there is a considerable lapse of time between the two for contract goods (Alchian and Woodward 1988).

Exchange goods are standardized goods that can be stored. A buyer can inspect an exchange good before concluding the transaction. Not even a prefabricated house fits this description, since the interaction of the chosen land and the house remains unknown at the time of the transaction.

Contract goods are the normal case in construction, and they are based on mutual promises. The contract describes the construction good. The future conditions of the contract and the quality of the contractor's performance are never clear; the contract must therefore cover eventualities. The contractor promises a structure without faults and the owner payment according to contract terms. Between contract signature and handover lies the construction period, and during this time both contract parties can settle on opportunistic behavior. Williamson (1985, p. 47) defines economic opportunism as "*self-interest seeking with guile.*" In contrast to the homo economicus of neoclassical economics with complete information and rational behavior, NIE models man as displaying bounded rationality, having incomplete information with a strong tendency toward opportunism (Williamson 1985).

Goods possess different qualities in transactions. On the highest level, we can distinguish between search, experience, and trust qualities (Nelson 1974). Search qualities are those that we can experience before concluding a transaction, accordingly only exchange goods have search qualities. In construction, this applies to materials and equipment (e.g. we check the quality of cement, steel, or an excavator when buying). Concrete displays an experience quality; typically, a certain statistically distributed concrete strength after 28 days is the result of a specific concrete mix. When executing a management contract, we can never accurately confirm the quality of the service. We are facing a trust quality.

Construction contracts define future behavior. Search qualities at not existing at signature; this is an ex-ante perspective with regard to the transfer of property rights at handover. Ex-post buildings and infrastructure clearly have search qualities. We call goods where we can never evaluate the quality with certainty trust goods and those where we can measure quality only ex-post as quasi-trust goods.

In the case of relational contracts with repetitive performance we find experience qualities. However, relational contracts are not the norm in construction. Relational contracts cast the shadow of the future on today (Axelrod 1984). A contractor looking for an advantage today might forgo profits from future contracts (Table 13.3).

The definition of construction goods as quasi-trust goods entails many problems since there are many different choices available to the contractor and the owner during the construction period. They are characterized by asymmetric information and a principal/agent relationship (Arrow 1985). The owner as principal entrusts the contractor as agent to fulfil the contract. The contractor as agent is privy to information that is not available to the

Table 13.3 Typology of construction goods in new institutional economics.

	Exchange goods	Experience goods	Quasi-trust goods	Trust goods
Examples	Cement	Relational Contracts	Building	Project management
Ex-ante search qualities	Yes	No	No	No
Ex-post search qualities	Yes	Yes	Yes	No
Experience qualities	Maybe	Yes	No	No
Trust qualities	Maybe	Maybe	Yes	Yes

Table 13.4 Problems in principal/agent relationships.

	Ex-post (after contract signature)	
Ex-ante (before contract signature).	**Principal can observe the actions of the agent.**	**Principal cannot observe the actions of the agent.**
Actions of the agent are fixed.	• Hidden qualities • Adverse selection	
Actions of the agent are flexible.	• Hidden intentions • Hold-up	• Hidden action • Moral hazard

principal. NIE posits that under such circumstances we must expect opportunistic behavior (Williamson 1985, p. 47): "*...opportunism refers to the incomplete or distorted disclosure of information, especially to calculated efforts to mislead, distort, disguise, obfuscate, or otherwise confuse.*" The principal faces a number of problems (Table 13.4). I will later show that before signature the contractor takes the role as principal and the owner as agent (i.e., the roles are not fixed during the transaction).

The contract terms and monitoring of the contract execution must provide means to safeguard the interests of the owner as principal. Hidden qualities occur in connection with search qualities, be it ex-ante or ex-post. The maintenance period provides some protection; however, hidden qualities are not only possible with regard to the structure but also to the construction process.

Hidden intentions and hold-up describe similar phenomena. Contractors can use gaps in the contract to their own advantage, even if this contradicts the contractual intentions (Alchian and Woodward 1988). Hidden intentions result from past considerations; hold-up is enacted in the present. Principals can only react ex-post to hidden qualities, hidden intentions, and adverse selection. They must counter a hold-up immediately.

When incentives are weak to stand in for the interests of the owner because opportunistic behavior promises advantages, then contractors can use hidden actions or exploit situations of moral hazard (Varian 2014). In this situation, a contractor makes the assumption that the owner will not notice.

Table 13.5 Consequences of opportunistic behavior.

	Hidden qualities	Hidden intentions	Hidden actions
Point of time	Before transaction	During transaction	During transaction
Asymmetric information	Agent has better knowledge of market and product	Actions of the agent that the principal cannot observe	Intentions of the agent that the principal cannot observe
Consequences for principal	Adverse selection	Hold-up	Moral hazard

Table 13.6 Possibilities of opportunistic behavior in the case of contract goods.

	Exchange goods	Contract goods
Hidden qualities	Possible	Possible
Hidden intentions	Not possible	Possible
Hidden action	Not possible	Possible

Hidden qualities, hidden intentions, and hidden actions arise at different times and from asymmetric information; they can trigger adverse selection, hold-up, and moral hazard (Table 13.5).

Contract goods provide for several possibilities to deceive the partner because of the characteristics of the transaction which are not possible in the case of exchange goods. The main reason is the time difference between the promise of performance at contract signature and the transfer of property rights at handover (Table 13.6).

13.1.4 Construction Goods as Investment

Most structures are investments goods, among them all public works, industrial buildings, and rental housing. Important characteristics of investment goods are their specificity and business relationship. Specificity describes the use of a good; if it can only serve one purpose, then it is very specific. Business relationships range from single transactions to continuous cooperation. The construction market relies heavily on single transactions or those with a low frequency. A low intensity of relationships and a high degree of owner integration are typical for buildings, civil, and plant construction.

Using the concept of NIE (contract goods), investment goods in construction are characterized by a high degree of trust qualities ex-ante and a high degree of search qualities expost with a transition during construction. Experience qualities are of little importance (Figure 13.3) because of the low frequency of transactions between the same parties.

13.1.5 Construction Goods as Services

Contractors generate services in at least two separate locations – in the head office and on site. Benkenstein and Güthoff (1996) developed a five-dimensional service typology based

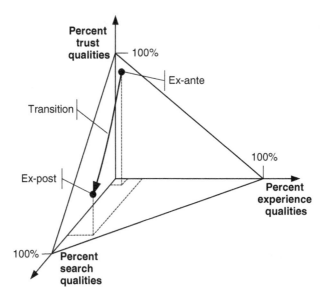

Figure 13.3 Characterization of construction investment goods.

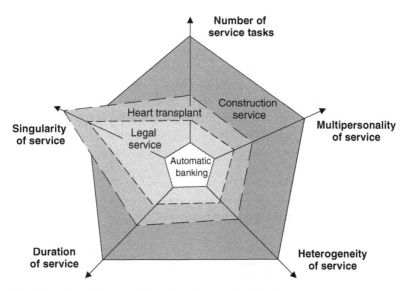

Figure 13.4 Five-dimensional typology of services and applications.

on the construct of complexity. The dimensions are: (i) number of different tasks, (ii) heterogeneity, (iii) multi-personality, (iv) duration, and (v) singularity of service.

This typology allows the characterization of construction services well. Even in a one-family home, the number of tasks and their heterogeneity are large, and the same holds true for the number of persons involved in the services during construction and its duration. Only the singularity of construction services is not outstanding. Figure 13.4 compares some services (construction, legal services, automated banking, and heart transplantation) with regard to these five dimensions. It shows that overall, construction services exhibit a high degree of complexity.

Benkenstein and Güthoff also consider two characteristics of customers (owners): The perceived risk and the involvement in the service process (owner integration). High risk and high involvement increase the service complexity. The perceived risk in construction is high for most owners and involvement almost extreme. The characteristics of construction services and those of the owners allow for great complexity.

13.1.6 Summary of the Characteristics of Construction Goods

Describing construction goods as quasi-trust goods is very helpful, especially as distinct from exchange goods. This statement holds true for consumer and investment goods as well as services in construction.

Some authors, such as Hillebrandt (2000), define construction goods as service. This is not tenable when we look at the process ex-post. A building is definitely a product; it is one of the longest-lasting products that humans create. It also makes no sense when we consider that a contractor spends less than 10% of the contract sum on pure services; the rest pays for labor and materials. It seems to be that the distinction between an ex-ante and an ex-post perspective provides a clearer idea opening the eyes for the transition period and its effects on the process, the quantity, and quality of the structure and the final price.

Describing construction goods as exchange goods also allows us to establish a framework for the cooperation between owner and contractor with the possibility of mutually opportunistic behavior. In the beginning, the owner uses his market information to lower the price; during construction, the contractor holds an information advantage and often uses it for aggressive claim management. Contract goods open up many opportunities for hidden qualities, adverse selection, hidden intentions, hold-ups, hidden action, and moral hazard.

The previous description helps to identify the major characteristics of construction goods. We have to differentiate between characteristics of the product and the process (Table 13.7).

Table 13.7 Comparative characteristics of construction goods.

	Characteristics of construction goods	
	Construction	Manufacturing
Product characteristics	Quasi trust good	Exchange good
	Ex-ante trust qualities	Ex-ante search qualities
	Ex-post search qualities	Ex-post search qualities
	Performance bundle	Good, service
	Transitional performance	Unchanged
	High specificity	Low specificity
	High complexity	Low complexity
Processes	Single transaction	Business relationship
	Opportunism possible	Opportunism not possible
	Integrative production	Autonomous production

Looking back at the arguments, I find that the best single description of construction goods would be as transitional contract goods. However, this might not be helpful or enlightening in some cases. The discussion up to now concentrated mostly on product characteristics. Therefore, it seems necessary to provide some thoughts on process characteristics in the following chapter.

13.2 Typology of Construction Goods

The goal for developing a typology is to provide a framework that allows us to explain and understand phenomena in the construction sector. Looking into a systematic order of construction projects is of a conceptual nature and it means advancing a theoretical contribution. Systematic orders are fundamental elements in the development of a scientific body of knowledge (McKelvey 1975). The lack of a theoretical background of existing orders such as the categorization in Table 13.1 makes this development necessary. A very simple example of an existing categorization is the well-known differentiation between residential, industrial, building, and civil engineering projects (Barrie and Paulson 1992; similar Langford and Male 2001). The group of building projects comprises offices, churches, sports arenas, opera houses, and universities. What do they have in common? Theoretical contributions need to identify building blocks (what), relations (how), and underlying dynamics (why). The reasoning (why) is of special importance because logic replaces data as the basis for an evaluation (Whetten 1989).

There are four forms of systematic orders on hand: nomenclatures, classifications, taxonomies, and typologies. Nomenclatures in science describe systems of designations. Examples are NACE (French: Nomenclature statistique des activités économiques dans la Communauté européenne), the statistical framework in the EU, or ISIC (International Standard Industrial Classification), as proposed by the United Nations. Nomenclature as a system of designation is nothing more than a series of nominalist definitions. Theory is not the base of nomenclatures – they have neither truth values nor empirical content but they can be practical.

We create classification systems by assigning objects or phenomena to specific classes according to predetermined criteria (similarity or relationships) and decision rules; a hierarchy (Doty and Glick 1994) characterizes them and they are precise, complete, and disjunctive (not overlapping).

If natural laws provide the basis of a classification, then it becomes a taxonomy. The most widely known is the systematic order of plants (Linnaean taxonomy).

Archetypes are the components of typologies. Categories of typologies are precise (however, the assignment is imprecise), complete but not clearly disjunctive. Unambiguous decision rules for the assignment are not possible. We can distinguish different characteristics for these four systematic orders (Table 13.8).

Clearly, either a classification or a taxonomy would be preferable, as they are most stringent. Unfortunately, construction projects do not follow the order of natural laws: establishing a taxonomy is impossible. A classification, on the other hand, poses problems when the population that it tries to capture is continuous. Project size measured in monetary units would result in placing two construction projects that differ by the amount of one dollar into different classes: Establishing a classification is undesirable. Some fuzziness

Table 13.8 Characteristics of systematic orders.

	Nomenclature	Classification	Taxonomy	Typology
Ontological base	Definitions	Rationality	Natural laws	Rationality
Assignment by	Definitions	Decision rules	Decision rules	Similarity with archetypes
Truth values	No	Yes	Yes	Yes
Empirical content	No	Yes	Yes	Yes
Precise	Yes	Yes	Yes	Yes
Complete	Yes	Yes	Yes	Yes
Disjunctive	No	Yes	Yes	No

about the categories will prove beneficial. In sum, a typology of construction projects promises the best results.

13.2.1 Approach to Developing a Typology

In a wider sense, the typology that I will develop is an explication. Carnap (1956) defined explications as the process of replacing unclear or defective pre-theoretical concepts (explanandum) with clear ones (explicatum). The method to achieve this comprises analysis and replacement. In the end, the explicatum must satisfy the conditions of adequacy (similarity with the explanandum, governance of rules, usefulness, and simplicity).

The existing pertinent literature helps to establish dimensions for the typology. These are the building blocks of the theory (what). In a second step, I will scrutinize the relationships between the dimensions (how). During the analysis, I will give reasons for the choices referring to the underlying concepts. Rules of use for the dimensions also follow from the analysis and from industry practice. This is clearly an exercise in theory building, and it does not include theory testing (Dubin 1969).

What can a typology achieve? In the best case, they are complex theories. Doty and Glick (1994) provide five guidelines for building a typological theory: (i) Make the underlying grand theoretical assertions explicit. (ii) Define the set of archetypes completely. (iii) Use the same set of dimensions when describing archetypes. (iv) Explain the constructs used as dimensions. (v) Operationalize and empirically test the typological theory.

13.2.2 Conceptualization

In construction, it is fundamental to differentiate between product and process. The client typically designs the product (building, structure) with the help of architects and engineers. The contractor develops the construction process. Product and process planning are separated; in the end, both contribute to the final result. This is not only true for cost and schedule but also for safety and quality.

13.2.2.1 Choice of Dimensions

Starting point for the selection of possible dimensions for the construction industry are publications analyzing organizations and their tasks. In this business context, authors distinguish between task and resource dimensions. Tasks can be difficult, variable, interdependent, complex, and novel, or easy, routine, independent, simple, and repetitive. Resources can be specific or generic.

Difficulty: It is not easy to analyze a difficult task. Algorithms for solutions are unknown (Van de Ven and Delbecq 1974). Perrow (1967) describes difficulty as the degree of complexity of the search process in performing the task, the amount of thinking time required, and the body of knowledge that provides guidelines for performing the tasks. The underlying theory is task contingency – a special form of contingency theory, i.e. the idea that tasks determine the structure of an organization (Fiedler 1964). This holds also true for the next two dimensions: variability and interdependency.

Variability: Tasks with a great variability demand use of a multitude of different approaches (Perrow 1967).

Interdependency: We can differentiate between interdependency within and between teams. It is high when either many members of a team or many teams need to interact for the solution of a problem. A high interdependency increases the demands for coordination (Tushman 1979).

Complexity: Complex tasks can be broken down into many parts, and there are many connections between the parts possible. A third component of complexity is the cross-impact of decisions, the question of consequences for the system. When many decisions entail important consequences, we face a high complexity. Multiple parts, interactions, and consequences create uncertainty when solving a problem (Tushman and Nadler 1978). Complexity belongs to a theory that interprets organizations mainly as information processing and decision-making entities. It also plays an important role in systems theory (Luhmann 2013).

Novelty: Novel tasks are the opposite of repetitive tasks; the degree of novelty to the actor defines the dimension (Puddicombe 2011). For their solution, we first need to develop new structures and then new algorithms. Repetitive tasks are already pre-structured. The construct of novelty also belongs into the realm of contingency theory.

Resource specificity: Of high importance are human, asset, location, and temporal specificity. High resource specificity describes high-quality demands in these areas (Williamson 1991). Using the concept of opportunity cost, we can state that resources are very specific if the difference to the second-best alternative is especially high (Williamson 1985). He introduced this term as part of the theory of governance costs.

I have now identified six dimensions of construction projects. Difficulty, complexity, novelty, and resource specificity are rather independent of each other. However, variability and interdependence are strongly affecting complexity with regard to the number of parts and their interrelationships. If we subsume these two dimensions under complexity, we get a scheme with four dimensions determining a construction project (Figure 13.5).

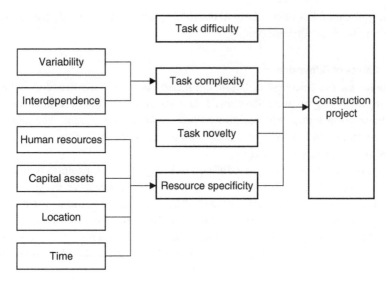

Figure 13.5 Conceptualization of construction projects.

Table 13.9 Existing combinations of dimensions.

	Difficulty	Specificity	Novelty	Complexity
Case 1	Easy	Generic	Routine	Simple
Case 2	Easy	Generic	Routine	Complex
Case 3	Difficult	Generic	Routine	Complex
Case 4	Difficult	Generic	Novel	Simple
Case 5	Difficult	Generic	Novel	Complex
Case 6	Difficult	Specific	Novel	Simple
Case 7	Difficult	Specific	Novel	Complex

13.2.2.2 Typical Cases

The four chosen dimensions represent each a continuum with values ranging from *high* to *low*. To simplify the typology, I will only consider the extreme values of the continuum. In this way, we can designate the variables by discrete attributes. Difficulty has the extremes of very difficult and very easy; complexity ranges from very complex to very simple; novelty spans from very novel to very routine; and resources can be very specific or very generic. An array of four dimensions with two possible values each generates 16 solutions. However, some of these are not possible. It is impossible to combine high difficulty with routine novelty and low complexity. We cannot solve highly novel tasks with ease. Routine tasks do not require specific resources. These practical exclusions reduce the possibilities to seven cases (Table 13.9).

13.2.2.3 Typology

Generic resources and routine tasks typify cases 1, 2, and 3; they differ only with regard to difficulty and complexity. Taking generic resources as defining dimension, we can

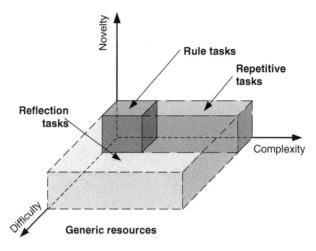

Figure 13.6 R-tasks with generic resources.

distinguish three types of tasks defining a group of projects. Rule tasks describe the simplest construction projects, traditional rules are sufficient for a solution (e.g., one-family home). Repetitive tasks differ from rule tasks by their increased complexity; managers and engineers must apply the rules repetitively (e.g., pipeline). If we increase difficulty as well as complexity, then we face reflection tasks, repetitive use of rules will not generate a solution (e.g., tunnel in a difficult geology). These three types form the R-tasks (Figure 13.6).

Small companies are well suited to implement rule tasks because such projects do not require a sophisticated overhead. The resources are available on the market, the implementation poses no unwarranted problems, and financing is no hurdle. Such boundary conditions allow for an easy market entry, and that again is a reason for the large number of companies in the sector.

For repetitive tasks, there is a need for an iterative approach to the problems and the observance of impacts of decisions on other parts in order to solve the structural complexity. New is the interdependence of partial solutions, not the general approach. Repetitive tasks provide the natural growth path for companies that master rule tasks.

A further horizontal task enlargement characterizes reflection tasks, and it entails the development of new solution algorithms in planning. This requires reflection based on previous experience. Planning becomes more important, and essential engineering resources must be available. There exists a possibility to create a niche market, based on the competence to solve difficult problems.

Generic resources can also appear together with high novelty and high difficulty, with complexity being the only variable. These tasks are innovation driven but need no specific resources except for highly qualified personnel.

Innovation tasks are novel and difficult to solve (e.g., Frank Gehry's Dancing House in Prague). This is the ideal setting for companies with a strong engineering background; they can be of smaller or larger size.

Integration tasks add to novelty and difficulty the dimension of complexity, thus increasing the size of the tasks (e.g., modern sports stadia). In such cases, expertise most likely

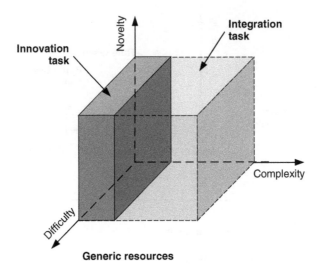

Figure 13.7 I-tasks with generic resources.

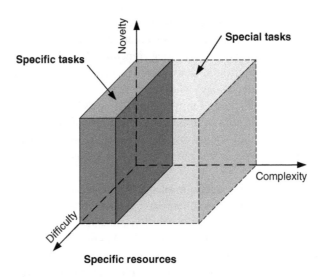

Figure 13.8 S-tasks with specific resources.

comes from different sources, and there must be a system leader to integrate all the efforts. The ideal company must have command of superior project management and engineering competences. This can still be a specialized company or a large one. Innovation and integrations tasks (I-tasks) are shown in Figure 13.7.

Special tasks (S-tasks) combine high degrees of novelty and difficulty with specific resources (Figure 13.8). Specific resources are expensive, so companies that attack such problems must possess the necessary capital.

Smaller high-tech buildings pose specific tasks (difficult, novel, specific). An example is the Sydney Opera House, where there were high demands on engineering and financing.

Megaprojects are examples of select tasks; they have the highest requirements in all four dimensions. An example is the Doha Metro Project with 350 km of metro lines and 100 stations. A company tackling such a task must have many highly qualified managers and engineers with megaproject experience. Access to financing is also of greatest importance, as shown in Figure 13.8.

I have developed a typology of construction projects by logical deduction. Starting from a review of the relevant business and construction literature, we are able to identify four independent dimensions. There exist seven possible combinations based on these four dimensions. A discussion of these seven cases leads to three project groups:

1) R-tasks or routine projects (derived from rule, repetitive, and reflection tasks)
2) I-tasks or innovation projects (driven by innovation and integration tasks)
3) S-tasks or special projects (results of specific and select tasks)

13.2.3 Applications

The starting point for the considerations above was the lack of existing systematic orders for construction projects. After establishing such a typology, there must be some proof how the new typology can be applied in a useful way. The following chapters analyze the impact of this typology for the understanding of different topics. I have chosen a number pertaining to contractors (market entry, optimum firm size, strategic planning). Other possible topics for contractors could be marketing or construction technology; one concerning the owner could be project organization. The derived theory is useful for enterprise and project planning, not so much for implementation.

13.2.3.1 Market Entry

In all most countries, the number of construction companies is large. However, small companies seldom compete with large ones for the same project. We can interpret the seven different project types of the typology as different markets that require different competences. Rule tasks need the least amount of capital and expertise; they are the typical entry projects for small companies. Special tasks, on the other hand, require considerable capital and a very high amount of expertise. Repetitive, reflection, and innovation tasks require approximately the same amount of capital but increasing and differentiated skills. Integration, specific and select tasks depend on ever-increasing amounts of capital and knowledge (Figure 13.9).

Typically, many companies will seek entry with little money and little know-how; these form the bulk of all construction companies. At other levels, a direct startup is still possible but mergers and acquisitions are more likely. For performing S-tasks either long-term growth or an acquisition requiring a high investment are options.

13.2.3.2 Optimum Firm Size

The optimum firm size is an economic concept. Williamson (1967) describes it as a management problem, but one can also interpret it as a microeconomic problem. When considering the management factor, the optimum size is determined on a continuum between hierarchical and market organization. R-tasks would rather rely on a market solution and S-tasks on a hierarchy.

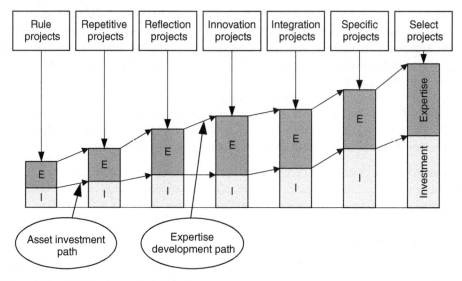

Figure 13.9 Market entry and required resources.

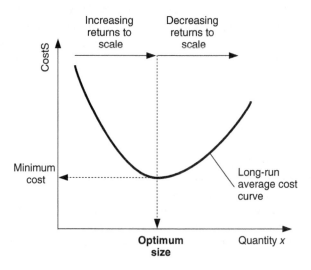

Figure 13.10 Optimum firm size based on long-run average cost curve.

Microeconomics defines the optimum firm size as the minimum of the long-term average cost curve given a certain amount of resources. The curve has a U-shape because of increasing returns to scale up to the minimum costs and decreasing returns to scale thereafter, which raise the costs. The use of all resources is efficient only at the point of the optimum firm size (Figure 13.10).

Since the construction industry is highly competitive, firms must produce close to minimum cost or they will fail. Looking at the seven different tasks and the associated levels of required capital in Figure 13.9, we can conclude that there are different optimum firm sizes in construction because different levels of capital are required. This separates the construction market into different submarkets. If we take a one-family home as an example, we face a rule task. A small company (with little capital) can get the work done efficiently while a larger company produces to the left of the optimum size and way above minimum costs.

13.2.3.3 Strategic Planning

Strategic planning often starts with a strengths, weaknesses, opportunities, and threats (SWOT) analysis and moves on to generating options; decision-making and implementation are the final steps. If we choose a resource-based approach to the problem, Figure 13.9 provides qualitative data on two types of resources: capital and expertise. We can plot these two resources in a two-dimensional graph (Figure 13.11).

A contractor can decide from his starting point where he wants to move and whether that means investing and/or hiring and an appreciation of the magnitude for both actions. Another option is, of course, to acquire expertise and capital assets by mergers and acquisition.

By building the typology, I have fulfilled most of the demands formulated by Doty and Glick (1994):

1) Explicit description of the underlying theoretical assertions when choosing dimensions for the construct construction project. Most belong to contingency theory.
2) Definition of a complete set of archetypes by R-, I-, and S-tasks.
3) Consistent use of the dimensions of complexity, difficulty, novelty, and specificity when describing the archetypes.
4) Explanation of the constructs used as dimensions.
5) I have offered examples to check plausibility.

The typology fulfills the conditions of adequacy as formulated by Carnap (1956) to a large degree. However, the developed terms are not similar to current custom. This was impossible because we do not have a similar systematic order available in construction. On the other hand, I was able to show the governance of rules, the usefulness for applications, and the simplicity of the use.

In practical terms, I have shown that the explanatory power of the typology is high. Market entry, optimum firm size, and strategic planning are just three applications

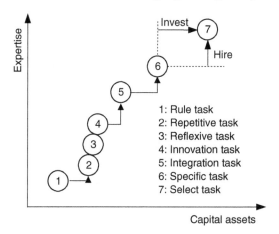

Figure 13.11 Strategic planning.

discussed; three others I have mentioned are marketing, construction technology, and project organization.

13.3 Summary

There is some extensive discussion in construction engineering management about the essence of construction goods. Few see it as a good, many as a service, and some as a mix between good and service. The idea of a contract goods holds the key to a deeper understanding by introducing a time frame: Contract goods draw attention to the fact that there is a production period between signing the contract and change of property rights at handover. Thus, construction goods change their characteristics from a pure service to a pure product. We can best describe them as transitional contract goods.

This allows to concentrate on the production period and the fact that the owner can assume a more passive or a very active role; in the latter case, he acts as employer for the construction project working with subcontractors in the same way as a management contractor could do it. Contract goods require the involvement of the customer. This involvement opens up new insights regarding construction goods. We can define them a quasi-trust goods and this opens a door for negative actions from both principal and agent.

Defining construction goods as transitional contract goods is useful for an analysis of the market interaction between owner and contractor but it does not help them understand the differences between projects. To this end, I have developed a typology of construction goods ranging from very simple projects to the most complex ones. This typology is theory-based and abstract. However, the applications discussed show that the framework is very helpful to both the contractor and the owner.

References

Alchian, A. and Woodward, S. (1988). The firm is dead; long live the firm: a review of Oliver E. Williamson's the economic institutions of capitalism. *Journal of Economic Literature* 26 (1): 65–79.

Arrow, K. (1985). The economics of agency. In: *Principals and Agents: The Structure of Business* (ed. J. Pratt and R. Zeckhauser), 37–51. Cambridge: Harvard University Press.

Axelrod, R. (1984). *The Evolution of Cooperation*. New York: Basic Books.

Barrie, D. and Paulson, B. (1992). *Professional Construction Management*. New York: McGraw-Hill.

Benkenstein, M. and Güthoff, J. (1996). Typologisierung von Dienstleistungen: Ein Ansatz auf der Grundlage system- und käuferverhaltenstheoretischer Überlegungen. *Journal of Business Economics* 66 (12): 1493–1510.

Carnap, R. (1956). *Meaning and Necessity: A Study in Semantics and Modal Logic*. Chicago: University of Chicago Press.

Doty, D. and Glick, W. (1994). Typologies as a unique form of theory building: toward improved understanding and modelling. *Academy of Management Review* 19 (2): 230–251.

Dubin, R. (1969). *Theory Building*. New York: The Free Press.

Fiedler, F. (1964). A contingency model of leadership effectiveness. *Advances in Experimental Social Psychology* 1: 149–190.

Hillebrandt, P. (2000). *Economic Theory and the Construction Industry*. Basingstoke: Macmillan.

Hirshleifer, J., Glazer, A., and Hirshleifer, D. (2005). *Price Theory and Applications: Decisions, Markets, and Information*. New York: Cambridge University Press.

Langford, D. and Male, S. (2001). *Strategic Management in Construction*. Osney-Mead: Blackwell Science.

Luhmann, N. (2013). *Introduction to Systems Theory*. Cambridge: Polity.

McKelvey, B. (1975). Guidelines for the empirical classification of organizations. *Administrative Science Quarterly* 20 (4): 509–521.

Nelson, P. (1974). Advertising as information. *Journal of Political Economy* 82 (4): 729–754.

Perrow, C. (1967). A framework for the comparative analysis of organizations. *American Sociological Review* 32 (2): 194–208.

Puddicombe, M. (2011). Novelty and technical complexity: critical constructs in capital projects. *Journal of Construction Engineering and Management* 138 (5): 613–620.

Tushman, M. (1979). Work characteristics and subunit communication structure: a contingency analysis. *Administrative Science Quarterly* 24 (1): 82–98.

Tushman, M. and Nadler, D. (1978). Information processing as an integrated concept in organizational design. *Academy of Management Review* 3 (3): 613–624.

Van de Ven, A. and Delbecq, A. (1974). A task-contingent model of work-unit structure. *Administrative Science Quarterly* 19 (2): 183–195.

Varian, H. (2014). *Intermediate Microeconomics with Calculus: A Modern Approach*. New York: W. W. Norton.

Whetten, D. (1989). What constitutes a theoretical contribution? *The Academy of Management Review* 14 (4): 490–495.

Williamson, O. (1967). Hierarchical control and optimum firm size. *Journal of Political Economy* 75 (2): 123–138.

Williamson, O. (1985). *The Economic Institutions of Capitalism*. New York: Free Press.

Williamson, O. (1991). Comparative economic organization: the analysis of discrete structural alternatives. *Administrative Science Quarterly* 36 (2): 269–296.

14

Construction Markets

Buyers and sellers meet in markets to conclude their transactions. These are voluntary, and accordingly each party must see a benefit in the exchange if concluded. Scarcity combined with necessity and market power can lead to forced exchanges. An example is someone who looks for an apartment while having no other place to stay. In such a case, the seller has market power. The inverse case is given when a fashion retailer has a full stock at the end of a season. Here, the buyer has market power. Market power is also evident in case of monopolies (monopsonies) when only one seller provides a good (or one buyer demands a good). Less power is in the hands of sellers in oligopolies, when few suppliers face a multitude of buyers. Table 14.1 shows typical market configurations.

I will look at markets in general in Section 14.1 before moving to the particularities of construction markets in Section 14.2. Section 14.3 follows with an analysis of such markets, focusing on homogeneity, theory, organization, structure, specialization, and legal aspects. Sections 14.4 and 14.5 concentrate again on owners and contractors respectively. Section 14.6 rounds of the descriptions with a look at the geography of construction markets before Section 14.7 summarizes the findings.

14.1 Characteristics of Markets

The following discussions are a concise repetition of Chapter 5 on perfect and Chapter 6 on imperfect markets. For most economic discussions, perfect competition is the case in point as we have already discussed. This serves to develop basic economic models that can later be adjusted to models of monopoly (monopsony) or oligopoly (oligopsony). The most important characteristic of perfectly competitive markets is that both, sellers and buyers, are price takers. They must accept the market prices that arise from supply and demand. In all other cases, either the seller or the buyer can influence the price to a certain degree. More common than perfectly competitive markets are oligopolies or monopolistic competition. The latter term describes a situation where many sellers and buyers are active in a market and where the sellers try successfully to differentiate their products. This is one of the goals in branding. An adherent to a specific brand faces a single seller (i.e., if you like

Construction Microeconomics, First Edition. Christian Brockmann.
© 2023 John Wiley & Sons Ltd. Published 2023 by John Wiley & Sons Ltd.

Table 14.1 Typical market configurations.

Buyers	Sellers		
	Many	**Few**	**One**
Many	Perfect competition	Oligopoly	Monopoly
Few	Oligopsony	Two-sided oligopoly	Limited monopoly
One	Monopsony	Limited monopsony	Two-sided monopoly

sneakers from Adidas, then Nike is not a competition). In sum, perfect competition is good for understanding theory but not so good for understanding real market situations.

The business cycle also influences market power in a given configuration. In a boom, typically demand exceeds supply with the consequence of rising prices. As long as this situation endures, sellers can reap higher profits. At the same time, sellers will expand their production possibilities and new entries to the market provide additional supply. After some time, supply and demand will balance and extra profits are no longer possible. This describes another characteristic of perfect competition: Market entry and exit (in a recession) are easy.

Tirole (2000, p. 6) describes a perfectly competitive market based on Arrow and Debreu as follows: *"The model starts with a fine description of available goods. An economic good is characterized by its physical properties, the date on which and the nature in which it is available, its location, and so forth. Consumers are perfectly informed about all goods' properties and have preferences over bundles of goods. Producers (firms), which are owned by consumers, are endowed with production possibility sets. A paradigm of market organization is added. All agents are price takers. The consumers maximize their welfare given that the expenditures must not exceed their income. . . This gives rise to demand functions. . . Producers maximize profits over their technological possibilities, giving rise to supply functions. . . A competitive equilibrium is a set of prices, with associated demands and supplies, such that all markets (one for each good) clear (i.e. total demand does not exceed total supply)".*

This definition contains much that I have already discussed to some detail plus some new elements:

- Goods are characterized by
  - Physical properties
  - Date of availability
  - Nature of availability
  - Location of availability
- Consumers (owners)
  - Are perfectly informed
  - Have clear preferences
  - Maximize their welfare (utility)
  - Demand goods

- – Face a limited budget
- – Are price takers
- Producers (contractors)
  - – Maximize profits
  - – Have production possibilities
  - – Have technological possibilities
  - – Supply goods
  - – Are price takers
- Markets
  - – Have an organizational form (Table 14.1)
  - – Allow the meeting of supply and demand
  - – Reach an equilibrium price by supply and demand
  - – Reach an equilibrium quantity by supply and demand
  - – Clear – i.e. all goods offered are sold; there is no surplus

We have already seen that many aspects of the archetypical perfect market do not apply to construction. By just picking on the last point, we can ask how a market can clear when we look at contract goods which do not exist physically ex-ante? A quick characterization of construction as – for example – a perfectly competitive market with all the characteristics above must fail. I have defined construction goods neither as good nor service but as a transitional contract goods. We have to expect similar complications for a market definition.

14.2 Particularities of Construction Markets

There is nothing like the construction market except in very abstract terms. A first distinction is between building and infrastructure and there are many more. We can describe markets in construction as highly diversified. This is, for example, different for mobile phones. At a given time, buyers face in many countries around the world the same goods, albeit with different prices. The question arises, what are the differences between markets for manufactured and constructed goods according to the description of Arrow/Debreu? Since the description pertains to a perfectly competitive market, the following discussion will only refer to this configuration.

14.2.1 Goods

I provided a fine description of available construction goods in Chapter 13. Owners buy some type of building or infrastructure; for this they have to sign a contract and then organize (endure?) the construction period. The time of availability stretches at least from signing of the construction contract to the end of the maintenance period. It is even more plausible to define it from the beginning of the product development (design) to the end of the maintenance period (construction sector). Of course, finally a construction good becomes available to the owner at handover. This is legally the point of transfer of property rights. If we would accept this also as the point in time of economical availability, we face a pure good and everything that happens before handover is irrelevant. Alas, it is highly relevant!

I have previously described the nature of availability by the term contract good. When concluding a construction contract, the good is available in the form of an intangible service; at handover, the contractor has transformed it into a tangible good.

The location of availability for most goods is a market (e.g., store, farmers' market, stock exchange, internet). It does not matter whether the market is real or virtual. However, in construction this is the owner's land. Even if a contract is signed in a posh conference room of a hotel, all services and production are tied to this singular plot.

Goods in perfectly competitive markets are homogenous. Construction goods are heterogenous. Homogeneity and heterogeneity are not exclusive ends on a spectrum; rather there exist all kinds of transitions. Mass produced goods tend to concentrate around pure homogeneity and tailor-made goods around pure heterogeneity. Pure heterogeneity means that every good is in an important way different from all others. This is true to a large degree for most construction projects. Design and production conditions differ in non-negligible ways.

14.2.2 Owners

Owners typically are not perfectly informed, mainly because of the complexity of construction goods. They need outside help by professionals. The drawings, specifications, and contracts that these professionals create together with the owner are in general incomplete, betraying missing information. While the product quality is therefore incomplete, the process quality lies in the future when the owner and contractor jointly implement the market deal. I have described this previously as the quasi-trust quality of contract goods.

Many authors have described the lack of clear preferences of owners. Again, the complexity of most construction projects makes it impossible for owners (and professionals) to understand the full impact of all the decisions they make and the possibilities that are available.

The assumptions of utility maximization, demand, and a limited budget without doubt not only apply to consumers but also to owners. In case of investment goods, owners bring a derived demand to the market, e.g. they need a factory for the purpose of production of apparel. The most intricate concept is that of utility maximization. However, all three will be discussed in following chapters.

A big question is whether owners are price takers. The typical case of a submission of bids as transaction process (this is a form of an auction) does not result in one equilibrium price but in many different prices. These prices never approach an equilibrium. In the moment, it suffices to state the difference, but I will analyze this in depth in a later chapter.

14.2.3 Markets

Markets in construction certainly have a specific organizational configuration, but how that can be described is the interesting part. The literature provides different views. Cooke (1996) and Hillebrandt (2000) observe many attributes of competition and oligopolies in construction markets. Gruneberg (1997) finds oligopolies characteristic. Myers (2017) ascribes to contractors some control over price-making (i.e., there exists an oligopoly).

Runeson (2000) sees the construction market as perfectly competitive. Ofori (1990), on the contrary, can detect only few features of perfect competition. Does this seem confusing? Yes, at least to me. I will provide my analysis and answer in the following chapters.

Markets have an organizational form. This might be a direct exchange of good versus money or a purchase contract with delayed delivery. The transaction might be conducted face-to-face or via the internet. Payments might be in cash, by credit card, or by invoicing. The buyer might have to pick up the good or the seller delivers it to his home. The market in construction is most often enacted as a sealed-bid auction. In this case, all contractors submit their bids including the price in a sealed envelope which the owner opens at a specific time on submission date. The contractors learn about the price of competitors only after submitting a binding offer. This is very different from other markets, where a seller might walk from stand to stand to learn about his competitors' prices or by looking into a catalogue or an online platform.

Certainly, in construction markets supply and demand meet. However, there is no equilibrium price as contractors submit a range of prices for each project: There is an array of prices. With regard to the quantity, we know that it is determined by the design of the owner for each project. Even if we create virtually an overall quantity of construction demand in one region for one-family homes, then this is the sum of all one-family homes. If the quantity of one is determined by the owner, then this must apply as well to this virtual sum. This means that in construction markets, quantity is not directly determined by supply and demand. Of course, there is an indirect off-market link. An architect, engineer, or quantity surveyor will check the cost of a design to adjust it to the budget of the owner. They will decide based on historic market costs.

It is a logic necessity that construction markets clear in the sense that there is no surplus. Construction starts only after concluding the market transaction. This statement is not true for real estate markets. At times, some cities have an oversupply in office space.

14.2.4 Summary

Table 14.2 provides a summary of the discussion by comparing the characteristics of a market under perfect competition against the construction market.

Looking at the comparison, it is impossible to describe the construction market as perfectly competitive. The different opinions expressed by the above-mentioned authors opinions stem mostly from the fact that they took the number of players in a predefined market to determine the competition. However, the description of construction markets needs consideration of all aspects from Table 14.2.

14.3 Analysis of Construction Markets

Markets are not physical objects like gravity or electro-magnetism; markets are social phenomena. Roth (2015), a winner of the Nobel Memorial Prize in Economic Sciences in 2012, distinguishes between commodity and matching markets. Prices (as well as supply and demand) govern commodity markets while matching markets require different mechanisms.

Table 14.2 Characteristics of perfectly competitive and construction market.

Characteristic	Perfectly competitive market	Construction market
	Goods	
Nature	Exchange good	Contract good
	Material good	Immaterial transformed to material
	Homogenous	Heterogenous
Time of availability	Given point in time	Construction period
Place of availability	Market	Owner's land
	Consumer	Owner
Information	Perfectly informed	Imperfectly informed
Preferences	Clear	Unclear
Utility maximization	Yes	Yes
Limited budget	Yes	Yes
Price takers	Yes	Maybe not
	Producer	Contractor
Profit maximization	Yes	Yes
Price taker	Yes	Yes
Possession of production factors	Completely	Partially
Technological possibilities	Yes	Yes
Supply	Goods	Performance/service/good
	Market	
Organization	Exchange	Sealed-bid auction
Supply and demand	Yes	Yes
Equilibrium price	Yes	No
Equilibrium quantity	Yes	No (fixed before market exchange)
Surplus	No	No

In the times of barter economies, all markets were matching markets. If someone wanted a pair of shoes and had meat to offer, it was necessary to find exactly such a match. Money made commodity markets possible and historically rules developed around these. Typical matching markets today are labor markets where talent, skills, and commitment are often more important than price. A move in the direction of matching markets includes partnering contracts in construction. Another clearly designed matching market used to be the one for architects and engineers.

According to Roth (2015) markets should be thick. That is, there is plenty of choice, without being congested. They should also be flexible, reliable, safe, and simple.

Construction markets were typically matching markets: Owners always had special needs and a master builder used to provide special skills. Roman roads and aqueducts as well as medieval churches and palaces are the result. When construction became more ubiquitous with ordinary people being able to afford more elaborate homes or entrepreneurs offices and buildings, it became customary to split design and execution. The design market remained to a certain degree a matching market, but construction became a commodity market. Two steps were necessary. Design separation homogenized the works. Instead of 10 contractors offering different designs, there was only 1, and in addition, owners took the mental step to standardize the execution by disregarding the influence of process quality.

There are all kinds of different procurement approaches (market designs). Design/build contracts untie the homogenization of design, partnering introduces the possibility of different process qualities provided by contractors. The same holds true for performance-based contracting. However, in all cases, it is the owner who creates a market design when drafting the construction contract.

All the possibilities in construction obfuscate the general functioning of markets. Therefore, the following discussions are based on the most widespread contract type used: competitive bidding with a design/bid/build sequence. In many countries, public owners are required to follow this path (at least most of the time) and private owners as often follow in line. The ensuing market has the following characteristics:

- There is a sequential process with design (architects and engineers), bid (quantity surveyors, project managers), and build (contractors) with three interfaces between design and bid, bid and build as well as build and operate.
- There is a single (homogenized) design.
- Contractors submit bids in sealed-bid auctions.
- Owners regard the process qualities offered by different contractors as exact substitutes.
- Profit or utility of owners is maximized by choosing the lowest bidder.

Evaluating such a market design against the criteria listed by Roth brings the result that it is thick, uncongested, flexible, reliable, safe, and simple. Generally, there are many contractors interested in submitting an offer. This might be different in strong booms when the capacities are stretched. In strong recessions, the market might become congested; I know of projects where the owner had to evaluate more than 50 offers; then, the owner can choose selective bidding to avoid congestion. In this case, most owners ask for offers from 5 to 10 contractors, which makes for a thick market (in the eyes of the owner). Design/bid/build contracts are very flexible, reliable, safe, and simple when concluding the contract. This might be and often is different during execution with all the problems caused by asymmetric information.

We have seen that perfectly competitive markets are paradigmatic in economics. Therefore, I will discuss next the characteristics of perfectly competitive markets. Perfectly competitive markets deal with homogenous goods, many buyers and sellers both acting as price takers, complete information and no entry or exit barriers. How does the typical construction market measure up to those benchmarks?

14.3.1 Heterogeneity

There are few dissenting voices when it comes to the heterogeneity of construction output. However, differences arise in the evaluation whether contractors erect products called structures (Gruneberg and Ive 2000) or provide services (Hillebrandt 2000; Runeson 2000). While the structures are heterogenous, the services are deemed to be homogenous.

The argument for providing a management service and thus for homogeneity of supply is an argument to somehow define the construction market in terms of perfect competition for which homogeneity is a condition sine qua non. It is also an argument that seems counterintuitive when looking at a construction site with all its material, workers, engineers, and equipment. It looks very much like a production process.

What else speaks against it? I can think of observation, theory, organization, structure, specialization, and law. In theory and practice, construction activities on site are predominantly aimed at production, and therefore construction inputs and outputs are highly heterogenous. The arguments follow in the next sections.

14.3.1.1 Observation

It is easy to walk by a construction site and take a look. You will see workers using all kinds of tools. Others are driving equipment, cranes are carrying loads, concrete pumps deliver concrete, trucks arrive loaded with prefabricated elements. There are also supervisors giving information and checking execution.

It is not as easy, but it is possible to arrange a visit to an automobile production plant. Workers use tools at fixed stations along a production line. Robots carry out tasks and supervisors control quality.

What is the difference? Different technologies are used but both produce some defined goods. In the case of automobiles, you can observe mass production and in the case of construction unit production.

14.3.1.2 Theory

An important theoretical concept is lean construction. Lean construction sees production at the center of all construction (Koskela 2000). The pertaining literature is full of complaints that not enough attention is paid to production. Surely, production must also be managed.

Microeconomic literature describes the technology constraints of firms, the resulting cost curves, and the ways to maximize profits. This applies to construction as well as to all other sectors of manufacturing, although with some differences.

14.3.1.3 Organization

The structure of construction firms does not support the view of construction as a service: it is dominated by workers. Table 14.3 provides an overview of the structure of German contractors (Hauptverband der Deutschen Bauindustrie 2021).

There is no other explanation for the fact that there are roughly 6.5 times as many workers on site as supervisors but the fact that something is produced on a construction site and not managed. While production is predominant on sites, management is the prime activity in the contractors' headquarters.

14.3.1.4 Structure

A similar picture emerges when looking at the inputs in construction compared to the automobile industry (Table 14.4).

Table 14.3 Structure of construction firms in Germany 2008, 2020.

		2008	2020
Headquarter	Owners	55,781	56,115
	Employees	140,010	188,099
Construction sites	Site Management	65,412	86,430
	Workers	453,845	563,170
	Ratio	1:6.9	1:6.5

Table 14.4 Input factors in structural works, finishing works and automobile industry.

Input	Structural works	Finishing works	Automobile industry
Wages	57.2%	47.6%	18.4%
Own	26.7%	32.4%	14.0%
Others	30.5%	15.2%	4.4%
Material	25.3%	33.2%	68.8%
Material, energy	24.9%	32.0%	48.7%
Products	0.4%	1.2%	20.1%
Depreciation	2.1%	1.3%	2.8%
Others	15.4%	17.9%	10.0%

Management costs in construction and finishing works are part of others and make up less than 15.4% or 17.9% of the total inputs, while more than 80% are in both cases production costs. The overall picture is similar to the automobile industry, with automatization lowering the share of wages and increasing the share of material in comparison to construction and finishing works.

14.3.1.5 Specialization

There is some serious discussion in the industry whether a construction project manager must be first of all a manager or an engineer. The answer depends on the project size. There are many specialized positions in very large projects so that the project manager must concentrate on coordination. In normal projects, the top person on site must solve many engineering problems and not only coordination problems. There are project managers for structural works, for finishing works and others for bridges, tunnels, roads, or dams. The idea of homogeneity would require that a manager can run all of these projects. This is not the case. Construction firms specialize their project managers and owners demand project managers with specialized knowledge. This is the case because production causes more problems than coordination.

14.3.1.6 Law

Construction contracts require the contractor to deliver a structure according to the contract without defects. This is checked at handover by comparing plans and the physical structure. Whatever the contract arrangement, even a managing contractor who subcontracts 100% of the works is responsible for the production result.

A service contract restricts the liability of the service provider to due diligence. The FIDIC White Book describes the responsibilities of a consultant in clause 3.3.1: *"The Consultant shall have no other responsibility than to exercise reasonable skill, care and diligence in the performance of his obligations under the Agreement."* Due diligence is characteristic of services, delivery without defects for products.

14.4 Owners

The definition of construction goods as transitional contract goods entails incomplete information on the part of the owners. They cannot observe the final building or infrastructure while they are in the process of bidding out contracts. The complexity of construction goods with its myriads of possible components and combinations makes the development of clear preferences impossible.

We have already looked at the market power of the owners, which is very different from that of consumers in markets for manufactured goods. In the extreme, the owner becomes the employer, i.e. the management contractor who organizes the work of subcontractors. The question arises whether this market power allows the owner to assume a position of a price maker instead of a price taker. I will forward the argument in Chapter 15 on contracting.

14.5 Contractors

After discussing the theory of the contractor in Chapter 12, I will now turn to the role that contractors play on construction markets.

14.5.1 Supply

Contractors typically supply structures only once they have received a contract, supply completely depends on articulated demand. This fact is captured by the description of construction supply as contract goods. While there are other forms of supply such as prefabricated elements produced and stored before a sales contract, these are by far not the norm. The following discussion will consider the norm: construction contract goods.

Assuming a contractor is successful on the market, then he might have construction works at hand in a given year as shown in Figure 14.1.

We can use estimating information to determine output per month for a contract. This is shown in Figure 14.2 for contract 1 by summarizing quantity finished times contract prices per month. Most contractors are allowed to submit an invoice per month and this is how they do it: actual quantities times contract prices. Invoices are the monetarily evaluated market supply by a contractor.

Figure 14.2 contains the first pieces of information that we need to build the individual supply curve. Remember that a supply curve pits price against quantity in one period. The line of thought brings us to a significant problem: What is the total quantity produced by a general contractor, the industry, or the sector? In a given month, it might be a

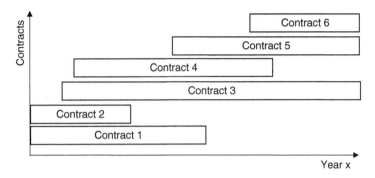

Figure 14.1 Construction supply of a single contractor over a year.

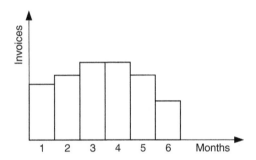

Figure 14.2 Turnover of one contractor from one contract.

certain mixture of doors, windows, beams, concrete, rebars, or steel, for example. An aggregate level may be square meters of floor. This would separate the supply into different markets. A square meter of a one-family home differs from a square meter of storehouse and a square meter of office building and a square meter of bridge deck. Can tunnels or dams be sensibly measured in square meters? Definitely not! The construction sector is not providing a set of homogenous goods and this make it impossible to define a supply curve for the sector. Varian (2014) defines an industry as the set of firms that produce goods which are seen as close substitutes by the buyer. Beauty lies in the eyes of the beholder and buyers define an industry, but there are few substitutes in the different domains of construction.

No public owner would see a bridge in place B as a substitute for one in place A, nor one built in 2018 as a substitute for one built in 2019. The bridge height and width would make them useful in one place and not in another. Similar considerations lead to the same consequence for most structures, there are no real substitutes. The industry boils down to a number of firms willing to build this one bridge in this place and at this time. The offers of these willing firms are complete substitutes in the eyes of the owner. The industry supply curve applies only to a single project.

14.5.2 Information

Perfectly competitive markets provide full information to buyers and sellers. This assumption does not hold in the construction market.

Overall, information is plentiful. Owners as well as contractors have access to detailed price and quality information through data banks. In addition, structures are visible to everyone. Every interested person can see what is going on.

It becomes more complicated when looking at a specific project. The owner and his aides (architects and engineers) have full product information and they must share them with the contractors in order to get what they have planned through the contract. Runeson (2000) concludes that full information is available to both parties in construction markets.

This is not tenable on closer inspection. As discussed before, there are numerous problems due to asymmetric information prevalent with contract goods. Possibilities exist for owners as well as contractors to deceive the other party.

The largest information disparity, however, exists during contract negotiations. The owner has at this time information from all bidders while each bidder can access only his own information. During negotiations, many owners use this privy information pressuring the contractors to lower their price offer or to increase their quality offer.

This type of privy information can be seen as market power. German public procurement law thus forbids public owners to discuss prices during negotiations. Private owners are not restricted in such a way.

In sum, two major information asymmetries exist in construction markets, one due to the nature of construction goods as contract goods and the other due to the procurement process by tendering. These two aspects are so important that construction markets cannot be seen as acting with full information.

14.6 Geography of Construction Markets

Contractors have principally the choice to enter five geographical markets: regional, national, international, multinational, and global. For each geographical market, different characteristics can be identified regarding the company's orientation, its behavior, the segments for which it produces goods, and for the type or organization. Table 14.5 provides a synopsis.

Firms that expand into ever-wider markets remain rooted in the most basic markets, the regional market and national markets. Two examples can illustrate this statement, Skanska from Sweden (Skanska 2021) and Strabag from Austria (Strabag 2021). Both have a rather small home market as Sweden has a population of 10 million and Austria of 9 million. Both firms also have strong traditional ties to the neighboring countries. These are the Scandinavian countries for Skanska and Germany for Strabag. The Engineering News Record (ENR 2021) publishes several top lists. One of them is the top list of global contractors with the worldwide turnover and the top list of international contractors with worldwide turnover minus home country turnover. Given these definitions, one should expect both countries to rank higher internationally because they are losing a comparatively small turnover from their home countries. The data in Table 14.6 allow us to categorize both firms as multinational contractors. The table does not contain enough information to rank them as global contractors, this remains an open question. It becomes also clear that both companies follow a different strategy with regard to market location. Strabag concentrates on the European market and Skanska on overseas markets.

Table 14.5 Characteristics of firms in different geographical markets.

Geographical market	Orientation	Behavior	Segments	Organization
Regional	Local network	Local	Few segments, specialized	Local headquarters
National	Many local networks within the home country	National	Many segments, diversified	National headquarter, local subsidiaries
International	Many local and some international networks	Ethnocentric	Many national segments, few international segments, diversified at home, specialized abroad	National headquarter, local subsidiaries, overseas department
Multinational	Many local and many international networks	Polycentric	Many national segments, few international segments, diversified at home, specialized abroad	National headquarter, local subsidiaries, overseas department, international acquisitions
Global	Many local and many international networks, megaprojects	Polycentric	Many national segments, few international segments, diversified at home, specialized abroad, megaprojects	National headquarter, local subsidiaries, overseas department, international acquisitions, megaproject centers

Table 14.6 Geographic distribution of turnover for Skanska and Strabag.

	Turnover Skanska (%)	Turnover Strabag (%)
Sweden	24%	
Austria		16%
Scandinavia	16%	
Germany		47%
Rest of Europe	17%	32%
Rest of the world	43%	5%
Total	100%	100%
ENR global rank	20	14
ENR international rank	9	6

14.6.1 Regional Markets

The fact that we cannot transport construction goods (with the exception of prefabrication) implies that construction activities have to take place on site. A firm that has not only one project must consequently become at least active on a regional level. A multitude of interactions between owner and contractor require their presence on site. Materials are bulky and transports are expensive so that a network of suppliers around the site is a competitive advantage. Construction activities are concentrated around civic or economic centers. Therefore, a small firm must serve at least one of these centers. Bowdich and Runeson (1985) report about a market area with a radius of 75–100 km in Australia and Lansley (1981) similar areas for the United Kingdom. This gives us a first idea about the size of regional markets. A given area will most likely depend on the center around which is located. It will be larger around London than around Oxford.

Brockmann and Nolte (2007) present more detailed findings in a study of the city of Bremen in Germany as a regional market. Bremen is not only a city but with some surrounding areas a metropolitan area. Together with Bremerhaven it is the smallest of Germany's federal states. As every state in Germany collects statistical data, we can find several data bases for construction activities. The city area of Bremen is 38 km long and 16 km wide. A contractor with offices in the center can reach any point within a radius of 38/2 = 19 km.

Typical segments of this market are general building, general civil works, harbor construction, and foundation works. A more detailed analysis will find even more submarkets. The concentration of works around an economic center can be seen in Figure 14.3, which displays empirical data for a representative set of construction companies. For this market around 60–70% of the activities take place within a radius of 25 km. In times of the deepening recession in Germany from 1995 to 2000 and further to 2005, markets expanded in geographic extension. Since the home market does not offer enough work, companies venture further away despite additional transportation costs and unfamiliar markets.

Market extension for the submarkets in harbor and foundation construction is also larger. The higher capital investment in these sectors cannot be supported on a small market, there simply is not enough work. The data above look only at the supply side of the construction market. However, data from several large public owners (demand side) support the notion of a concentrated regional market. Very small and large companies are more concentrated than mid-size companies. While the small ones are lacking information

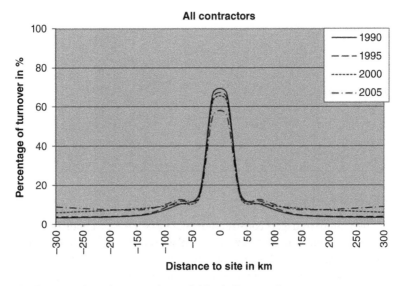

Figure 14.3 Concentration of construction activities in Bremen, Germany.

about and contacts in other markets, the large companies delimit the activities of their subsidiaries through internal boundaries as a matter of organization.

In Germany there were in 2000 (in the middle of the recession) roughly 80,000 construction companies – 78,500 of these had fewer than 100 employees. If we accept this figure to be descriptive of regional companies, then 98% of all companies on the national market are regional companies. The reasoning for this cut-off point (100 employees) is as follows: A company must sustain somewhat substantive construction activities in a number of places to be called active on a national market. Ten regional markets with 10 employees in each market seem to be a minimum expectation for a nationally active company (and it is a highly unlikely form of organization).

Especially fragmented is the market in the UK with a total of more than 164,000 companies in 1999. Out of these, approximately 750 had more than 100 employees, i.e. more than 99.5% of the companies had fewer than 100 employees. In the USA, there are more than 700,000 construction companies in the market, and this market is also highly fragmented. So, what applies to Germany also is true in the UK and the USA.

The fragmentation of regional markets can be explained by the demand. In 1999, 47.3% of all new build work in the UK were orders with a magnitude of less than £100,000, yet only about 50% of all contracts were new build work. Contracts for repairs and maintenance will be even smaller (Morton 2002). Every industry sector with substantive numbers of contracts ranging from a few thousand euros to megaprojects worth several billion euros will show similar patterns of fragmentation.

In general, competition in regional construction markets is high and dynamic due to the fragmentation. Table 14.7 shows data for the number of building permits and general contractors in housing and building in Bremen, Germany, for a number of years covering a business cycle (1993 boom; 2001 recession). The interpretation is clear: on such a market there is all the evidence of perfect competition (Bauindustrieverband Bremen-Nordniedersachsen 1997) on the supply and demand side at this macroeconomic level.

Collusion is always a possibility if there exists an oligopoly of bidders for specific types of tenders. Table 14.8 gives again data for Bremen. In the least competitive group 1 with 27%

Table 14.7 Supply and demand for general contractors in Bremen, Germany.

Year	1993	1994	1995	1996	1998	1999	2000	2001
Building permits housing	739	746	482	565	817	922	925	751
Building permits buildings	142	116	126	119	211	209	203	151
Contractors	~130	~130	~130	~130	~130	~120	~120	~120
Ratio owners vs. contractors	7:1	7:1	5:1	5:1	8:1	9:1	9:1	8:1

Table 14.8 Percentage of identical competitors in submissions.

	Group 1	Group 2	Group 3	Group 4	Group 5
% of recurring competitors	~ 80%	~ 60%	~ 40%	~ 20%	> 20%
% of all submissions	27%	33%	30%	6%	4%

of all submissions there are still 20% of new competitors. In other word: in 73% of all submissions, there are 40% of changing competitors; these are definitely not numbers that suggest a general oligopsony (few contractors). These numbers provide the necessary dynamics to make naturally caused collusion impossible (Nolte 2006).

What these data do not prove is the absence of niche markets with very limited competition and the possibility of collusion. However, those markets can only be few in number and small in size.

If we extend the information from the regional market in Germany, especially keeping Figure 14.3 in mind, then we see a mountainous and hilly landscape of regional construction activities. Around centers there are more activities than in between and at larger centers the activities are also more important than at lesser centers. The contractors serving a specific center are to a very large degree located in that center.

14.6.2 National Markets

Once a company has established a regional presence, it can add additional networks in other areas. At the end of such a process the company becomes a national competitor by being active in most of the important regional markets of a country. Since activities on the national market are the sum of regional activities, there can be no national strategy without regional activities. Since construction projects cannot guarantee continuous employment, most national companies diversify in order to ensure continuous production. National construction markets are defined by the building rules and regulations, the economy as well as by the national language and culture of the host country. These form a barrier to entry for contractors from other countries. The barriers are cumbersome but not insurmountable.

National companies have headquarters and many local subsidiaries. Intra-company services are concentrated at the headquarters, the contact with the owner is mostly organized on the regional level although there are also some national clients. Other than the regional company, the national company has gained the competence to grow into other markets and create separate networks, albeit this is limited within the set of national culture and regulations.

While there are no data on how many national companies are active on such a market, it can be said with the same reasoning as above, that it should be fewer than 2% of all companies. There will be variations to this order of magnitude from country to country, yet the picture will not change dramatically. This does not mean that competition in the national market is limited. Regional and national companies fight on the same markets, as was explained with regard to the market in Bremen. Large companies do not dominate the national market. The top five contractors in national markets do not share more than 10% of the market in industrialized countries, with Japan being the exception at ca. 11% (Bollinger 1996). A way to measure concentration in a market is the Herfindahl–Hirschman Index (HHI). The HHI is the sum of the squared market shares α_i^2 in percent. With i $(1, \ldots, n)$ firms in a sector it must by definition be:

$$\sum_{i=1}^{n} \alpha_i = 1 \qquad (14.1)$$

Thus, the HHI is defined as:

$$HHI \equiv \sum_{i=1}^{n} \alpha_i^2 \qquad\qquad (14.2)$$

If there is just one supplier, this adds up to 10,000 ($100\%^2$); if there are a thousand with a share of 0.1% each, then the HHI equals 10. Values below 1500 signify no concentration in a market (Krugman and Wells 2018). Using data from the German Monopoly Commission in 2001, it is possible to determine an approximate value for the HHI in Germany (Monopolkommission 2003). Only an approximation is possible since the data are given for classes of companies and not for single companies (Table 14.9).

The HHI has a value 12.83 for construction in Germany, signifying a very competitive sector with no determinant players. The top 10 contractors just have a market share of 9.1%. Including more than 100 companies will increase the HHI slightly. The tobacco industry might serve as a comparative industry. Here the HHI is 808.86 in Germany. Regional and national markets in construction are fragmented, and market share is not a competitive advantage. The ensuing question is, of course, why do construction firms not build up a larger market share? The Profit Impact of Marketing Strategies (PIMS) study claims that relative market share is one of the major reasons for higher profits (Buzzell and Gale 1987). I will discuss this topic in more detail in Section 14.7, and the argument will be that the maelstrom of the business cycle prohibits any firm from accumulating a commanding position based on market share.

 Observation = (obs.) A CEO of a leading German construction firm once told me that someone striving for market share in this business is a criminal. Given the uncertainties of construction estimates, striving for market share means to introduce an estimating bias toward underbidding competitors. It is easy to win a bid with a low price. However, before a firm could reach a commanding market position it would go bankrupt because of all the losses on the way. The upfront investment required for such a strategy is forbiddingly high and success is not guaranteed.

Table 14.9 Approximated Herfindahl–Hirschman Index for the German construction industry.

Number of firms in the class	Market share of the class	Average market share of one company	Squared market shares	Sum of squared market shares
3	4.8	1.60	2.56	7.68
3	2.0	0.66	0.44	1.33
4	2.3	0.58	0.33	1.32
15	4.7	0.31	0.10	1.47
25	3.9	0.16	0.02	0.61
50	4.6	0.09	0.01	0.42
			Approximated HHI:	12.83

Table 14.10 Market forms based on centration ratios.

	Market share top 5 firms	Market share top 15 firms	
High competition	<10%	<15%	Approaching perfect competition
Competition	10–18%	16–35%	Generally competitive markets
Low competition	19–34%	36–45%	Imperfect competition
Oligopoly	>34%	>45%	Monopolistic or oligopolistic markets

Table 14.11 Market forms in construction and supplying industries.

Industry	Input share	Top 5	Top 15	Evaluation
Construction	48%	5%	9%	Perfect competition
Materials, components	23%	23%	38%	Low competition
Professional consultancy	9%	11%	18%	General competition
Plant, equipment	9%	13%	26%	General competition
Real estate	5%	n/a	n/a	Oligopoly
Transport, services	4%	27%	40%	Low competition
Energy, supplies	2%	46%	68%	Oligopoly
Sum:	100%			

Lowe (2011) advances a very strong argument as proof of the very strong competition in the UK construction market. He uses the concentration rate of the top 5 and top 15 firms and the Table 14.10 for evaluation.

The data for the construction sector as the sum of construction, material, equipment, and services are given in Table 14.11, together with an evaluation according to the framework of Table 14.10.

Construction includes contractors and subcontractors, and both markets are highly competitive. This allows drawing the summary of Figure 14.5. The percentages per group of sectors shows that only 48% of construction output are sourced from perfectly competitive markets, 18% from competitive markets and 34% from markets with low competition or even oligopolistic structures. The statements describe only the supply side and they do not extend to the demand side (owners).

This allows us to draw up the direct influences on the owner (Figure 14.4). The owner deals in two markets, one for professional services and another for contractors. In both cases, he can profit from competitive prices resulting from highly competitive markets. There are two other markets where the conditions are not so beneficial to owners, i.e. the market for land and the financial markets.

Similar consideration for the markets of contractors put them in a disadvantage as they have to buy partially from markets which are not competitive while being in a market that

Figure 14.4 Competitive forces in owner markets.

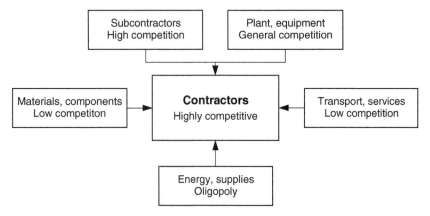

Figure 14.5 Competitive forces in contractor markets.

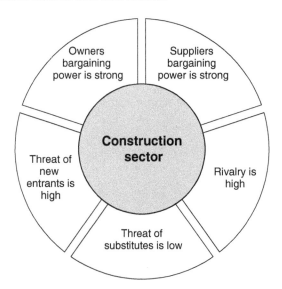

Figure 14.6 Porter's five competitive forces forming the construction sector.

is highly competitive by itself (Figure 14.5). This makes it difficult to pass on high price to the owner.

The analysis of the owner in Chapter 11 showed that the owner wields strong market power; substitutes to construction goods are few; market entry is easy (Section 14.7). These considerations allow to draw up Porter's five competitive forces in Figure 14.6. Three of the five competitive forces have a negative impact on profits: The bargaining power of the owner, possible new entrants, and rivalry among contractors. Mixed is the influence of the factors of construction. While material and energy sectors impact profits negatively, contractors can alleviate their own burden by passing it along to competitive subcontractors. The only other protection that contractors have is the lack of substitutes.

We know from the supply and demand model that demand changes will first lead to price changes and then to supply changes in perfectly competitive markets. Data from the years 1995 (end of a boom in Germany) and 1996 (beginning of a recession) show a quantity reduction of 3% and a price reduction of 1.3%. The insolvencies increased by almost 30% (decrease on the supply side). Theory of perfect competition on the supply side and empirical data match (Figure 14.7).

Figure 14.8 shows the data set of several years from 1995 (end of the boom) to 2006 (end of recession). In most years, a quantity reduction is followed by a price reduction. The data

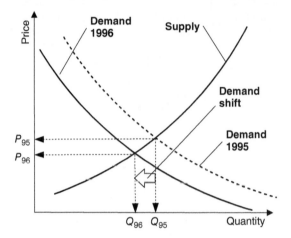

Figure 14.7 Shift in construction demand in perfectly competitive construction sector.

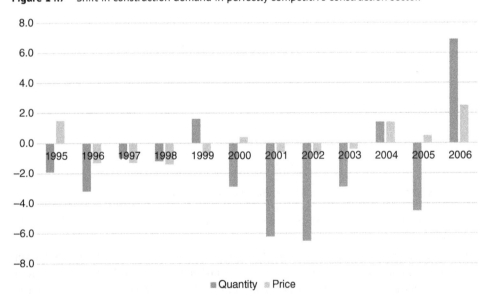

Figure 14.8 Quantity and price changes in the German construction market 1995–2006.

provide strong support for describing the market behavior on the supply side as perfectly competitive. The price reaction is sometimes delayed by a year.

14.6.3 International Markets

An international market strategy is characterized by a large share of the revenues being generated in the home market and some additional activities in selected foreign markets. Construction performances offered on foreign markets are specialized and limited to few contracts. The number of countries served is also small and typically they are adjacent to the national market (border hopping). The behavior of the company is ethnocentric with a limited amount of knowledge of foreign markets. The international activities are coordinated from the national headquarters through an international department. Despite all limitations, international companies must have the know-how to form and manage the networks around their foreign sites responding to differences in culture and regulations. There is no scientific way of determining what means a "small" number of countries. For reasons of clarity, I propose here a cut-off point of less than 10 countries.

In 1993, 249 of all German contractors were active in international markets, which is equivalent to 0.3%. 35 companies (or 14% of 249) had a share of 90% of this market. The volume of international contracts was a mere 1.5% of the volume of the German market. The conclusion is that very few of all regional construction companies are interested and capable of leaving their home market. Many of them do the odd job just across a borderline, and some take a deeper interest in international markets (Russig et al. 1996). Although the data are quite old, there is no reason to assume that they have changed considerably. They still portray the market.

There are only three German firms among the top 250 international contractors (Engineering News Record 2021). The top firm is much larger than the next two (Table 14.12). Again, the picture emerges that international construction is not the bread-and-butter business of contractors but rather a niche for very few.

Having provided data for the German construction market raises the question how this market compares with others. Table 14.13 provides data for the top international nations and the structure of their construction business. With the exception of China and Spain, Germany is a rather typical example of the international involvement of national companies.

Table 14.12 Top five German international contractors in 2019.

ENR rank 2019	Firm	Int'l revenue 2019 (million USD)	Int'l revenue 2005 (million USD)
2	Hochtief	29 303	17 599
—	Bilfinger	Left the market	6553
76	Züblin	1217	730
115	Bauer	542	442
—	SKE Group	—	366
—	Heberger Bau	102	201

Table 14.13 Number of international firms and revenue per country 2019.

Country	Percent of Int'l revenue	# of firms
China	25.4%	74
Spain	14.9%	11
France	9.9%	4
Germany	6.6%	3
USA	5.2%	35
Korea	5.2%	12
Turkey	4.6%	44
UK	4.2%	3
Japan	4.1%	13

It is the norm in international markets to do some work outside the national borders. However, it is also possible to become almost independent of the home market. Good examples are Hochtief (Germany) which produces 97% of its revenue on international markets or ACS (Spain) with 87%. The most international of the Chinese firms, China Communications Constructions, generates only 26% internationally (Tulacz and Reina 2020).

14.6.4 Multinational Markets

If the activities in foreign countries contribute a larger share to a company's revenues and if these contributions come from many different foreign markets (10 or more), then the company can be called multinational. International and multinational markets cannot be distinguished by groups of contractors. While contractors have either an international or a multinational strategy, they compete on the same physical markets.

The companies that are multinational differ quite significantly from those who are international. Ethnocentricity is replaced be a polycentric behavior with strong orientations toward the different host countries. This, of course, implies a decentralized organization of the company with many top managers in the host countries having a local origin. Besides coordination of the international activities through an international department, the company also acquires foreign companies. A multinational company has many foreign subsidiaries. An example is Actividades de Construcción y Servicios (ACS, Table 14.14).

The ability to work in foreign environments and to build up networks in a score of different cultures is the main competence of the multinational company. Through its polycentric behavior such a company is prepared for and at ease in foreign environments.

However, much of ACS's international revenue is produced locally. Turner was in 2021 the largest contractor in the USA, adding up most of its revenue in local projects all over the USA. These projects were managed and staffed by Americans; the owners were Americans as well as the suppliers and the equipment producers. Technology and culture were also American (Adolphus 2021). Since Turner is owned by ACS all these revenues count as international revenue for ACS. The rankings of the Engineering News Record for the

Table 14.14 Subsidiaries of Actividades de Construcción y Servicios (ACS).

Company	Sector	Country
ACS, Actividades de Construcción y Servicios	Corporation	Spain
ACS, Servicios Comunicaciones y Energía	Industrial services	Spain
ACS, Servicios y Concesiones	Services	Spain
CIMIC Group Limited	Construction	Australia
CLARK BUILDERS	Construction	Canada
CLECE S.A.	Services	Spain
GRUPO COBRA	Industrial services	Spain
COGESA	Construction	Spain
CONTROL Y MONTAJES INDUSTRIALES	Industrial services	Spain
CPB CONTRACTORS PTY LIMITED	Construction	Australia
HOCHTIEF Aktiengesellschaft	Construction	Germany
IMESAPI	Industrial Services	Spain
INITEC ENERGÍA	Industrial Services	Spain
INTECSA INGENIERÍA INDUSTRIAL	Industrial Services	Spain
IRIDIUM CONCESIONES DE INFRAESTRUCTURAS	Construction	Spain
J. F. WHITE Contracting	Construction	USA
JOHN PICONE	Construction	USA
LEIGHTON ASIA, INDIA AND OFFSHORE	Construction	China
LEIGHTON PROPERTIES PTY LIMITED	Construction	Australia
MAESSA	Industrial Services	Spain
MAETEL	Industrial Services	Spain
MAKIBER	Industrial Services	Spain
MASA	Industrial Services	Spain
PACIFIC PARTNERSHIPS	Construction	Australia
POL-AQUA	Construction	Poland
PRINCE	Construction	USA
PULICE	Construction	USA
SCHIAVONE	Construction	USA
SEMI	Industrial Services	Spain
SICE TECNOLOGÍA Y SISTEMAS	Industrial Services	Spain

biggest international contractors are based on the international revenue as shown in the books of the holding. Most of it is regionally produced with no foreign influence except the transfer of profits and an occasional meeting of top management representatives to determine the strategy.

The total value of construction spending has been estimated to be around $11.5 trillion US dollars for the year 2020 (Business Research Company 2021). The top 250 international

Table 14.15 Revenue of the top 15 international contractors.

#	Firm	Country	Int'l revenue (billion USD)	Total revenue (billion USD)	Int'l work
1	ACS	Spain	38,950	45,016	87%
2	Hochtief	Germany	29,303	30,243	97%
3	Vinci	France	24,499	54,574	45%
4	China communications	China	23,303	89,506	26%
5	Bouygues	France	17,142	33,225	52%
6	Strabag	Austria	15,659	18,668	84%
7	Power Construction	China	14,715	57,009	26%
8	China State	China	14,143	180,355	8%
9	Skanska	Sweden	12,881	16,116	80%
10	Technip	UK	12,852	13,409	96%
11	Ferrovial	Spain	12,065	15,714	77%
12	China Railway Constr.	China	8,205	123,427	7%
13	China Railway Group	China	6,572	154,905	4%
14	Hyundai	Korea	6,300	15,166	42%
15	China Energy	China	5,325	27,001	20%
		Total international revenue:	473 000		

contractors had in 2019 a revenue of 473 billion US dollars from international operations (ENR 2021 - See the above Table 14.15), which is equivalent to 4% of total construction spending. While nobody knows how much revenue is generated by smaller international companies, it cannot be a big sum. The last 20 firms in that list all have only an international revenue of less than 100 million USD each.

The interpretation of these data is clear: About 95% of the global construction spending is allocated on local markets to local and national contractors. It definitely shows that the construction market is not a global market. Whatever happens in China is of very little interest to most construction companies in the US.

Table 14.16 lists the 15 firms with the largest international revenue in 2019. Based on the data, the top 5 international contractors have a market share of 28%. This is considered as low competition. The top 15 international contractors have a market share of 51% which signifies oligopolistic markets. Unfortunately, the reasoning is not correct because the national competitors are not included in the analysis. The Chinese firms, for example, are only included with their international revenue while they compete in this market with their national revenue and there will be many more competitors which are not in the list of the top 250 international firms and which are still be capable of participating in large-scale projects.

The concentration data above are the absolute maximum when not considering all competitors. They do not allow to categorize the multinational construction market as oligopolistic.

Table 14.16 Ranking of the top nine German construction firms 1998 and 2019.

Rank	Firms 1998	Firms 2019
1	Philipp Holzmann AG (6.100 Mrd. Euro), insolvent	Goldbeck (2.900 Mrd. Euro)
2	Hochtief, acquired by ACS, Spain	Zechbau
3	BilfingerBerger, Construction sector sold	Max Bögl
4	Strabag, acquired by Bau Holding, Austria	Bauer
5	Dyckerhoff & Widmann, insolvent	Leonard Weiss
6	Walter Bau, insolvent	Köster
7	Züblin, acquired by Bau Holding, Austria	Wolff & Müller
8	Heilit & Woerner AG, insolvent	Papenburg
9	Wayss & Freytag, acquired by BAM, Netherland	Bremerbau

14.6.5 Global Players and Global Markets

Global players producing exchange goods (consumer goods) are not simply active in more foreign countries than the multinational company but they use standardization of their products to achieve the highest possible economies of scale. They profit from a network spanning the globe to organize production. Managers come from countries around the globe, factors of production are bought where cheapest, and production is set up where labor costs are low and the institutional environment reliable. The orientation of the company is not tied to any one national culture, it is global, the behavior is polycentric. The headquarter coordinates the activities of a multitude of affiliated and owned companies around the world. The important point for this strategy to work is, however, that the products can be standardized. An example of a global player is Apple. Some companies also follow a transglobal strategy where some parts of the product are global and others are national. This is the case for Coca-Cola which uses a global marketing approach but different tastes of their products in different countries.

It is out of question that the products of the construction sector can be standardized around the world. Global markets seem to be no option for the sector. Whether there are global players in construction remains an open question that I will analyze in the following paragraphs.

The five described markets – from regional to global – form a hierarchy. Any multinational company is still rooted in a multitude of reginal markets, it is active in a number of national markets. One of these is the home market, the others are international markets. Projects and companies in many foreign countries as well as the home country characterize the multinational company. Companies need time to develop from a local contractor to a multinational contractor or a global player. For each step on the market ladder upwards, additional resources are required (Section 12.3.3).

Megaprojects differ from international projects. Miller and Lessard (2000) have studied megaprojects around the world and the average contract size in their sample is 985 million US dollars. As such, they form another subgroup of all international projects. In addition to size, they require cutting-edge technology. Structures included in the sample are hydroelectric projects, thermal and nuclear power projects, urban transport, roads, tunnels, bridges, oil projects, and technology projects. Not included are buildings and manufacturing structures. Besides technological know-how, the companies undertaking megaprojects need the ability to deal with the extreme complexity of such projects which includes the ability to establish the necessary network around the project in any part of the world. Such networks need to incorporate a large number of stakeholders. Besides the typical project participants shown in Figures 14.4 and 14.5, we need to include governmental ministries, public works authorities, utility authorities, banks, lawyers, insurers, press, police, and the public.

Such projects are typically contracted by international construction joint ventures (ICJV) with the multinational contractor responsible for know-how and financial resources and the local contractor in charge of local contacts. The project itself will be a multi-organizational enterprise (MOE). We can call these multinational contractors of megaproject ICJVs global contractors. They must possess certain experiences that are standardized on a high level. In this way, standardized inputs characterize global contractors and not standardized outputs. Most notable required experiences for global contractors are (Brockmann 2021):

- They know how to deal with the overwhelming task, social, cultural, cognitive, and operative complexity of megaprojects.
- They must have the competence to organize a megaproject: A decentralized organization is required.
- Megaprojects are mostly built by ICJVs, so the global contractor must have experience with international joint ventures. Especially since the local partner might never before have worked with an international partner and the same might hold true for the owner: A cultural competence is required.
- The global contractor must organize and manage the international network: A social competence is required.
- The global contractor (at the side of the local partner) must deal with a foreign owner, often with high-ranking government employees: A diplomatic competence is required.
- The global contractor must offer the owner and the ICJV partner a worldwide reputation as hostage against possibilities to resort to opportunism. Since the international contractor does not reside in the foreign country he can rather easily leave it. Once he behaves opportunistic, the owner and partner can only retaliate by hurting his reputation: A worldwide reputation is required.
- The global contractor must provide cutting-edge technology: A technological competence is required.
- The global contractor must possess sufficient financial resources for the tendering, the negotiation phase and for the start-up of the megaproject: Large financial resources are required.
- During the negotiation phase the global contractor must understand the reversal of the principal/agent relationship. He must find ways to withstand the pressures and to counter the asymmetric information favoring the client: A negotiation competence is required.
- The international contractor must be able to price projects of a kind never built before: a megaproject pricing competence is required.

14.7   Entry and Exit Barriers

There are two different views on the function of barriers. Bain (1956) describes advantages of established sellers in an industry over those outside the industry as reason for a barrier and Stigler (1983) refers to costs that firms have to bear in order to enter an industry. Runeson (2000) defines the construction market as perfectly competitive and therefore assumes that there are no barriers to entry or exit, i.e., insiders do not enjoy advantages over outsiders or outsiders do not have to bear significant costs for entry or exit. De Valence (2011) sees the construction market as a three-tiered system made up of small, medium, and large contractors with no barriers to enter the group of small firms but significant ones for the other two levels. He ascribes oligopolistic competition to the top level and monopolistic competition to the middle level. As the number of firms increases from the upper level downward, this model resembles a pyramid (Figure 14.9).

This model is the most sophisticated one for the construction industry. However, it does not address exit barriers, and it is static. For an appreciation of the market structures, we need to also consider time-dependent effects caused by the business cycle. This approach demands using data from upturns and downturns and the passage from one to the other. It will become clear that the maelstrom of the business cycle has an all-important influence on the market structure and on entry as well as exit barriers. The result is the model of Figure 14.10, which shall serve as a hypothesis for the ensuing discussion.

The model in Figure 14.10 postulates that during upturns the market is attractive and demand grows often faster than the supply by existing firms. These firms expand production and enjoy economic profits. As these profits do not go unnoticed, new firms enter the market, mostly as small-size firms while the expanding firms grow to mid-size or large-size firms. At the end of this suction effect, supply and demand find again an equilibrium and economic profits approach zero. During downturns, the competition increases because supply exceeds demand. Instead of profits, companies face losses and bankruptcies increase. The easiest strategy for companies is to downsize. As there are no growth barriers between the differently sized companies, so there are no barriers for shrinking firms. Exit barriers exit nonetheless in the form of loss of prestige. Especially owners of small-size firms invest sizable amounts of private money to keep their firms afloat. Once the size of

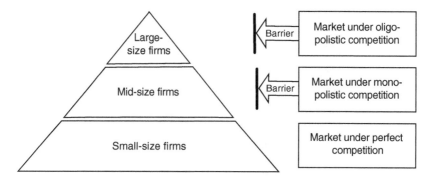

Figure 14.9   Model of entry barriers (de Valence).

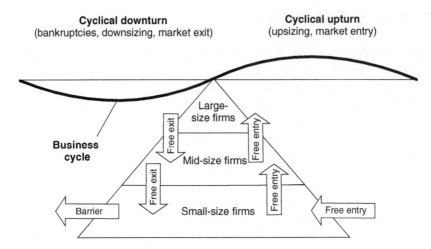

Figure 14.10 Entry and exit barriers.

the construction firms has adjusted to the low demand, another increase in demand will again trigger a new round of growth. The free entry at the level of small-sized firms and the exit barriers at the same level keep some pressure on the construction sector, it is not a free-flowing system.

Are there any data in support of this model? This we will see in the next chapters.

14.7.1 Effects of the Business Cycle

The construction business cycle in Germany is shown in Figure 14.11; the data are based on constant prices and the development is in percent with 1990 as base year. In 1990, the German unification was completed and for the first time, data from the former East Germany were incorporated. It can be seen that there was an upturn in construction investment from 1990 to 1994 followed by a sharp downturn until 2005. From 2005 to 2009,

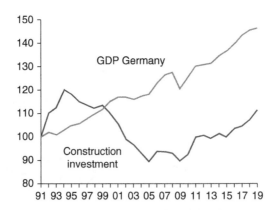

Figure 14.11 Construction investment and GDP, Germany 1990–2019. *Source:* Data from BAUWIRTSCHAFT IM ZAHLENBILD 2021

investment was rather stable. Next followed an ongoing upturn. It can also be seen that construction investment started lagging behind the general economic development as measured by the gross domestic product (GDP) around 1999 (Hauptverband der Deutschen Bauindustrie 2020).

An entry/exit model based on the idea of no significant barriers suggests an increasing number of firms during an upturn and a decreasing one during a downturn.

Statistical data for the construction sector are not always consistent. Often the statistical population is different between two sets of data. This holds true for developed countries as well as for developing ones. Therefore, it is necessary to carefully avoid drawing conclusions from different populations.

The term construction investment refers to all expenses borne by the owners. Thus, the figures not only include value-added by the construction sector but also services by architects and engineers, material, and equipment. The construction sector can be split into structural works including all the firms working on the structure and finishing works referring to finishing trades as well as plumbing, mechanical, and electrical works.

Figure 14.12 shows the number of firms in the category structural works as total and split into firm sizes above and below 20 employees. The data are given in percent with 1991 as base year (100%). The business cycle is shown as the number of orders per year. This is more direct than construction investment as in Figure 14.11 as payments (investments) follow orders. However, basically the run of the business cycle is very similar between the two representations. In the upturn until 1994 the number of firms with more than 20 employees increases stronger than the total numbers of firms. The same holds true for the upturn from 2011 onward. This is a consolidation process in good times. The number of small firms (<20) increases in the first upturn (1991–1994), decreases in the following downturn to increase again during the continuing recession. This is process of fragmentation. The number of larger firms fluctuates considerably more than that of small firms.

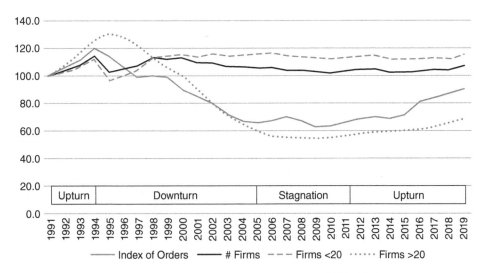

Figure 14.12 Number of firms for construction works in Germany, 1991–2019.

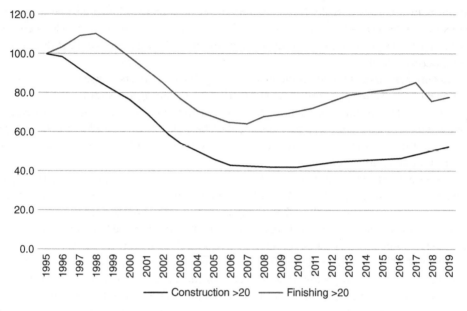

Figure 14.13   Number of firms >20 employees for construction and finishing works, Germany.

This time series explains that the construction sector is and remains fragmented. During the end of the first upturn (from 1991 onwards) companies of all sizes grow faster than orders, a sign of no significant barriers to entry. At the beginning of the second upturn orders grow faster than the number of firms; there is some reluctance.

Figure 14.13 shows the number of firms split into structural and finishing works and for firms with more than 20 employees. The data are normalized with 1995 as base year (there are no data available before 1995). While the run of the curves differs, the overall message is the same. Downturns decimate the number of firms with more than 20 employees. Any exit barriers cannot be strong. The market in up- as well as downturns is volatile with regard to the number of lager firms.

 Observation = (obs.) Behind the data there is a story explaining the mechanism. In a downturn, all firms come under pressure. Many face losses or as one CEO explained, construction firms envy the other companies' losses and submit an even lower bid at the next submission. If a firm of 200 employees goes bankrupt, it might have three project managers. One goes into early retirement and the other two set up two new construction firms by taking each the most efficient 15 workers along. Some 170 employees lose their job but two new firms enter competition. They buy the equipment from their old company for very little money, nobody else makes an offer. Instead of a headquarters, the business is conducted at the living room and kitchen tables. The Mercedes of the old boss is replaced by a well-used Toyota Corolla. To cut a long story short, fixed costs are cut drastically and productivity increases.

During the first upturn (1991–1994) in Germany, competition stayed strong while capacity was stretched; accordingly, the labor productivity defined as output per labor hours increased; the same continued during the recession; slack was cut and costs reduced. A year after the end of the recession in 2006, the trend broke off. Life became easier and profits rose without increasing the productivity (Hauptverband der Deutschen Bauindustrie 2018; Figure 14.14).

Figure 14.15 provides the theoretical explanation of the fragmentation/consolidation process. A larger firm requires greater investments and can then produce at lower variable costs. Small firms have little fixed costs but higher variable costs. There is a breakeven point

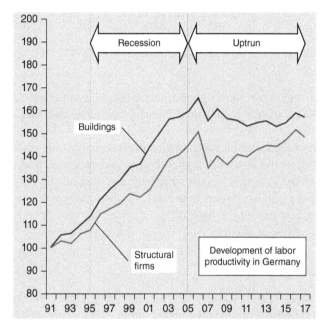

Figure 14.14 Changes of labor productivity during the business cycle in Germany. *Source:* Data from Zahlen & Fakten

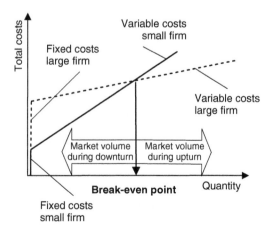

Figure 14.15 Functioning of the fragmentation/consolidation process.

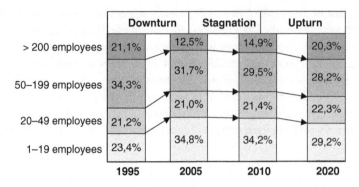

Figure 14.16 Business cycle, firm size, and construction volume in Germany.

depending on construction volume (quantity). Downturns lead to quantities to the left of the break-even point favoring small firms. Upturns then allow cost advantages for larger firms.

The effects of the fragmentation/consolidation process are not only visible when looking at the number of firms but also when comparing the percentage of industry turnover per company size. Figure 14.16 clearly shows the fragmentation during the downturn (construction volume of small-sized firms increases), almost stable conditions during stagnation and consolidation during the upturn. The number of small firms increases during downturns and decreases in upturns.

14.7.2 Number of Exits and Entries

Data are also available for insolvencies; however, they are generated from a different population than the number of firms in construction (appr. 75,000 firms in 2019), being from a statistic on turnover (about 110,000 in 2019). Important is the trend and it shows that insolvencies increase strongly during a downturn (Figure 14.17). If we take insolvencies as an

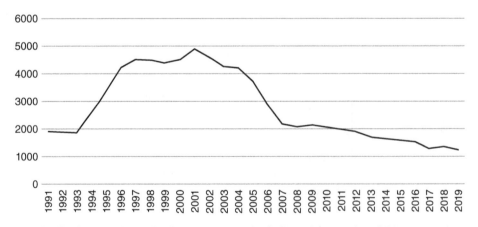

Figure 14.17 Insolvencies in the German construction industry (structural works).

indicator of competition, then a strong competition built up during the first upturn; insolvencies jump up in 1993, two years before the end of the boom. The consolidation process lasted 10 years until 2005, the end of the recession.

The overall level of exits is rather high. Manufacturing might serve as a benchmark for insolvencies in construction. Figure 14.18 shows that the insolvency ratio in construction per 10,000 firms always surpasses the ratio for manufacturing with the exception of 1995 (reorganization of East Germany with a heavy toll on manufacturing). In 2000, the figure is more than four times higher. The construction market functions well in the sense that firms which are not competitive leave the market in large numbers. While there are data of insolvencies, there are none on firm closures. Insolvencies as involuntary exits and closures as voluntary ones certainly show that exit barriers have little importance.

The number of start-ups in the construction sector (structural and finishing works) is given in Figure 14.19. The data refer again to a different population with approximately

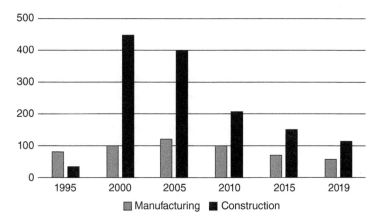

Figure 14.18   Comparison of insolvencies in manufacturing and construction per 10,000 firms.

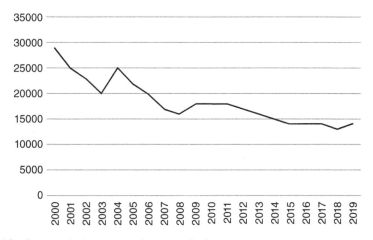

Figure 14.19   Startups in the construction sector in Germany.

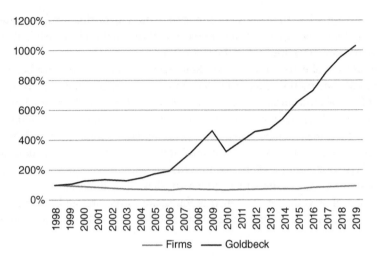

Figure 14.20 Comparison of orders of all structural contractors and turnover of Goldbeck.

300,000 firms. There are no continuous data available starting in 1991, but in the first reces-sion until 2005 the number of startups decreased due to a lack of market attraction.

A final group of firms of interest consists of the top nine firms in the German construc-tion sector. The nine largest firms controlled by German majority owners are listed in Table 14.16. Not a single top-nine contractor survived the long recession from 1995 to 2005 unscathed. All firms in the group of the top nine in 2019 have grown into these top posi-tions, and the turnover of the largest firm is substantially lower than in 1998. The turnover of Philipp Holzman was twice as large (6 billion euros, not inflation adjusted) as that of Goldbeck with 3 billion. Overall, the large contractors have become smaller; consolidation starts at a low level.

Goldbeck provides a good example of the consolidation process. Besides being the most successful German contractor over the past years, turnover grew strongly during the stag-nation period and exploded during the upturn. Figure 14.20 shows turnover for Goldbeck and orders for the German structural firms, showing that the end of market stagnation provides great chances for those well prepared.

14.8 Summary

Without a clear and comprehensive definition of the term market, it becomes difficult to discuss the prevailing market type in construction. I have introduced the definition by Arrow/Debreu for this purpose.

An analysis of construction goods proves that they are heterogenous. This means that there exist a multitude of markets with the owners yielding considerable market power. These markets are most of the time competitively contested by contractors. There are no barriers to protect market share and profit of firms in the market. In downturns of the market, low costs and high productivity guarantee survival. Companies also downsize. In

Table 14.17 Characteristics of construction markets.

Perfectly competitive market	Monopolistic market	Monopolistic competitive markets	Oligopolistic market	Construction market
Homogenous goods	Homogenous good	Differentiated goods	Homogenous goods	Heterogenous goods
Many buyers as price takers	Many buyers as price takers	One owner as price setter	Many buyers as price takers	One owner as price setter
Many sellers as price takers	One seller as price setter	Many contractors a price takers	Some sellers as limited price setters	Many contractors a price takers
Complete information	Asymmetric information	Asymmetric information	Asymmetric information	Asymmetric information
No barriers	Strong barriers	No barriers	Some barriers	No barriers

upturns, companies again grow in size. The succession of up- and downturns plus the single project markets keeps the industry fragmented.

The German construction sector is similar to others markets in western countries. For these, the data tell us that a large number of firms characterize the national construction market and the exit and entries are easy; barriers are not strong. Thus, the market shows the attributes of perfect competition in this regard. There seems to be no chance of an increasing market concentration as the business cycle triggers the fragmentation/consolidation process. Large firms are especially susceptible to the business cycle.

In downturns, there is a downward movement in the pyramid of firm sizes and the pyramid gets flatter. Growth in upturns happens mostly from within the market. While there are exit barriers to enter the top of the pyramid from the outside, there are none to achieve this from the inside.

The results of the previous discussion are summarized in Table 14.17, and they show that construction markets are neither perfectly competitive, monopolistic, monopolistic competitive nor oligopolistic.

References

Adolphus, E. (2021). The pandemic's second act: many ENR Top 400 contractors plan a return to normal, but will the COVID-19 vaccine be a shot in the arm or a shot in the dark? *Engineering News Record* (May 24/31).

Bain, J.S. (1956). *Barriers to New Competition*. Cambridge: Harvard University Press.

Bauindustrieverband Bremen-Nordniedersachsen (ed.) (1997). *Öffentliche Vortragsveranstaltung*. Bremen: Bauindustrieverband Bremen-Nordniedersachsen.

Bollinger, R. (1996). Auslandsbau. In: *Handbuch der strategischen und taktischen Bauunternehmensführung* (ed. C. Diederichs), 567–586. Wiesbaden: Bauverlag.

Bowdich, G. and Runeson, G. (1985). Unpublished survey of the building industry in the Wollongong area.

Brockmann, C. (2021). *Advanced Construction Project Management: The Complexity of Megaprojects.* Hoboken: Wiley-Blackwell.

Brockmann, C. and Nolte, L. (2007). Regional construction markets: The example of Bremen, Germany. In: *Proceedings of the CIB World Congress 2007*, 3146–3158. Cape Town: CIB.

Business Research Company (ed.) (2021). *Construction Global Market Report 2021: COVID-19 Impact and Recovery to 2030.* London: Business Research Group.

Buzzell, R. and Gale, B. (1987). *The PIMS Principle: Linking Strategy to Performance.* New York: Free Press.

Cooke, A. (1996). *Economics and Construction.* Basingstoke: Macmillan.

De Valence, G. (2011). Market types and construction markets. In: *Modern Construction Economics* (ed. G. de Valence), 171–190. London: Spon Press.

Engineering News Record (ed.) (2021). *Top global contractors.* Troy: ENR.

Gruneberg, S. (1997). *Construction Economics: An Introduction.* Basingstoke: Macmillan.

Gruneberg, S. and Ive, G. (2000). *The Economics of the Modern Construction Firm.* Basingstoke: Macmillan.

Hauptverband der Deutschen Bauindustrie (2018). *Kapazitätssituation im deutschen Bauhauptgewerbe: Schwerpunkt Wohnungsbau.* Berlin: Hauptverband der Deutschen Bauindustrie.

Hauptverband der Deutschen Bauindustrie (2020). *Die Bauwirtschaft im Zahlenbild 2019.* Berlin: Hauptverband der Deutschen Bauindustrie.

Hauptverband der Deutschen Bauindustrie (2021). *Die Bauwirtschaft im Zahlenbild 2020.* Berlin: Hauptverband der Deutschen Bauindustrie.

Hillebrandt, P. (2000). *Economic Theory and the Construction Industry.* Basingstoke: Macmillan.

Koskela, L. (2000). *An Exploration Towards Production Theory and Its Application to Construction.* Espoo: VTT.

Krugman, P. and Wells, R. (2018). *Microeconomics.* New York: Worth.

Lansley, P. (1981). *Maintaining the Company's Workload in a Changing Market.* London: The Chartered Institute of Building.

Lowe, J. (2011). Concentration in the UK construction sector. *Journal of Financial Management of Property and Construction* 16 (3): 232–248.

Miller, R. and Lessard, D. (ed.) (2000). *Strategic Management of Large Engineering Projects: Shaping, Institutions, Risks, and Governance.* Harvard: MIT Press.

Monopolkommission (ed.) (2003). *Hauptgutachten 2000/2001.* Baden-Baden: Nomos.

Morton, R. (2002). *Construction UK: Introduction to the Industry.* Oxford: Blackwell.

Myers, D. (2017). *Construction Economics: A New Approach.* Abingdon: Routledge.

Nolte, L. (2006). *Baumärkte als regionale Baumärkte: Nachweis am Beispiel Bremen.* Unpublished master thesis. Bremen: Hochschule Bremen.

Ofori, G. (1990). *The Construction Industry: Aspects of Its Economics and Management.* Singapore: Singapore University Press.

Roth, A. (2015). *Who Gets What – And why.* Boston: Houghton Mifflin Harcourt.

Runeson, G. (2000). *Building Economics.* Victoria: Deakin University Press.

Russig, V., Deutsch, S., and Spillner, A. (1996). *Branchenbild Bauwirtschaft.* Berlin: Duncker & Humblot.

Skanska (ed.) (2021). *Annual and Sustainability Report 2020.* Stockholm: Skanska.

Stigler, G. (1983). *The Organization of Industry.* Chicago: Chicago University Press.

Strabag (ed.) (2021). *Annual Report 2020*. Vienna: Strabag.

Tirole, J. (2000). *The Theory of Industrial Organization*. Cambridge: MIT Press.

Tulacz, G. and Reina, P. (2020). Struggling with Covid-19: Rocked by the worldwide pandemic and plunging oil prices, the global construction market attempts to cope. *Engineering News Record* (August 17/24).

Varian, H. (2014). *Intermediate Microeconomics with Calculus: A Modern Approach*. New York: W. W. Norton.

15

Contracting

Contracting describes what actually happens in construction markets when owners and contractors meet to sign a contract. While Chapter 14 on construction markets often took a macroeconomic perspective, now I will focus on microeconomics when describing contracting. This is a lengthy process. The owner must prepare a design and a contract before starting the process with the owner's notice to the contractors. Next, interested contractors will pick up the tender documents (contract, design, bill of quantities), evaluate and price them. The result is a bill of quantities with unit prices or a lump sum offer. Tendering ends with the submission of the offers by the contractors. The owner will in turn have to evaluate all offers received. Typically, there is a need for contract negotiations to clarify the offer and to discuss the price. This stage ends with the signature of the contract by both parties. Construction contracts are prime examples of incomplete contracts because there will always be amendments, additions, and changes to the contract due to planning mistakes or unforeseeable events. The changes require renegotiations of the contract impacting scope, quality, time, and price. Construction formally ends with handing over the completed structure from the contractor to the owner. Most of the time, there is a punch list of defects that require correction before the contractor can leave the site. The maintenance period follows where a contractor must rectify all new defects. Contracting ends only at the end of the maintenance period (Figure 15.1).

Contracting is only possible, when dealing with contract goods. Therefore, I will recapitulate the main characteristics of contract goods defined as construction goods (Section 15.1). Another repetition regards the construction markets on which owners and contractors engage in contracting. The market transaction lasts as long as the contracting period; the two are identical. It starts with the first contact and ends when all obligations from the contract are fulfilled. Personal face-to-face interaction characterizes all phases of contracting in construction markets (Section 15.2).

Owners bring a demand to the construction market and contractors a supply. Owners define their basic demand in the design before contracting starts (Section 15.3). They also define the conditions of the exchange by the contract (Section 15.4). Contractors then evaluate the demand and prices it by estimating (Section 15.5). The contracting market design takes typically the form of sealed-bid auctions. This has a rather specific impact on the market outcome and needs close analysis (Section 15.6). Estimating a contract price causes considerable problems for contractors who must determine a price

Construction Microeconomics, First Edition. Christian Brockmann.
© 2023 John Wiley & Sons Ltd. Published 2023 by John Wiley & Sons Ltd.

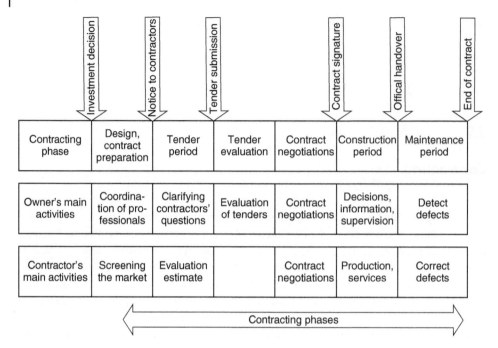

Figure 15.1 Phases of contracting.

for future activities without the benefit of repeating an experience (Section 15.7). When both sides have completed their tasks, supply and demand finally meet at submission date (Section 15.8). However, owners are often not content with the market outcome and start a phase of negotiating and bargaining. At the end of this process, owner and contractor will sign their agreement (Section 15.9). Due to the incompleteness of the contract, there will be changes that require renegotiations of scope, quality, time, and price (Section 15.10). The warranty period might bring additional changes to the market outcome (Section 15.11). I will conclude all these processes in a summary (Section 15.12).

15.1 Construction Goods

I do not know how often I have already repeated that construction goods with all their heterogeneity have one thing in common: They are contract goods. The only justification for this repetition lies in the fact that no other monography in construction economics uses this definition from new institutional economics. Figure 15.2 shows the development and transaction of an exchange good and Figure 15.3 the same for a contract good.

In the development, production, and sale of an exchange good we find a clear separation of the activities between the producer who controls all phases until delivery to the market and the consumer. Both join their activities at a specific moment in time to undertake the market transaction. Many authors describe these phases in detail (e.g., Kotler and

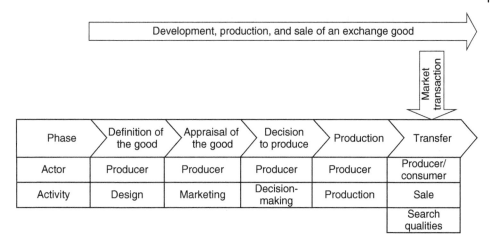

Figure 15.2 Development, production, and sale of an exchange good.

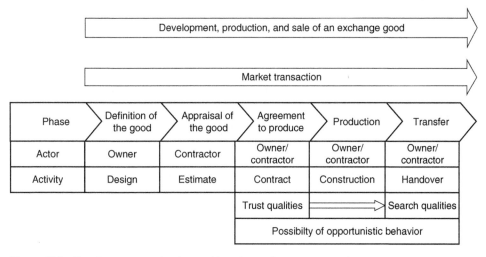

Figure 15.3 Development, production, and handover of a contract good.

Armstrong 2017). I have left out the role of retailing. It is important but not for this analysis (Levy et al. 2018). The main characteristics of exchange goods are therefore:

- Development, design, and production of an exchange good take place before the sale.
- The producer has full and exclusive control of development, design, and production.
- The consumer is not involved before the market transaction (buying).
- The market transaction takes place at a moment in time.
- The consumer can find information about the product (experience quality) because it is mass produced.
- The consumer can test the product (search quality).

When developing, producing, and handing over a construction good, the actions of the owner and contractor overlap. The contractor is an external factor from the owner's point of view, and from the contractor's point of view it is just the opposite. The market transaction takes a long time with many possibilities for opportunistic behavior. The main characteristics of construction goods are:

- Development and design are the responsibility of the owner.
- Contracting (the sale) takes place before production.
- The owner and contractor exchange promises when signing the contract, i.e., the legal obligations to fulfil the contract.
- Owner and contractor cooperate during production (team production).
- The market transaction takes a long time.
- Experience qualities are very limited due to the singularity of structures and the multitude of contractors.
- When signing the contract, the owner must rely on trust qualities.
- Different from other contract goods such as haircuts, the structure has search qualities at handover.

15.2 Construction Markets

I have discussed that owners and contractors trade one-family houses for consumption and factories or offices as investments. To find out whether we are facing a market for consumptive goods or investment goods, we just need to look at the structure of construction spending. I introduced already the idea, that the difference between consumption and investment depends on the use of a building. Infrastructure projects are always investments. A one-family home can be both. If a family builds a house for self-occupancy, then the use will be consumptive. If the same family builds a house to rent it to someone, then it is used as investment. Only residential buildings can be used as consumption goods.

Table 15.1 provides data on the importance of different subsectors of construction with the goal to distinguish between consumption and investment. The data are from the US Census Bureau for April 2021 (US Census Bureau 2022).

Table 15.1 Subsectors of US construction spending.

	Billion USD	Percentage overall (%)	Percentage detailed (%)
Total construction spending	1524	100	
Total private construction spending	1181	77	
Nonresidential	452		30
Residential total	729		47
New single family	396		26
New multifamily	98		6
Improvements	235		15
Total public construction spending	343	23	

Consumptive use is only possible for single family homes and improvements (maximum 41%). However, we have to deduct the homes put on the real estate market for sale. Builders are producing investment goods which families buy for consumptive purposes. In Germany the percentage of custom-designed and self-contracted homes is at 10% and this is most likely higher than in the US. Combining these figures, we find for this group for new-built and improvements a percentage of 4.1% of total construction ($26\% \times 0.1 + 15\% \times 0.1$). In sum, the overwhelming amount of construction spending is an investment.

The investor is nothing but the producer of residential, nonresidential, and public construction. Accordingly, the owner is the producer who works with contractors to achieve his goals. The very important takeaway here is that we are looking at a factor market. On factor markets, producers buy inputs to achieve their own productive goals of providing goods or services.

If we go back to the market definition by Debreu and Arrow (Tirole 2000), then we find that locality, time, and the product determine the market. Typical are regional markets with regard to location but it is also possible that a construction good is traded on a global market. Locality puts few limits on the market. Time constraints the market much stronger: A one-family standard home in Mumbai built in 2010 is traded in a different market then the same home in the same place two years later. The most important separation of the relevant market comes about by the individual design of a structure.

In the end, each individually designed structure creates a separate market. A bridge in Warsaw is no substitute for one in Reykjavik. A chip factory built in 2020 is no substitute for a factory built in 2019. A one-family home with three bedrooms and two baths in South Bend, Indiana, under construction in 2022 is no substitute for a one-family home with two bedrooms and one bath built at the same time in the same place.

Each construction project with a minimum of singularity creates its own market, whether it is used for consumption or investment.

15.3 Owner's Demand

When the owner produces a unique design with the help of architects and engineers, the quantity later demanded on the market becomes fixed; the owner also fixes the quality, which will decisively influence the price.

The simple supply and demand model has only two variables, quantity and price. In the following chapters, I will introduce further influences. However, the model in Figure 15.4 concentrates only on these quantities.

Other than in the supply and demand model, the quantity in contracting is not a variable; it is a constant determined by the owner's design. Most owners have a fixed price limit but they are willing to accept a lower price. There will also be a lowest acceptable price, which will depend on the circumstances. Owners will accept an unexpected low price until the point where they become convinced that it is not possible to deliver the expected structure based on such a price level.

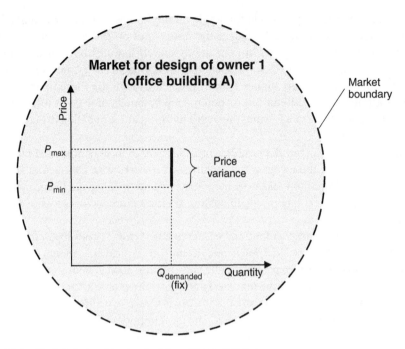

Figure 15.4 Market demand by owner 1 for office building A.

While it is conceivable that the same owner has a demand for two identical buildings in the same location and at the same time, it is also highly unlikely: These two buildings would constitute one market.

15.4 Contractor's Supply

The nature of contract goods makes it impossible to have an independent supply. Supply on construction markets depends on demand. Contractors offer a possible performance in construction markets until they engage in bidding for a concrete project. Then, they offer a specific performance.

To take an example, we can imagine a mid-size contractor (20–49 employees) who negotiates signing a contract for a $2 million building project. This is the contractor's specific supply on this project market. The contractor is also preparing to bid on another 10 contracts. Thus, he is looking at a potential supply from these projects. By experience, the contractor expects to win 1 out of 10 bids; the probability of the market supply is accordingly 10%. If all of the other 10 projects also have a value of $2 million, the contractor can expect another supply volume of $2 million. This is the expectancy value, which is the product of probability and contract values ($0.1 \times 10 \times \$2,000,000 = \$2,000,000$).

While owners bring their demand only to the project-specific market, contractors bring their supply to several actual or potential markets (Figure 15.5).

Figure 15.5 Actual and potential market supply contractor X.

15.5 Construction Contracts

I have stressed repeatedly that time plays an important role in construction. Elsner et al. (2015, p. 14) provide the following statement: *"Behaviors and institutions are especially important when exchanges are concerned that are not executed simultaneously, but where some agents have to get active first, and have to rely on others to follow suit and honor commitments they may have entered into in order for long-term projects to become viable. . . Reciprocity and trust become necessary ingredients in order to bridge over time. The necessity to put trust into the interaction partners is most relevant as written contracts are necessarily incomplete."* Construction contracts are prime examples of incomplete contracts. However, before I turn to the theory of incomplete contracts, I wish to provide a general overview.

Property rights define the degree to which an individual has control over an object. This can be absolute or restricted. Property rights to land are, for example, restricted by zoning laws. Economists call the actual property rights residual control rights. We trade such residual control rights in markets. Each market transaction has accordingly a physical and a legal aspect (Kolmar 2017). Contract goods rely in addition on an exchange of promises at the beginning of the transaction and the physical exchange becomes possible only at the end, months or years later. Contracts detail the applicable rules for a specific transaction, and contract laws the rules governing the contracts.

For Adam Smith and his idea of capitalism, the freedom of contract was of absolute importance. This is no longer the case to the same degree today. Contracts are regulated by contract law, which applies to a lesser degree to private than to public owners. Public owners in many countries have to follow tendering laws that do not apply to private ones. Tendering laws have two objectives: (i) The tendering process shall be transparent to eliminate corruption. (ii) The tendering process should be fair to both parties. Public owners are at least in some countries not allowed to negotiate the contract price after submission in order to avoid use of an information advantage on the side of the public owner. Tendering laws in the EU also have the objective to guarantee equal market access to construction firms from different EU countries. The World Bank goes one step further by mandating the use of FIDIC contracts. FIDIC (Fédération Internationale des Ingénieurs Conseils) is the International Federation of Consulting Engineers and it publishes different types of standard contracts. Private owners enjoy a much larger degree of contractual freedom and exert more market power. Table 15.1 provides the information that private construction investments (77%) considerably exceed public investments (23%) in the US in 2020.

Contracts can be complete or incomplete, bilateral or multilateral, and they can be based on symmetric or asymmetric information. Construction contracts are most often bilateral (owner/contractor), incomplete because of project complexity, and based on asymmetric information. Asymmetric information gives rise to three models of contracts (Salanié 2005):

- *Adverse selection models.* The uninformed party does not have complete knowledge of the other party's characteristics and it moves first. Example: This model describes the situation in which an owner signs a construction contract or a contractor a subcontract. The owner does not know the contractor's intentions and neither does the contractor know those of the subcontractor.
- *Signaling models.* The situation is as before but the informed party moves first. Example: This model applies to contractors showing off reference models.
- *Moral hazard models.* The uninformed party does not completely know the actions of the informed party but moves first. Example: The contractor does not know fully the actions of the owner (with whom else is the owner negotiating what?). However, he is requested to move first, most likely by reducing a price.

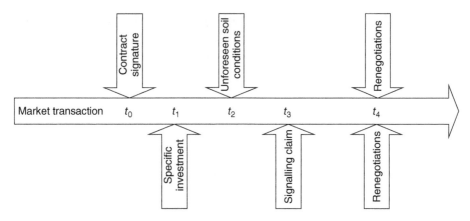

Figure 15.6 Consequences of incomplete contracts.

There are two major reasons for incomplete contracts, limited rationality and prohibitive costs. Since nobody can predict a future event, we are looking at an example of limited rationality if the future is involved. There is simply no way to get full information. Soil conditions are an example of prohibitive costs. It would be possible to extend soil investigations so that the owner will have complete knowledge. It might, however, not make economic sense since the incomplete information will cause problems only in some cases. Over several projects, it can well be less costly to pay for a claim than for a multitude of soil borings.

Figure 15.6 shows the effects of an incomplete contract (Erlei et al. 1999). Owner and contractor sign the construction contract (t_0) based on some soil investigations. The contractor must execute a large number of long piles for a deep foundation. For this purpose, the contractor buys a pile-driving rig with a capacity to drive piles to a depth of 40 m given the friction of the soil layers as investigated by the owner (t_1). While driving the piles in a certain area, the pile-bearing capacity cannot be reached because the assumed sand layer does not start at level -38 m but at level -50 m. The unexpected soil conditions are an unexpected event (t_2). The contractor must obtain a stronger pile-driving rig at extra cost and will certainly hand a claim to the owner (t_3). Construction continues while the parties renegotiate the contract (t_4). We can analyze the situation with the signaling model.

The example of Figure 15.6 assumes the typical case that the owner keeps the residual control right over the soil conditions. Sometimes owners choose to transfer these to the contractor during the construction period. However, this transfer of risks might come at a cost to the owner.

15.6 Contracting Market Design

How producers of investment goods with a need for a factory or a prospective owner of a one-family home procure their resources is a question of contracting market design, and this again is a special form of market design. We are looking at a matching market and not at an exchange market (Roth 2015). It does not matter in which supermarket we buy our toothpaste; a specific brand has always the same quality. The retailer has no influence on this quality. Contract goods can only be traded in matching markets exactly because what we get does not only depend on price but also on the supplier. Think of a haircut or a restaurant dinner. The choice matters even if you want just a crew cut or a pizza. There can be no doubt that different service providers deliver different qualities, and the same is true for contractors. This applies to product as well as process quality. Finding the right match should be of high importance to the owner in construction.

The owner's first decision is about the thickness of the market with a choice of open and selective tendering as well as single-sourcing (Masterman 2002). The second decision concerns the control that the owner retains by choosing between a separated or integrated approach (Greenhalgh and Squires 2011).

When choosing a separated procurement method, the next decision is about contracting each trade separately (owner coordinates the trades) or a general contractor who takes over the tasks of coordination. The separation lies in the elaboration of the design through architects and engineers and construction by contractors. This is called a design/bid/build approach.

Integrated procurement systems are characterized by an overlapping of design and construction because this task is given to one design/build contractor. An even stronger integration can be achieved through partnering where owner, designer, and contractor work together from the beginning as a team. The economic idea here is to cut down transaction costs and to improve innovation.

Other venues include BOT procurement (build/operate/transfer), where the contractor takes on the additional role of operator for a certain amount of time. This approach can have different variants – for example, design/build/operate/transfer (DBOT).

Authors of books on procurement systems provide in-depth details. From a microeconomic perspective, it is important that the owner has many possibilities to organize the market. Choices imply decision-making and decision-making considers incentives. Thus, it becomes necessary to analyze the main procurement methods regarding their incentive systems (Figure 15.7).

We can distinguish five major phases in traditional design/bid/build contracts: (i) design, (ii) tendering, (iii) construction, (iv) operation, and (v) demolition. The last phase is important because construction waste contributes almost 70% to all solid waste in many countries; minimizing this waste is a question of efficiency and economic as well as ecological sustainability. None of the procurement systems provides incentives to plan for demolition and to use recyclable materials. Different actors with different goals are active in each of the phases and it is the task of the owner to guide the process toward his own goals. Design decisions might have a negative impact on tendering. A typical case is the choice of minimum track radii in high-speed rail projects. If the radius chosen is small enough, then only one supplier of rolling stock remains in the world (i.e., the choice of a radius can establish a monopoly). Tendering decisions can also lead to adverse selection. Design and tendering decisions might rule out an optimum construction process. The finished structure might make operation less efficient than possible. The three most important transition points are between design/tendering and construction (transition 1), between construction and operation (transition 2), and between operation and demolition (transition 3).

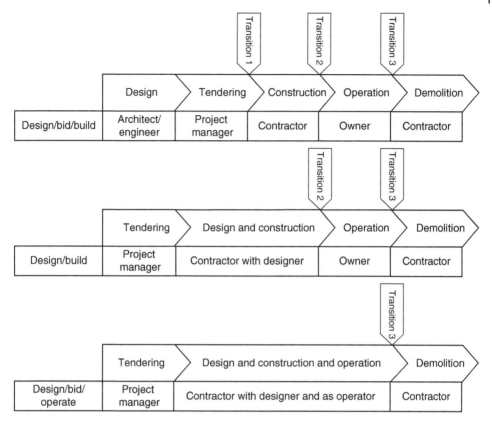

Figure 15.7 Incentives in procurement systems.

Design/build contracts allow contractors to optimize design and technology decisions. There is a considerable incentive for integration of product and process to reduce costs. The disadvantage to the owners is a loss of control.

In design/build/operate contracts, contractors have an incentive to care about life-cycle costs. Thus, they will not only try to minimize investment costs at the time of handover but also operation costs. Operation costs in a simple building are around 70% and investment costs 30%. Specialized buildings have operation costs of 90% and investment costs of 10%.

It is important to consider who is in the best position to determine the optimum approach besides incentives. Engineers are trained to minimize quantities, contractors to minimize costs based on a given design, and owners prefer to minimize overall costs including operation.

15.7 Pricing of Construction Contracts

There is a clear rule in microeconomic theory for the quantity to produce given a market price: Stop production when marginal costs equal market price. We need to consider whether this rule is also applicable to construction, and we will see that it is not.

Construction firms use markup pricing instead. This means that they do not adjust the quantity produced according to market price and internal costs but that instead they determine a price for a given quantity (Section 15.7.1).

The owner must organize a market and have a mechanism to choose on the market among different offers. Owners do this typically by using sealed-bid auctions and contract award to the low bidder (Section 15.7.2).

15.7.1 Marginal Cost Decisions Versus Markup Pricing

The rule in microeconomics for the producer is to extend quantity to the point where marginal costs equals price. This is one of the most important findings from theory. However, this result depends on several assumptions, among which the following are important for our further discussion:

- Continuous production of the same good, so that overall the last unit is marginal
- Decreasing rates of return
- Increasing marginal costs
- Knowledge of the market price
- Free choice of the produced quantity

If we apply these conditions to a construction firm, we find that contractors do not produce the same good in the firm over a certain period. Instead, the product portfolio of a general contractor can include residential and nonresidential buildings as well as infrastructure. The owners demand heterogenous products. This prohibits applying the marginal cost principle. As I have shown in Section 12.7.3, marginal costs of contractors are constantly horizontal and not increasing: If we now consider the rather hypothetical case of a one-product contractor such as a builder of one-family homes of a repetitive type, we find that even in this case the marginal costs do not constantly increase but they vary; sometimes they increase, sometimes they decrease. The reasons for this behavior are the different crews that the contractor employs. The first project might start with the most productive crew, the second with the second most productive crew, and so on. Here we can expect increasing marginal costs. However, once the first crew has finished its job and it moves to the next one, then the outcome for this project in the sequence of production are decreasing marginal costs. We simply cannot find evidence of cost curves in accordance with monotonously increasing marginal returns.

Returning to the typical case of a contractor with a variety of projects, we will most likely find that several projects are running simultaneously; there is no sequential production. What result would a thought experiment bring based on sequential production, i.e., the following project always starts at the beginning of the previous? Let us also assume that each project has the characteristics of increasing marginal costs. The difference between manufacturing (on which the marginal concept is based) and construction becomes evident. Each construction project has a different price quality, and for each one the marginal costs are also different (Figure 15.8).

The manufacturer of mass production goods knows the market price P^*, since it has sold this good many times before; it also knows the costs since it has produced the good many

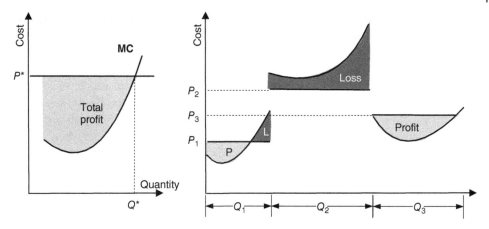

Figure 15.8 Marginal cost concept in manufacturing and construction projects.

times before. With some accounting, it can identify the marginal cost and thus determine the quantity Q^* at which the total profits are maximum. This is exactly the quantity where marginal cost MC equal equilibrium price P^*.

The contractor is in a different situation. He looks at project 1 with a fixed quantity Q_1, which he can find by making a take-off from the owner's design. The contractor is looking at a price based on an estimate; there is no prior knowledge of the cost except by experience from similar projects. Inputs are manifold and the output is one, so the contractor needs a proxy for quantity – such as square meters of usable space. The square meters of usable space are different but so is the quality of the projects and the resources required in production.

The contractor also does not know in advance the marginal cost (MC) curve. Let us assume for the three construction projects in Figure 15.8 that we have a learning effect at the beginning with decreasing marginal costs and schedule slips with overcrowding at the end with increasing marginal costs. This is just one possible run of the MC curve. Ideally, there would be learning effects and then constant marginal costs at the lowest level. At the beginning of the project, the contractor hopes to make a profit. The first project is profitable overall, but there are losses at the end. In this case, the contractor should stop construction once $MC = P$, but he is contractually bound to finish the project. The hope for profits vanished from the start for project 2; the whole project turns out to be losing money. The third project runs as planned and hoped for: the profit is maximum because $P = MC$ holds. In construction, this is the least likely case. Even with hindsight, contractors cannot draw up the MC curve because the required effort to determine the actual cost per each square meter is prohibitive. In addition, the distribution of fixed costs in accordance with differentiated input is impossible. A uniform distribution distorts the results beyond repair.

It should be clear by now that marginal pricing is not an option for contractors, not on the project level and not on the firm level. Contractors need a different approach. They take recourse to markup pricing. Starting from the given quantity by the owner's design they try to find the optimum production size and resource mix. Project 3 in Figure 15.8 shows how a positive project can be interpreted with hindsight. Many times, the contractor will not be successful when bidding for a project and sometimes he will lose money after winning a submission (project 2).

In markup pricing, the estimator evaluates the costs for the project by profiting from the law of large numbers: the estimate will be based on as many different items as possible without bias. The result is a normal distribution of mistakes where the mean is close to an achievable outcome. Next, he will add a markup for general risks and profits. Once this is complete, marketing considerations come into play: Does the situation require a reduction of the markup and maybe even the fixed costs? Does it allow a higher markup while keeping a good chance of winning the submission? The process runs through two stages: (i) intelligent number crunching with the balancing of the 11 partial plans and (ii) a strategic marketing decision.

Some other rules of microeconomic theory apply to the strategic marketing decision. The competitive nature of the construction does not allow for extra profits beyond earning the opportunity costs of all used resources, i.e. overall economic profit is zero in the long run. Some projects might be very profitable, others might be disasters. If the contractor cannot cover the opportunity costs in the long run, he should shut down the firm and invest in more profitable ventures. In the short run, it might be advisable not to include any fixed costs in the estimate, but all variable costs must be covered. In this case, the contractor is as bad off with or without the next project; in both cases, the contractor will be stuck with the fixed costs. With a project at hand, he can at least remain operational.

15.7.2 Auctioning

Sealed-bid auctions are only one of many pricing possibilities in procurement. There are other auction designs. However, most important in construction are sealed-bid auctions, and these pose several problems, which I will discuss in the following chapters.

15.7.2.1 Construction Goods and Auctions

Construction goods are very different from exchange goods. They are fabricated after signing a contract, and they are most often single units and of considerable complexity. This implies three major problems for estimating construction projects: (i) There is no repetitive production of the same good and thus no direct learning about pricing. When someone produces a million pencils, it is of little importance whether the initial price is correct, it can be adjusted within short time. In single-unit production the initial price cannot be changed because the contract is signed and binding before production starts. (ii) The inherent complexity of many construction projects makes it hard to consider and judge all relevant facts. (iii) There is no control over the production conditions; productivity is influenced by the environment as well as by the required contributions of the owner. As construction is a highly integrative process, owners are an important external factor of production. They must know what is required of them (process evidence). Thus, productivity depends not only on the contactor but also on the owner (Brockmann 2011). A more radical view is that of the owner as employer and the contactor as external factor.

Milgrom (1989, Nobel Prize winner 2020) discusses two premises in conjunction with pricing of complex contract goods: the private and the common values assumption. The private values assumption states that contractors can determine their cost correctly (labor, materials, equipment, subcontractors, indirect cost) and Milgrom argues that this assumption does not hold in construction. He assumes estimating errors by all bidders (ε_i) with a normal distribution about the mean (i.e. no bias). Detailed analyses of single estimates and the bid-spread of submissions support the statement. The estimating approach takes this into consideration and deals with the problem by detailing a structure into an extensive work breakdown schedule. Judgment mistakes occur for most items; however, they are not systematic (unbiased). Over a large number of items, they cancel each other out, and there is a tendency toward a mean value. In an example of a post-construction analysis of a construction project, the differences in single items reached almost 300% (planned vs. actual) while the overall difference was only 3%. The contractor was lucky in this case; he had overestimated the total cost (Birol 2009).

The second assumption is accepted by Milgrom for construction projects: All companies face approximately the same cost (C); the common values assumption holds. In different segments of the market, companies of equal size tend to compete against each other, therefore the purchasing power of the companies is the same. Short-term advantages of one competitor (e.g., use of cheap foreign labor) must be imitated by the others due to the competitiveness of the market. Another argument goes as follows: Since all contractors use the same subcontractors and many works are executed by subcontractors, construction prices can vary only due to the efficiency of the management processes (Drew 2011). This argument might be true in some countries, in others it is not. It is possible to define countries with a trading orientation (many Asian countries) and those with a crafts orientation (e.g., Germany). The value of subcontracting as percentage of the total production value has never exceeded 32% over the past 30 years in Germany. Yet, subcontracting also contributes in such a case to the tendency toward a common value.

With these considerations Milgrom formulates the value of an individual bid (X_i):

$$X_i = C + \mu_i \tag{15.1}$$

While the estimating error (ε_i) is unbiased overall, this is not true for the successful bid X_{min}. The lowest bid lies below the mean value and therefore below the market equilibrium price P^*, which equals the mean of all submitted bids X_i, because of the assumption of unbiased pricing (Figure 15.9).

The quantity demanded is fixed in Figure 15.9; this reflects the scope resulting from the owner's design. The demanded quality determines the general price level together with the scope and cannot be shown in Figure 15.9. There are only four offers here, for simplicity. In reality, there are often more, and they are in theory uniformly distributed because of the common values assumption and an unbiased (although erroneous) approach to estimating. The low bidder's offer is accepted by the owner with a positive price difference between equilibrium price and contract price, a gain for the owner. The price difference is negative for the contractor. If the markup for profit and risk is larger than the difference, the contractor will still make an accounting profit; otherwise, the contractor will face a loss. The story so far assumes that the contractor with the biggest estimating error will win the bid. This cannot be the end of it – we still have to look at innovativeness of contractors.

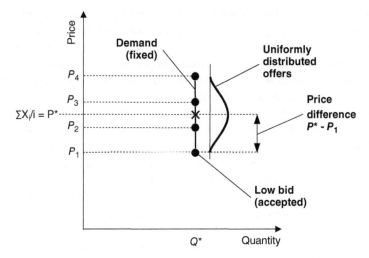

Figure 15.9 Result of auctions in construction markets.

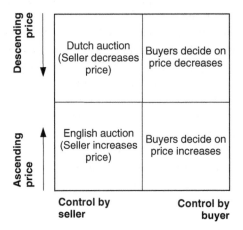

Figure 15.10 Auction designs for selling a good.

15.7.2.2 Auction Designs

There are a number of different auction designs (Leitzinger 1988). The first question is whether someone wants to sell or to buy a good using an auction design. Sotheby's typically sells goods but owners buy goods and services. A second distinction is whether the buyer or the seller controls the auction. A third one is whether the bidding process is open or sealed. Drew (2011) makes the following statement: *"Sealed bidding is the conventional mechanism used in the construction industry for allocating construction work to willing contractors."*

Therefore, I will concentrate my analysis on sealed-bid auctions. A short description of other types of auctions allows us to evaluate the outcome of sealed-bid auctions. The other auction designs can serve as points of reference.

Selling auctions are the best-known examples of all auction designs, as we can sometimes follow one in movies or on the news, such as the auctioning of a painting by Van Gogh. Four principal options are shown in Figure 15.10. If an auctioneer (seller) controls the auction, he will set the starting price and the price increases. If the buyers control the auction, then they will determine at least the bid changes. Important selling auctions are the Dutch and the English auction.

The auctioneer in a Dutch auction starts with a high price and gradually lowers this price until the first interested bidder agrees to the demanded price. Thus, the highest bidder wins and pays exactly his bid price in this descending bid auction. If all bidders know their private values of the good on offer, the bidder with the highest private value will obtain the good at approximately this price. Differences to the exact private value are due to the steps by which the price is reduced. If, for example, the second-last price was $1,000 and the next offer comes at $950, a bidder with a private value of $990 will have a gain of $40.

In an English auction, the auctioneer starts at a low price and slowly increases the price. Each bidder knows his private value. Once the price exceeds this private value, this bidder will drop out of the auction and cannot enter it again. The winning bidder in this ascending price auction will be the one with the highest private value and will have to pay a price close to this, depending on the steps by which the price increases.

It is also possible that bidders determine the price increases or decreases. The winning high bid will again be close to the highest private value, maybe a bit closer because the bidder can choose very small increments.

Designs similar to Dutch and English auctions can also be used for buying goods and services. To differentiate buying and selling situations, these designs are called licitations when buying as the auctioneer wants to elicit the lowest price. Most common in buying situations are sealed-bid auctions; award can go to the winning bidder at the lowest (first-price) or second lowest price (second-price). Figure 15.11 shows the different options.

In a Dutch licitation, the auctioneer is in a position of a monopsonist and will lower the price until no bidder is willing to offer a lower bid. Award is made to the lowest bid price. This should be close to the bidder with the lowest private value. In construction, this would be the contractor with the lowest estimate. Licitations were at least in Germany used in construction many years ago. The contractors asked to change the practice because they felt that they sometimes got carried away by the emotions similar to gamblers. Many thought a

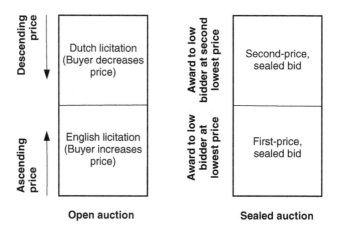

Figure 15.11 Auction designs for buying a good.

lower and even lower price could be possible. The English licitation is working upward from a low value. The auctioneer will increase the buying offer in increments until the first bidder accepts. In this case, that makes the buyer a monopsonist.

One of the important differences between open and sealed-bid auctions is the information that bidders and sellers receive. In open auctions, pricing behavior is visible to all; the information is complete and symmetric. By this information, participants can learn to understand each other's behavior. Especially when the same participants regularly meet in auctions, repetitive behavior allows formulating an advance strategy. Bidders can use learning to adjust their own behavior, as each bidder can submit many bids in one auction. Bidders in sealed-bid auctions have only one chance; the offer enclosed in the sealed envelope is their only bid. As there is no bidding process with changes in behavior of the participants, bidders cannot learn from each other. Learning takes place in between bids since owners most often publish the results of their sealed-bid auctions. However, as random mistakes are the norm in construction bidding, the information does not allow formulating a strategy. If a contractor makes an offer below costs in a recession, learning will induce others to follow suit. The opposite is true in boom times. Sealed-bid auctions tend to strengthen the effects of business cycles. Empirical data support this statement: Price fluctuations in construction are stronger than in other industries.

Information in sealed-bid auctions is incomplete for the contractor while complete for the owner. We have a situation of asymmetric information. An auction also establishes a well-defined market where only one good is traded. The owner is by definition a monopsonist if he chooses a sealed-bid auction for procurement.

Table 15.2 shows the result of a descending price licitation (Dutch) versus an ascending price licitation (English). The first assumption is that the bidders know their private values; these are given in the first column for four firms. The auctioneer chooses in both cases rather large steps in reducing or raising the price; they are identical in both cases, i.e., $100,000. In the Dutch licitation, the owner will have to pay $1,199,000 and the contractor will have a gain over the private value of $199,000 USD. In the English licitation, the owner pays $1,099,000 and the gain of the contractor is $99,000.

What can we learn from this example? Auctions are zero-sum games. Gains of one party are the other party's losses. Results in auctions depend on design and circumstances

Table 15.2 Result of licitations.

Private value (USD)	Dutch licitation			English licitation	
A : 1,000,000	1,499,000	↓	D drops out	↑	
B : 1,200,000	1,399,000				
C : 1,300,000	1,299,000		C drops out		
D : 1,500,000	1,199,000		B drops out		
	1,199,000		Award to A	1,099,000	A accepts
				999,000	No interest
				899,999	No interest

(private values). An auctioneer is well advised to change the offers in small steps. If the auctioneer would not proceed with discreet steps but use a monotonous function, the results would be identical and, in both cases, firm A would get the contract for a price equal to its private value ($1,000,000).

As contractors have no information about the bidding behavior of competitors in advance, their only strategy is to make an estimate as good as they can and then make a business decision about the markup depending on the general market condition. The markup will be different in recessions and booms. If the private values in Table 15.2 would include a markup, then the firms would typically offer their private values in the sealed envelope. In sealed-bid auctions, firms tend to disclose their private values and the owner will benefit from the lowest price offer available in the market; contractors cannot realize a gain in sealed-bid auctions. Making matters worse for contractors is the fact that they are unsure about their private values, as they depend on estimating accuracy.

Harris et al. (2021) show that contractors tend to lower their prices step-by-step before winning a contract with a low bid. As blind people try to feel their way ahead, so do contractors.

Second-price sealed bid auctions are used very seldom in construction. Jehle and Reny (2011) as well as Milgrom (2004) provide detailed information on this topic.

15.7.3 Sealed-Bid Auctions

The result of sealed-bid auctions is a monopsony market structure for any given project. Assuming the prevalence of Milgrom's formula $X_i = C + \varepsilon_i$, the individual bid X_i is unbiased and the estimating error ε_i is normally distributed. With this assumption, we can calculate the expectancy values of winning a bid depending on the number of bidders. The expectancy value of a bid $E(b)$ for a number of contractors (n) depends on this number n and is in all cases except for $n = 1$ below the price level of the equilibrium price P_0 in competitive markets (Leitzinger 1988). The equilibrium price is the price resulting from the interaction of demand and supply under the conditions of perfect competition, i.e. the ideal postulated under competition is good and it serves as benchmark. The larger the number of bidders is, the smaller are the chances to win an auction by submitting the equilibrium price. Winners are faced with a price below equilibrium in competitive markets (Table 15.3).

McCaffer (1976) provides some empirical evidence to this theoretical observation (Figure 15.12). He compares the result of sealed-bid auctions with the owner's estimate. These estimates are not exact, they serve only as a benchmark. The results show the difference between the mean of all bids versus the owner's estimate in percent as well as the lowest bid in comparison to the owner's estimate. I have added to the original diagram two trendlines to show the results clearer.

Table 15.3 Expectancy values in first-price sealed-bid auctions.

Number of contractors	1	2	3	4	5	6	7	8	9	10
Expectancy value $E(b)$	0	−056	−085	−103	−116	−127	−135	−142	−148	−154

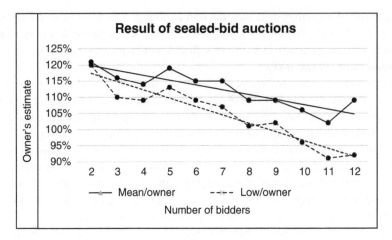

Figure 15.12 Outcomes of sealed-bid auctions and number of bidders.

The data give proof of two observations: (i) The larger the number of bidders, the lower the mean value of all bids. Bidders react to the decrease in expectancy values depending on the number of bidders. (ii) The larger the number of bidders, the higher is the difference between mean value and lowest bid.

15.7.3.1 Pricing in Sealed-Bid Auctions

Pricing in construction depends on the procurement method chosen by the owner as buyer. There is a large number of different procurement methods. To simplify the discussion, I will concentrate on the most common one, the traditional method (design/bid/build, Masterman 2002) in the form of a sealed-bid auction and award to the low bidder. In many countries, this is the prescribed procurement method for public owners. Sealed-bid auctions with award to the low bidder are characterized by a pricing bias (Section 15.7.3.2), an information bias (Section 15.7.3.3), and an uncertainty bias (Section 15.7.3.4). The first two are a result of the monopsonistic power of the client, the last one is an estimating bias for complex contract goods. Additionally, technology advance plays a role (Section 15.7.3.5).

15.7.3.2 Pricing bias

Milgrom's formula $X_i = C + \varepsilon_i$ rests on the assumption that the estimating error is normally distributed and unbiased. The award to the low bidder, however, favors the company with the largest estimating error (winner's curse). Assuming a normal distribution, we can calculate the expectancy value of winning a bid at the market equilibrium price as it would develop in competitive markets by the interaction of supply and demand. The values are shown in Table 15.3 and they depend on the number of bidders. Except for the case of just one bidder, the values are negative (i.e., the bidders have to expect a price below market equilibrium). The option of the buyer to use sealed-bid auctions puts the contractors at a clear disadvantage. While the theoretical reasoning might not be clear to the contractors, the results are all the clearer to them. They must make sure that the estimated profit is larger than the possible difference between market price (zero economic profit) and sealed-bid auction low bid price.

15.7.3.3 Information Bias

In many cases, private owners use their complete information of all unit prices from all different bidders to further negotiate the price downward. Theoretically we face a turnaround of the typical principal/agent relationship. In construction contracts, the owner acting as principal appoints the contractor as an agent to implement the project on his behalf. During the implementation the contractor (agent) gains a lot of information that is not available to the owner. This establishes an information asymmetry favoring the contractor. In many cases, contractors will use this asymmetry for their gain. This happens after signing the contract when the two parties have entered into a two-sided monopoly. Before signing the contract, however, the situation is reversed. The contractors as principals endow the owners as agents with the task to agree on a contract. They have no influence on the decision-making process once they have submitted their bids. Information asymmetry favors the owner. The actions of owners as free agent mean that there are no restraints on their behavior and the principal (each single contractor) cannot watch the actions. This allows clients to take hidden actions and to exploit a moral hazard situation. The latter situation arises when agents face no risks for acting in their own self-interest.

The asymmetric information, the moral hazard situation, and the possibilities for hidden action allow owners to play one bidder against the next. False information about the price of one bidder given to another one cannot be detected by either bidder during a round of simultaneous negotiations. Only at the end of the negotiation do the bidders exchange and check the owner's information. This is a legal strategy. Certainly, however, it will be seen as unfair by the contractors. It is also seen as unfair in many countries by legislation. In Germany, for example, public owners are not allowed to reduce the price during negotiations, thus making it impossible to use the asymmetric information advantage. However, private owners are free to negotiate the contract price. Making use of false information is often not even perceived as an ethical problem but as shrewd negotiation tactics.

The above paragraph describes the outcome of the monopsony situation on the project level. Yet the consequences of enacting a monopsony are as detrimental to the overall welfare on the national market level as is true for monopolies. Monopsonies are seen as the opposite of monopolies in textbooks.

Figure 15.13 explains the process of reaching the final contract price to be paid by the owner. The submitted price of the low bid is below market equilibrium price. Then, the owner engages in price negotiations. The assumption in the figure is that the original low bidder remains the low bidder after negotiations. This is not necessarily the outcome in a specific case. It can also be that the low bidder is undercut during negotiations by another bidder.

Owner and contractor sign a contract at the end of the negotiation phase based on the contractual price. The contractual price is typically well below the low-bid price during submission. As a rule of thumb, we are looking at a reduction of 5%. The contractor enjoys the reversal of roles during construction and has no reason to shy away from aggressive

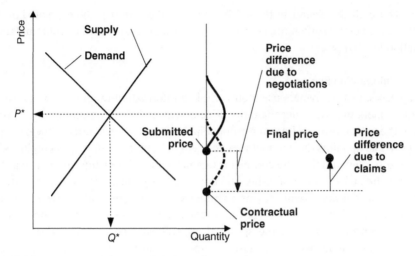

Figure 15.13 Influence of asymmetric information on price.

pricing of claims if the owner used an aggressive negotiation approach. An adversarial approach is a customary component of a sealed-bid auction with an award to the low-bidders and subsequent price negotiations. The power to reduce the price during contract negotiations as well as the power to raise the price during claim negotiations is based on asymmetric information. Please keep in mind that the supply and demand lines in Figure 15.13 belong to the model of a perfectly competitive market (not a construction market); the are shown only to anchor the uniformly distributed bids around the equilibrium price.

15.7.3.4 Uncertainty Bias

The uncertainty bias of estimating should not be confounded with Milgrom's estimating error ε_i. It is closer to what Flyvbjerg et al. (2003) describe as optimism bias on the owner's side. According to these authors, owners tend to underestimate costs and to overestimate benefits of megaprojects.

Something similar happens when estimating complex projects. I use the term uncertainty bias to differentiate the phenomenon from the above, although it is also driven by optimism. One-of-a-kind projects are hard to estimate because there exists no experience relating directly to the problems they pose. Most problematic are the evaluation of productivity rates for labor, the performance rates for equipment, and the completeness of the estimate. Textbooks typically give values ranging from 100% to 200% for productivity rates (e.g., for concrete pouring a range from 0.4 to 0.8 h/m³). In a highly competitive bid, estimators know that they have to produce a low price and develop a tendency to use values on the lower side for labor productivity and equipment performance. I explained before that estimated values are never correct but that mistakes are unbiased, thus having in tendency to balance out. The strain of perceived competitiveness can introduce a bias to choose low values and thus bias the estimated values toward optimistic assumptions. The outcome will be a dangerously low submission price.

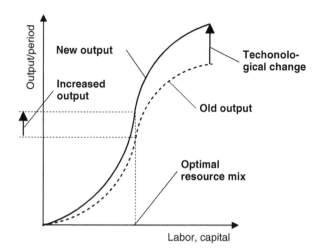

Figure 15.14 Technological change for a production function of Type A.

15.7.3.5 Technology Advance

Advances in technology allow to produce more output with the same amount of inputs. The idea is most easily conveyed with a Cobb–Douglas production function but it also works with a Type A production function (Figure 15.14) as it applies to a construction project. A (positive) technological change shifts the production function upwards.

We can observe that it is possible to produce more output with the same inputs. The point of the optimal resource mix coincides with the maximum slope of the tangent to the curve which is at the point of inflection; this is the maximum production speed. The figure shows also the increased output for the optimal resource mix. However, since the output quantity is fixed by the owner's design, contractors can reduce production costs. The contractor with the most advanced technology can produce at the lowest cost. Whether this translates into a low bid is not sure because of likely estimating mistakes.

Contractors can decide how to use the advantage if the cost reductions by technological change are large enough; they can lower the bid price, improve their profit, or somehow split the difference. Technologically advanced contractors will most likely enjoy higher profits. The high competition in the construction sector will force the other contractors to close the technological gap as quickly as possible.

15.8 Supply and Demand in Construction

Initial perfect competition in construction markets combined with sealed-bid auctions and followed by a monopsony situation with an asymmetric information advantage assures the market power of the owner. The contractor is confronted with four factors: (i) Sealed-bid auctions as institutions are rewarding estimating errors, driving the low-bid award price below equilibrium price. (ii) Sealed-bid auctions as institutions provide asymmetric information, driving the low-bid award price below equilibrium price. This holds especially true in a two-phase award process, when the auction is followed by price negotiations. (iii) Sealed-bid auctions as institutions further uncertainty, resulting in overly optimistic assumptions and driving the low-bid award price below equilibrium price. (iv) Sealed-bid auctions as institutions exploit technological advances by firms, driving the low-bid award price below equilibrium price. All effects will overlap and aggregate. In the worst case, the technologically most advanced contractor commits the biggest estimating error by using overoptimistic assumptions, and is being taken advantage of during a negotiation phase.

Accordingly, contractors feel continuously pressed into an unfair pricing system in comparison to competitive market structures. Their only chance to counter the asymmetric information advantage of the owner is through collusive cooperation. In terms of game theory, it can be stated that the payoffs for a noncollusive outcome of a sealed-bid auction are negative in comparison with the equilibrium price that is accepted as being fair. The incentives in the auction game are not set in a way to keep the contractors interested in keeping the rules. Anti-trust laws are required to keep them in line. However, these are not always successful.

15.9 The Owner as Monopsonist

If we stay with the definition of a market in construction as consisting mostly of separate markets for each project, then one owner faces in this market several contractors competing against each other; the owner wields market power as monopsonist. Differing from monopolists, who exert their market power through quantity sold (the price follows then from the demand curve), monopsonists do this through quantity purchased. While monopolists typically face anonymous buyers, owners additionally enter into face-to-face contract negotiations with the contractors and can employ this asymmetric information advantage.

We have seen that a firm in a competitive factor market is exposed to a flat supply curve. Monopsonists, however, are confronted by an upward-sloping supply curve. The monopsonist's profit maximization takes the following form:

$$\max_{x} pf(x) - c(x)x \tag{15.2}$$

The monopsonist can solve the problem by setting the quantity at that point where marginal revenue for the factor equals marginal cost. In Figure 15.15 there is an upward-sloping supply curve with a slope of b. Similar to a monopoly, the slope of the marginal cost curve has double the slope (2b). The marginal revenue is a downward sloping curve. The monopsonist will choose the quantity x where $MC = MR$, at Q^*. This is less than the volume in a perfectly competitive market and the price is lower (Blair and Harrison 2010) (Figure 15.15).

Comparing the described situation of a typical monopsonist and an owner acting as a mono-psonist by using sealed-bid auctions for a self-designed structure yields three differences:

i) The contractors are not completely depended on the offered contract, they have other options. However, those options have a much lesser value, maybe a probability of 10%.

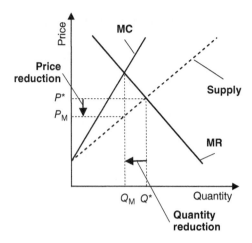

Figure 15.15   Typical monopsony.

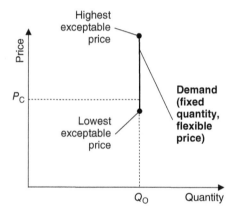

Figure 15.16   Monopsony outcome in construction.

Then it is critical how many such options a contractor has at a moment in time and how pressing the need for a new contract is to keep employment.

ii) There are no diminishing returns to construction; the marginal cost curve of the contractor is horizontal.

iii) The demand is not sloping downward but instead vertical with a fixed quantity.

The horizontal marginal cost curve means the contractor can expand supply endlessly as long as he can get resources on the factor market at constant costs. More important is that the negative economic market outcome from a typical monopsony (lower quantity at a lower price) cannot come into play. Horizontal supply and vertical demand provide a market price for the fixed quantity. This is shown in Figure 15.16 for a single contractor offering a price P_C and a single owner demanding a quantity Q_O. We get a graph that is similar to Figure 15.9.

The points discussed above about the price pressures on contractors (i.e., especially the low-bid award and the price negotiations together with asymmetric information) remain valid: There is a high competitive pressure on pricing for contractors unless a boom and demand above production capacity allows to escape this pressure.

15.10   Bargaining for the Contract Price

Once the owner has opened the bids, the owner must choose a contractor. Often, public owners are not allowed to discuss prices, but they still might want to clarify technical details with one or more bidders. The situation for private owners is different; there are no restrictions on their further actions and most enter into negotiations on price and content of the offer. The following discussions are specifically pertinent to this situation.

To characterize the context of the negotiations, it is helpful to look at the existing power structure. We can differentiate between five types of power: (i) power to withhold, (ii) bargaining power, (iii) coercive power, (iv) institutional power, (v) cultural power. All of these forms can impact negotiations in construction, some more so, some less.

The owners' power to withhold is strong. It is up to them to award the contract. It is stronger for private owners, as public owners have to follow the procurement rules. In many countries, it is not easy for public owners to reject the lowest bidder. The owner's power also extends into the future: They can decide to never again ask a certain contractor for a quotation. Public owners have, for example, blacklists of noncompliant companies. The threat by a contractor to back away from negotiations is not convincing; there are others to step up. Definitely, the owner has more power to withhold.

Bargaining power depends on how much more one party needs an agreement than the other party. The one with the lesser need has more bargaining power. The strength of this power accordingly depends on the difference in need. Few will be the contractors who have contract negotiations with more than one owner at the same time. This depends, of course, on the size of the contracting firm. The statement does not apply to the very large international contractors, but those are few. Typically, the owner will have several offers with sometimes small differences in price. An additional component of bargaining power is information, and this also tilts the balance in favor of the owner. A third impact can be attributed to the business cycle. In a recession, the owner will receive many bids and the contractors have few chances of landing a contract. The bargaining power of the owner will be considerable. During boom times, there will be fewer offers to the owner and more possibilities for the contractors. The bargaining power of the owner will decrease, and sometimes the power will shift to the contractor. This is especially the case when an owner needs a project and there is only one contractor offering to take on the job – a situation that is definitely not the norm. The British economist Beveridge developed a curve matching up unemployment against job vacancies (Beveridge curve). This curve can also illustrate the bargaining power in construction, depending on the business cycle (Figure 15.17). The horizontal axis shows the number of contractors looking for work at a given moment (the unemployed) and the vertical axis the number of owners offering contracts (vacant jobs). Point A reflects a situation where there are more owners looking for contractors and C shows a situation where contractors are chasing work. In both cases, one party is more in need of a contract than the other (i.e., one party possesses bargaining power).

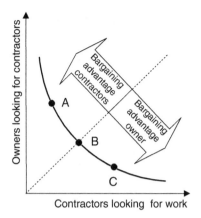

Figure 15.17 Matching owners and contractors.

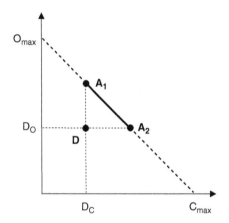

Figure 15.18 Negotiation possibilities with default option.

Only at point B there is a balance, and bargaining power vanishes. It should be clear that there is no balance for specific construction project (one owner, few contractors).

Coercive power exists only for owners. Think of an owner who repeatedly offers contracts, a professional owner. This owner can threaten contractors with the shadow of the future. The threat consists of taking away all chances for future work (and profits).

Institutional power exists when a public owner is more or less in a monopsonistic position. This is in some countries the case for rail and road infrastructure.

A simple bargaining model for construction comprises two parties, the owner (O) and the contractor (C). They can either agree or not agree. In case of disagreement, both sides have other options, and each one knows the economic value of these fallback options. A fixed sum is to be distributed between the two parties and each on tries to maximize profits. In terms of game theory, this is a zero-sum game that is not played repeatedly. Figure 15.18 shows a graphic representation of such a game.

If the complete sum of money available for bargaining goes to the owner, the owner will receive O_{max}; in the opposite case, the contractor will receive C_{max}. Whatever one party gets, the other loses; thus, the slope of the agreement curve equals −1. Both have a default option D and they can determine the monetary value D_O and D_C, respectively. The default option for the owner is the second-lowest bid; the default option for the contractor is the expected monetary value of a future project. Any point on the agreement curve below D_O does not make sense to the owner, he would be better off with the default option. The same is true for every point to the left of D_C for the contractors. Therefore, there are two limits to a mutually acceptable option, A_1 and A_2. This segment of the agreement curve shows all the possible options for striking a bargain.

There is no definite solution to the problem. Nash (Nobel Prize 1994) proposed a simple solution: The maximum product of each side's benefits. Following this rule, the parties would split the difference (Figure 15.19). The solution is designated by S^*.

Up to now, there was no leverage for either party in the bargaining. To be more realistic, we can imagine a project where the contractor C_1 has submitted the lowest bid at 9.7 million USD. The second-lowest bid is from contractor C_2 at 9.9 million USD. This is the default option for the owner, who has full information; he knows his private value. The contractor C_1 is working on two further submissions at 11.4 and 17.1 million USD. Typically, the contractor wins 1 out of 10 submissions. The contractor will have to work with the expected monetary value of the two bids, which amount to: $0.1 \times 11.4 + 0.1 \times 17.1 = 2.85$ million USD, and this is the default option. The contractor does not know his private values. The difference for the owner is maximum 0.2 million USD, but the difference for the contractor is 6.85 million USD ($9.7 − 2.85$). As can be seen from Figure 15.20, this shifts the bargaining advantage decisively in the direction of the owner, and it reflects the information that each party has at the moment of bargaining.

Up to now, we have not considered the influence of information. The information available is asymmetrically shared between the owner and the contractor. The owner has full information, while the contractor cannot even be sure about the correctness of his own estimate. Additionally, the contractor does not know the other bid prices, nor about the true fallback position. This information asymmetry favors the owner, so a bargaining solution will not be halfway between A_1 and A_2 in Figure 15.21 but closer to A_1.

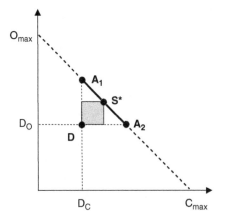

Figure 15.19 Nash solution to the zero-sum game.

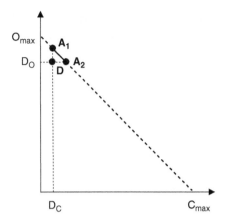

Figure 15.20 Bargaining solution in construction contract negotiations.

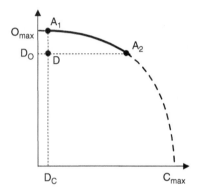

Figure 15.21 Bargaining possibilities through value engineering.

A last possibility is easing the restriction on a fixed sum. If the negotiations do not only focus on price, it is possible to create a larger sum by value engineering. An example is a construction pit. Most owners do not care the least bit about a hole in the ground; they care about the floors that fill this construction pit. In such a case, the contractor could propose alternative construction methods that are cheaper and serve the same purpose. This is a simple example but value engineering is not confined to such simplicity. Value engineering allows solutions to the upper right of the negotiation line. Instead of a straight line, the solutions can now be found on a concave curve. Figure 15.21 demonstrates that there is a good chance for the contractor to improve his bargaining result.

15.11 Change Orders and Claims

Construction contracts are examples of incomplete contracts. Instead of trying to regulate all future possibilities during the construction period, they provide for mechanisms to deal with those problems when they arise. Change orders are one option through which the owner can demand changes to quantity or quality while keeping the responsibility to reimburse the contractor. Claims are mostly initiated by contractors for work they have to do without these works being part of the contract. Change orders and claims together with profit-maximizing behavior leads to the adversarial nature of the construction process.

During the construction period, there are only two parties that can try to solve the problems by negotiation. This is in economic terms a two-sided monopoly. Asymmetric information now favors the contractor; however, the amount that a contractor can claim is hemmed in by the necessity to provide proof.

In the case of a change order, the contractor must propose a solution and provide an estimate of the costs. The owner does not have to reward the works of the change order to the contractor; principally, he could also get quotes from the market. However, the contractor on site has a big advantage over outside competitors (not having to factor in fixed costs). In the end, the contractor already owns the complete site installation from offices to workshops, from equipment to labor. The contractor wields power during the negotiations and does not have to pay full attention to the market price. The owner's threat with competition is not very credible.

The owner is not helpless, though. There is plenty of detailed information in the estimate, and the contractor has to observe all prices submitted with the contract. Thus, the strategic options for the contractor are limited. In the worst case, the owner with the help of quantity surveyors can assemble a price by picking other items apart and putting them together to reflect the change at hand.

It becomes especially difficult if the change order impacts the production process. This can affect the productivity of the workers, and the impact of productivity changes is difficult to determine.

Claims by the contractor will face harder opposition from the owner than change orders, as many times they arise from different interpretations of contractual obligations.

In some countries there is a tendency in these circumstances to award the contractor with a time extension resulting from a claim but without any monetary compensation. This is economic nonsense, as costs depend on time and available resources.

The press is often reporting cost overruns, especially on public projects. These cost overruns are necessarily due to actions by the owner. If the responsibility for the problems were on the contractor's side, there would be no need for increased compensation. The profit or loss of a contractor in a contract with a cost overrun of 150% remains unknown. It is fiction that the contractor always makes big money on changes and claims.

In sum, the contractor will very often be in a position to improve the overall price quality of the contract through change orders and claims, but this is not a given and always limited.

15.12 Summary

The main lines of the argument in this Chapter 15 on contracting are as follows:

- The predominant procurement method worldwide uses a design/bid/build approach with sealed-bid, first-price auctions.
- The owner acts like a monopsonist. However, the contractual work being offering does not have exclusivity. Contractors as buyers have other options to maximize their profits.

- The long-run marginal cost curve of contractors on an industry level is horizontal. In this case, the market structure appears to be monopsonistic but the owner as single buyer does not yield monopsonistic power (Pauly 1998).
- Since the owner has no monopoly power, the negative effects of monopsonies, i.e. loss of welfare, low prices and low quantity, do not emerge.
- The owner still yields much power by fixing the quantity by a particular design, by developing the contract conditions, by determining the market design, and often by an asymmetric information advantage.
- With the exception of boom times when contractors are pushing their capacity, construction markets are highly competitive and long-run profits are zero. Unimpeded entry and small exit barriers with a mobility between the layers of different sized companies allows only short-term economic profits, while periods of economic losses can be longer.
- Sealed-bid auctions frequently favor the contractor with the greatest estimating error. This entails adversarial relationships and low quality.
- The horizontal supply curve and the heterogeneity prohibit the use of a marginal cost analysis. Contractors are bound to markup pricing.

References

Birol, F. (2009) Nachkalkulation – dargestellt am Beispiel einer Industriehalle. Unpublished bachelor thesis at the University of Applied Sciences Bremen, Germany.

Blair, R. and Harrison, J. (2010). *Monopsony in Law and Economics*. Cambridge: Cambridge University Press.

Brockmann, C. (2011). Collusion and corruption in the construction sector. In: *Modern Construction Economics – Theory and Application* (ed. G. de Valence), 29–62. Abingdon: Spon Press.

Drew, D. (2011). Competing in construction auctions: a theoretical perspective. In: *Modern Construction Economics – Theory and Application* (ed. G. de Valence), 63–79. Abingdon: Spon Press.

Elsner, W., Heinrich, T., and Schwardt, H. (2015). *The Microeconomics of Complex Economies: Evolutionary, Institutional, and Complexity Perspectives*. Amsterdam: Academic Press.

Erlei, M., Laschke, M., and Sauerland, D. (1999). *Neue Institutionenökonomik*. Stuttgart: Schäffer-Poeschel.

Flyvbjerg, B., Bruzelius, N., and Rothengatter, W. (2003). *Megaprojects and Risk: An Anatomy of Ambition*. Cambridge: Cambridge University Press.

Greenhalgh, B. and Squires, G. (2011). *Introduction to Building Procurement*. London: Spon Press.

Harris, F., McCaffer, R., Baldwin, A., and Edum-Fotwe, F. (2021). *Modern Construction Management*. Hoboken: Wiley-Blackwell.

Jehle, G. and Reny, P. (2011). *Advanced Microeconomic Theory*. Harlow: Pearson.

Kolmar, M. (2017). *Principles of Microeconomics: An Integrative Approach*. Berlin: Springer.

Kotler, P. and Armstrong, G. (2017). *Principles of Marketing*. Harlow: Pearson.

Leitzinger, H. (1988). *Submission und Preisbildung: Mechanik und ökonomische Effekte der Preisbildung in Bieterverfahren*. Cologne: Heymanns.

Levy, M., Weitz, B., and Grewal, D. (2018). *Retailing Management*. New York: McGraw-Hill Education.

Masterman, J. (2002). *Building Procurement Systems*. London: Spon Press.

McCaffer, R. (1976). Contractors' Bidding Behavior and Tender Price Prediction. PHD thesis. Loughborough: Loughborough University of Technology.

Milgrom, P. (1989). Auctions and bidding: a primer. *Journal of Economic Perspectives* 3 (3): 3–22.

Milgrom, P. (2004). *Putting Auction Theory to Work*. Cambridge: Cambridge University Press.

Pauly, M. (1998). Managed care, market power, and monopsony. *Health Services Research* 33 (5): 1439–1460.

Roth, A. (2015). *Who Gets What and Why*. Boston: Houghton Mifflin Harcourt.

Salanié, B. (2005). *The Economics of Contracts*. Cambridge: MIT-Press.

Tirole, J. (2000). *The Theory of Industrial Organization*. Cambridge: MIT Press.

US Census Bureau (2022). *Monthly Construction Spending November 2021*. Suitland: US Census Bureau.

16

Market Imperfections

Market imperfections or failures have several causes: (i) market power in monopolies (monopsonies) and oligopolies (oligopsonies), (ii) principal–agent problems, (iii) information asymmetries, (iv) externalities, (v) corruption and collusion, and (vi) public goods. We find all of these in construction markets.

An an example of an externality are smoke emissions. Often, proucers are allowed use the air without paying for this resource. This leads to an overuse with negative effects on the air quality. While everybody in the vicinity has to live with the polluted air, the producer enjoys the right to pollute. Social responsibility is often not strong enough an incentive to deal with the problem.

We have seen in Chapter 5 the ideal workings of perfectly competitive markets. A deviation from this ideal are imperfect markets (monopolies, oligopolies, etc.) and the reason for the imperfection is market power (Chapter 6). Other market imperfections can exert negative influences in perfect as well as imperfect markets – externalities are one example.

Imperfect (incomplete or asymmetric) information can distort market outcomes and lead to inefficiencies (Chapter 16.1). Principal-agent problems arise from differences in incentives and asymmetric information, we already have looked into this in various places. Externalities originate from factors of production, which are free of charge. A well-known example is the use of the environment by firms as a sink. When firms emit exhaust fumes, they are using the clean air as a factor of production. The same happens when firms discharge used water into a river. The price for the economic activities is paid by the social and ecological environment (Section 16.2).

The construction industry is also prone to violations of competitive markets. Sometimes firms resort to price-fixing (collusion) or corruption (Section 16.3). I will discuss the problems of public goods in a separate chapter (Chapter 18).

16.1 Imperfect Information

Imperfect information describes a state where someone has not all the information required to make a rational decision. One of the assumptions of the perfectly competitive market is

Construction Microeconomics, First Edition. Christian Brockmann.

that both sides of the market have full information. This assumption is a stretch, as we all know by our own experience. The degree to which there is a lack of information on one or the other side of the market remains hidden by the term incomplete information.

If one side has an information advantage, we speak about asymmetric information and this can be exploited by the side with the information advantage. A person applying for a job knows more about his or her abilities than the potential employer. Someone receiving a loan knows better whether he is in a position to pay it back. A manufacturer knows more about his products than a buyer.

Regulations determine whether someone is obliged to disclose full information. Warranties also serve the purpose of safeguarding the buyer against undisclosed defects. However, for other purposes the motto remains caveat emptor, i.e., let the buyer beware.

Very seldom the focus is on the opposite, the case where the buyer has more information than the seller. I have given several examples of information favoring the owner and disfavoring the contractor (Chapters 11 and 12). In such a case, the motto that applies is caveat vendor – let the seller beware. I have argued that this is to a large degree pervasive in contracting when the owner buys the production of his structure.

Discussing asymmetric information requires reference to the famous example of the market for lemons by the Nobel Prize winner of 2001, Akerlof. In a used car market with good cars (peaches) and bad cars (lemons), buyers will form an idea of an average price that considers the offers for both types of cars, higher for the good ones and lower for the bad ones with hidden defects (Akerlof 1970). Since the average price is lower than the price for good cars, nobody will buy them as they seem to be too expensive. In the end, the good cars will be withdrawn from the market and only bad ones remain. The market breaks down because the buyers do not have enough information to separate the peaches from the lemons.

This can happen in construction markets as well. One example is the selection of subcontractors. Some of them choose a business model of low-quality factors of production (equipment, skills of workers). These are the lemons and, in the end, they might drive out the peaches. What applies to subcontractors also applies to contractors. Are we in a market for lemons? There are enough complaints about botched jobs in construction because of the owners' tendency to award low bidders. The difference between bidding in construction and used car markets is that a sizable number of owners and contractors repeat the experience often enough to collect information about the typical behavior of a contractor or subcontractor. They play a repeated game in terms of game theory. While a good reputation might not suffice to sign a contract when the price is not the lowest, a bad one can exclude contractors from participating in selective tendering.

Individuals in the used car market have less experience than professional owners or contractors.

The essence of a contract good is incomplete information as it deals with the future. I have discussed the consequences in Chapter 15.

I also want to repeat the arguments of incomplete information for the contractor's estimate. The task is to determine future costs for a production that has never been done before; there is no complete experience. An estimate might include the use of excavator y on a given day of the future as well as the output of worker x. As the future is unknown, at least to us humans, this is an example of incomplete information about output rates based on a guess informed by past similar but not the same experience. I already explained that the solution to the problem is breaking down the project into many activities, an unbiased estimate of each activity, and the fact that the estimating errors tend to be uniformly distributed. When the number of activities is large enough, the overall estimating error will be close to the mean. With all the problems contractors are facing when estimating, they still have more information than owners: this is a case of asymmetric incomplete information.

16.2 Externalities

Kolmar (2017, p. 104) provides the following definition: *"An institution is inefficient if not all interdependencies caused by individuals are internalized. The noninternalized interdependencies are called externalities or external effects."*

The problem of external effects often results from incompletely specified property rights. A solution could be the assignment of such rights to the firm (e.g., right to emit) or the residents (right to prohibit emissions).

There are not only negative but also positive externalities (Table 16.1). However, negative production externalities are the most important ones.

Externalities are factors of production for which there is no market price. The beneficiary uses this factor without paying for it and, while doing so, harms someone else in the case of negative production externalities. This other person bears the costs of the economic activity. A typical example are emissions from a factory with negative effects for the neighborhood. The owners of the factory use the air as a sink without paying and the general populace might suffer from trees without leaves.

Use of the air or water as a sink is principally not different from using a disposal area for solid waste. The only difference lies in the attached property rights. These are assigned to an owner in the case of the disposal area and they are not always assigned to somebody in the case of air and water. Assigning complete property rights is the economic solution to the problem according to the Coase-theorem by the 1991 Nobel Prize winner (Coase 1960). It postulates that an efficient solution to externalities lies in the assignment of complete property rights and complete information. This efficient solution is independent of who

Table 16.1 Classification of externalities.

	Negative externality	Positive externality
Production	Noninternalized interdependencies whose internalization increases opportunity costs of production (Pollution)	Noninternalized interdependencies whose internalization decreases opportunity costs of production. (Research, education, pollination)
Consumption	Noninternalized interdependencies whose internalization decreases the consumption of others (Vaccination)	Noninternalized interdependencies whose internalization increases the consumption of others. (Maintenance of property for the benefit of the neighborhood)

owns the property rights – the producer or the public in the example above. Of course, there is another aspect to the problem. Whoever receives the property rights profits from an endowment.

One way to assign property rights was enacted in 2005 through the European Union Emissions Trading System (EU ETS). The property rights were initially accorded to the producers, and after this, the rule was cap and trade (European Commission 2021). Every recipient of a property right has a cap and can sell certificates if it stays below the cap. Exceeding the cap means that the producer has to buy certificates with attached property rights to emit larger quantities. The externalities are internalized in this way.

Important emissions in construction are noise and dust pollution. In addition, construction contributes the biggest share of all sectors to solid waste disposals in most countries.

 Observations = (obs.) A contractor operated a separation plant for tunnel muck in a populated neighborhood. After measuring the noise immission level, the neighbors complained and pointed to the fact that allowable limits were exceeded. The contractor was forced to encase the separation plant with a noise reducing shell. The considerable costs were not part of the estimate.

A contractor in Thailand trucked 27 m long prefabricated bridge segments over a well-used highway from the precast factory to the points of delivery on site. In order not to add to the traffic problems, the contractor was allowed to ship only during six hours at night. While the contractor enjoyed the right of using the highway free of charge, he might cause harm to others stuck in the ensuing traffic jam.

Internalization in these two cases was brought about by regulation. The contractors as beneficiaries had to pay a price. This will reduce negative effects but not lead to an economically efficient outcome.

The economic analysis of the impact of externalities such as pollution starts with the discussion of social benefits and costs. The social benefits are those things society could buy by accepting more pollution. Social costs are the damages done by pollution to the air we breathe, the water we drink, and fauna and flora we enjoy. Marginal costs are those added by one more unit of pollution, be the effects positive or negative. We have seen that the demand curve represents the marginal utility (social benefit) and the supply curve the marginal costs. Therefore, the socially optimal solution would be at the intersection of the marginal social benefit and the marginal social cost curve. This result arises from internalizing the externality (Figure 16.1).

The result of Figure 16.1 will not occur without assignment of property rights. Without this, the producer will emit a maximum amount of polluting emissions. This is the situation in case of an externality that is not internalized (Figure 16.2).

The producer in Figure 16.2 is free to use the air for emissions, and therefore social costs play no role. As long as there is a marginal social benefit to society, the producer will extend production (and pollution) to this point. The resulting quantity of pollution is Q_{market}.

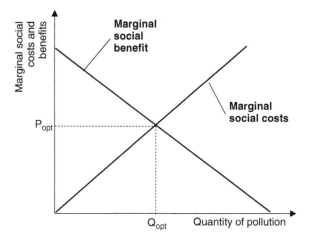

Figure 16.1 Optimal solution to externalities with assigned property right.

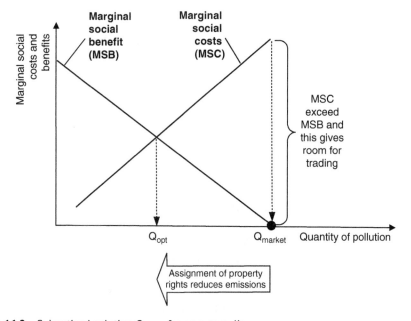

Figure 16.2 Suboptimal solution Q_{market} for an externality.

The producer and society can both be better off by finding a solution where society pays the polluter for the reduction of the amount of emissions. This is, of course, a solution if the producer is allowed to use the environment as a sink. Should this right rest with the society, then the producer would have to pay in order to emit pollution. Society would be willing to do so until again the optimum amount of emissions (Q_{opt}) would be reached. This amount of pollution is Pareto-efficient because nobody can be made better off without someone else becoming worse off. The solution with an emission level of Q_{market} is accordingly not efficient.

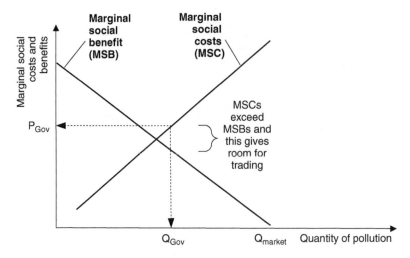

Figure 16.3 Solution with government limit.

Figure 16.3 shows an example where the government set a limit for noise emissions from contractors. The quality of the solution depends on how close the limit is to the optimal amount. Imagine a construction site of a supermarket in a populated area. The set noise limit is above the socially optimal solution. The noise will impact fewer people than the group, which will benefit from the coming shopping possibilities. For the group of affected people, it is very difficult to reach an agreement with the contractor because it is not a coherent group and everyone has a temptation to shirk the responsibility. Thus, the situation with a high noise level will persist. Under such circumstances, even a suboptimal government solution might still be preferable.

16.3 Collusion and Corruption

Collusion and corruption are two different phenomena of market imperfections. Both are ways to prevent the results from otherwise perfect competition from the supply side. Collusion or price-rigging requires contractors to sit together and to determine jointly the results for a submission.

Transparency International (2021a) defines corruption as: "... *the abuse of entrusted power for private gain.*" People on the owner's side are entrusted with power to act on behalf of the owner (principal–agent relationship) and they have the opportunity to abuse this power. Providing a contractor with an information advantage over competitors in exchange for a bribe to the owner's representative can serve as example. Transparency International describes the consequences of corruption: "*Corruption erodes trust, weakens democracy, hampers economic development and further exacerbates inequality, poverty, social division and the environmental crisis.*"

16.3.1 Collusion

Collusion is the illegal act of producers to meet and take coordinated action. The first step is to exchange information about prices and quantities. To this purpose, producers form cartels. While collusion always refers to an illegal act, cartels can be legal. The best-known example is the Organization of the Petroleum Exporting Countries (OPEC). By limiting the oil supply, prices increased in 1973 fourfold. Consequentially, the global economy went into a strong recession. Even legal cartels can be very powerful and detrimental to overall welfare.

Monopolists have no need of a cartel, and in oligopolies there are often legal ways of signaling prices. A case in point are gas station operators. The prices are visible to everybody; what remains unclear are the quantities of each supplier. However, we have seen that a price war in oligopolies has negative effects on all producers. If one reduces its price, that producer will increase its sales. The other producers must follow suit, and everybody is caught in a downward price spiral. Nevertheless, there are examples of cartels in oligopolies: dredging in construction (Netherlands) or beer prices (Germany).

Collusion in a duopoly is the typical case used in textbooks to introduce this practice. While duopolies are practically nonexistent in the construction industry, the example allows us to understand the basic mechanics of the process. The following theoretical case is from Taylor (1995) and uses game theory for analysis.

Two companies called Bageldum and Bageldee produce rather homogenous products, bagels. They have a choice of charging the competitive price where they will earn no economic profit as marginal cost equals price or they could collude and charge the monopoly price making a profit of $2 million each (Table 16.2). There is also an incentive to defect from the collusion by undercutting the monopoly price just slightly (thus becoming competitive with a price above marginal cost) and by selling a large volume of bagels with a comfortable profit ($4 million in the example). The other company then will make a loss equal to fixed cost (−$1 million). The payoff matrix shown in Table 16.2 resembles that of the well-known prisoner's dilemma.

Without communication, both companies will end up with a competitive price and no profit. With successful communication, both will reap a profit of $2 million, and this is a strong incentive to cooperate.

In the prisoner's dilemma, communication is physically impossible, while in the case of a duopoly price communication is illegal but possible. However, the incentive to defect is large, and a naive Bageldum might choose this option. Bageldee would in such a case have no other choice but to follow in reducing the price – otherwise, it will be wiped out. Thus, both arrive at the competitive price. Bagels are sold continuously and the game is repeated over and over, again contrary to the prisoner's dilemma that is played just once. Bageldum and Bageldee will get the idea sooner or later and might want to collude to charge the monopoly price. If the game is played often enough, there is not even need for explicit collusion. Understanding the mechanics, both companies will converge toward the monopoly

Table 16.2 Payoff matrix for a collusive duopoly.

		Bageldee (A)	
		Competitive price	Monopoly price
Bageldum (B)	Competitive price	A: $0/B: $0	A: −$1 mil./$4 mil.
	Monopoly price	A: $4 mil./−$1 mil.	A: $2 mil./$2 mil.

Figure 16.4 Tacit collusion in the beggar's market. *Source:* Gunston | Strandvik

price by tacit collusion that is not illegal (Figure 16.4). It is well established that the results of monopoly pricing are quantities supplied below equilibrium quantity Q_o and prices charged above equilibrium price P_o, definitely a suboptimal outcome with regard to overall welfare.

The difference between the prisoner's dilemma and duopolistic collusion is due to two facts: communication is possible and the game is repeated in the case of duopolies. As a duopoly is highly unlikely to be found in the construction industry, we need a model of market structures for the construction industry to advance the argument.

Most tempting is collusion in markets with strong competition, as we find it in the construction sector. When a collusive ring is successful, they will increase the price and reduce the supplied quantity. The following ideas rely heavily on those developed by Brockmann (2011).

16.3.1.1 Naturally Caused Collusion

Oligopolies in construction exist because there are some natural factors limiting competition. One possibility is a limited regional oligopoly; another is a long-term oligopoly in a niche of the construction sector. Deep-water dredging is an example for the latter. Dredgers are undoubtedly required resources; they are visible, and the whole interested world knows who owns them. Competitors for large deep-water dredging contracts are thus known, and they can form a naturally caused oligopoly. Market entry is limited by the high investment for dredgers.

Tacit collusion is not possible in construction because there is not a large quantity of goods being supplied continuously to the market, as is the case for bagels (or refinery products such as gas etc.). Instead, the goods traded are defined by large, single-unit contracts being awarded by sealed-bid auctions. These games are not repeated often enough, and the work is not similar enough to establish market equilibrium at monopoly prices. The size of a single contract offers considerable incentives to defect from collusion and this is facilitated because the contract prices are always made public to all competitors at submission. Except for accepting the sealed-bid auction price, the competitors can only engage in explicit collusion.

The mechanics of the ensuing process are driven by two mechanisms. Firstly, the colluding contractors must agree on a selection mechanism and second, they must decide on a price setting mechanism. Third – but not necessarily – a profit distribution mechanism needs to be established. Bribery is not required to gain information, as the competitors are known by possession of the limiting factor (in the case above, by the dredgers). Selective tendering offers another possibility to know the participating players in a submission; contractors just need to bribe someone in the owner's organization to hand over the list of selected contractors. Knowledge of all participants is indispensable in collusive schemes.

The selection mechanism must allow determining whose turn it is. This can be based on statistical data, such as market share at the beginning of collusion, phantom bookkeeping, or on argumentation where a bundle of criteria might be considered.

The price-setting mechanism again depends on three options: Companies can generate their own profits once they have been chosen, they can give back part of the profit to the others, or the whole profit will be shared by all colluding contractors. The first case sets the stage for a two-phase game that is cooperative in the first phase and competitive in the second. Here, the chosen contractor wants to establish the highest reasonable price possible while all the others want to limit the contractor's profits, since it will still be a competitor in other areas or at other times. The price will shift from below equilibrium price upward. How much upward depends on the price effect of the collusion. In an older study (Bülow et al. 1977), the price effect was found to amount to 2.5% as part of return on turnover for all projects (competitively and collusively bid). Since the total return on turnover during the same period was smaller than 2.5%, prices were still below equilibrium with collusion. This is a very strange result!

The second and third cases also bring about a two-phase game, but both phases are cooperative. Since all companies are interested in the profit from the focal transaction, they tend to charge the highest price possible, which is the monopoly price. The monopoly price decreases welfare due to the overall deadweight loss; it is not a desirable result (Varian 1999).

The distinction between these two types of scenarios is of utmost importance for an analysis of the collusive outcome on social welfare.

16.3.1.2 Artificially Caused Collusion

An example of a well published collusive scheme is that of the Dutch construction industry from 2002 (Dorée 2004). It was all pervasive and thus not a niche problem. All companies involved had a claim account that was recorded in phantom bookkeeping. During the collusive meeting, contractors could bid for the focal contract by offering a financial compensation to the other bidders. The high bidder would be the winner of the collusive part of the game. The price was decided on jointly and therefore competitively. This is also an example of the two-phase game, with the first phase being cooperative and the second competitive. The compensation to be paid also introduced some competitiveness into the first phase. No money changed hands, but money spent and received was recorded in the phantom accounts.

Dorée also discusses factors supporting the proliferation of collusion based on a literature review and the Dutch case study. Among the supporting factors are undifferentiated performances (as they exist on the construction market), price-oriented competition (sealed-bid auctions), similar cost functions (common values assumption), high rate of risk and uncertainty (uncertainty bias), high concentration of buyers (monopsonies on the project market), risk of the winner's curse (price bias), and a predictable selection process. The list includes other factors that cannot always be found in the construction industry, such as a high concentration of sellers. Some other factors are based on culture such as social homogeneity and therefore differ from country to country.

Owners basically have three options in arranging a market through procurement: (i) A perfect competition/monopsony by allowing all interested contractors to submit a bid. The number of contractors in the game is large and the participants are unknown to everybody (open bidding). (ii) A perfect competition/limited monopsony by preparing a list of bidders. The owner knows the number and names of the bidders (selective bidding). (iii) A two-sided monopoly by negotiating with just one contractor. In this case, information is symmetric (direct negotiation).

Case (i) does not provide enough information for collusion. In order to enter the game, there must not only be an incentive but also the knowledge of all participants. The Dutch case was an exception, as it was almost all-encompassing.

Case (ii) is the classical setup for collusion in a market that is generally in perfect competition on the supply side. In order to get the information contained in the list of bidders, contractors must bribe someone in the owner's organization. A principal–agent relationship is an absolute prerequisite for bribery. The agent in such relationships can profit at the expense of the principal. In a private firm, the owners are the principals and all employees are agents. Accordingly, all employees with knowledge of the list of bidders are possible targets for bribery. The taxpayer is the principal in public agencies, all employees are agents, and therefore all of them are possible addressees of bribes.

Case (iii) does not lend itself to collusion because of lack of players. Bribery is still a possibility to get access to information for the negotiation process and to create an asymmetric information situation.

Bribery in construction is facilitated by the large contract sums and the imprecise knowledge of prices. An add-on of $1 million for a contract of $10 million cannot be easily detected as being excessive. A bribe of $100,000 out of the extra million is in most cases enough to convince a morally weak agent.

For a collusive scheme to work, there must be repeated tenders, preferably an infinite number. Then and only then can the contractors play repeated collusive games among themselves. It is not necessary that all contractors are always invited. The group playing the repeated games can be larger than the bidders for one contract. The collusive arrangement must, however, include all contractors that have been or will be invited.

16.3.2 Corruption

Corruption is widespread in construction. Transparency International (2008, 2011) provides a bribe payers index (Table 16.3). The index reflects the supply side of bribes, and it is based on perception. The lower the index value, the more corrupt is a sector. The construction sector is the most corrupt, easily outpacing notorious sectors such as arms and defense. Real estate and property development take rank 3. A comparison between the values from 2008 and 2011 shows that construction and real estate have become more corrupt, while oil and gas, mining, heavy manufacturing, as well as arms and defense became less corrupt. Not only the absolute value but also the trend for construction and real estate is troubling!

Table 16.3 Ranking of sectors according to degree of corruption.

Ranking	Sector	Index 2008	Index 2011
1	Public works contracts and construction	5.6	5.3
2	Utilities	—	6.1
3	Real estate, property, legal and business services	5.9	6.1
4	Oil and gas	5.7	6.3
5	Mining	5.8	6.3
6	Power	—	6.4
7	Pharma industry and health care	—	6.4
8	Heavy manufacturing	6.1	6.5
9	Fisheries	—	6.6
10	Arms and defense	6.4	6.6

Table 16.4 Corruption perception index.

Ranking	Country	Index 2020
1–9	Denmark, New Zealand, Finland, Singapore, Sweden, Switzerland, Norway, Netherlands, Germany, Luxemburg	88–80
11–22	Australia, Canada, Hong Kong, United Kingdom, Austria, Belgium, Estonia, Iceland, Japan, Ireland, United Arab Emirates, Uruguay	79–71
25	United States	67
78	China	42
129	Russia	30
170–180	Democratic Republic of the Congo, Haiti, North Korea, Libya, Equatorial Guinea, Sudan, Venezuela, Yemen, Syria, Somalia, South Sudan	18–12

Corruption is also a cultural problem: There are significant differences between countries. Transparency International (2021b) publishes yearly a corruption perception index. Some selected data are provided in Table 16.4. The higher the value, the less corrupt is a country; a country with no corruption would score a value of 100. The least corrupt region is Europe and the most corrupt Africa. It is also notable that countries which speak up publicly against corruption with vehemence not necessarily follow up with action (e.g., USA).

Corruption seems to be a basic ingredient of action in the construction sector, in some countries to a larger degree than in others. It can be employed together with collusion to improve the situation of the contractors.

The form of power at play is the power to withhold (or give). We find this power in private firms as well as in public agencies. The situation requires two accounts: one into which the benefits go and one from which the costs are paid. These accounts must belong to different entities.

An example from a private sector is an employee who takes money from a collusive ring to hand over the list of bidders in selective tendering. It is a prerequisite for collusion to know the players of the game. Remember the contract with a competitive price of $10 million and a price increase of $1 million due to price fixing and you have $1 million that can be distributed between the fixed contractor and the provider of the information. There is quite an incentive for both parties even in small projects. The costs are borne by the owner. If the owner were a family, norms would be much stricter for someone to cheat his own family. If the owner were a single person, corruption would make no sense at all.

The public sector can arrange for price fixing at a larger scale (i.e., for all projects within one year if supervision is only cursory).

16.4   Mechanics or Ethics of Collusion

The presented argument has stressed the mechanisms leading to collusion. The model of man in new institutional economics on which the argument is based is not an ethical one. It supposes that all actors are opportunistic (acting with guile). While this model is basically sound, it remains ethically unsatisfying.

Zarkada-Fraser and Skitmore have presented a study looking at the ethical side of collusion (2000). They conclude: *"The results show that collusive tendering, in all its forms and variations, is a result of a decision with moral content, and generally perceived as necessarily unacceptable in Australia."* The problem with these findings is that the whole study has serious flaws. Estimators are the chosen focus group; however, they are not the ones taking part in collusion primarily. Collusion is a business decision, and the players are the business unit managers. This becomes clear when analyzing documents on the prosecution of perpetrators. Collusion is illegal and it carries serious punishment. Nobody will ever acknowledge taking part in it freely. To be against collusion is the socially accepted answer.

The case of the Dutch collusion scheme shows that collusion is not only a problem of individuals but of a large group. Therefore, group dynamics have also to be considered when discussing the problem. As has been shown, collusion is most of all a structural problem. These statements do not deny the responsibility of each individual involved, they have to decide whether they want to act according to law or whether they want to risk acting against it. The predicament is that an ethical decision by individuals will not change the structure. Individuals have only the chance to walk away from the game as this will continue to be played by others. I think that walking away is the right choice.

There are also the ethical problems on the other side of the table: Why is it ethically acceptable that owners have such market power? Why is it acceptable that they can use *"shrewd"* negotiation tactics? This often is just a euphemism for plain lying. Collusion will persist as long as the institutions of procurement are not changed, giving both market sides equal power and reinstituting perfect competition on the project market through regulation of the owner's behavior.

16.5   Conclusion

The line of the complete argument can be found in a condensed form in Figure 16.5. There are strong incentives in the construction sector to engage in collusion. The main argument is that widely used sealed-bid auctions with award to the low bidder produce outcomes below equilibrium price. This is unacceptable to the bidders and economically undesirable.

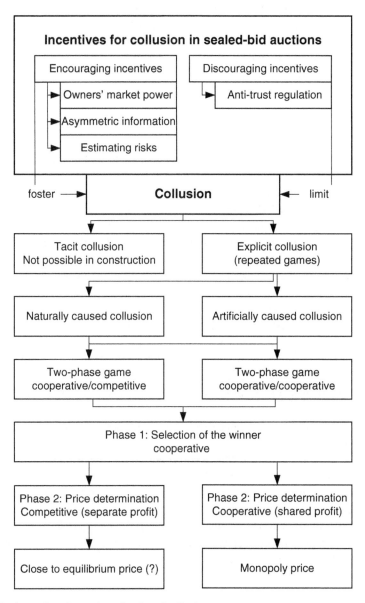

Figure 16.5   Incentives for and mechanics of collusion.

Depending on the mechanisms chosen in collusive games, the result will be monopoly pricing (economically undesirable) or a price not far away from the equilibrium price (economically desirable). The first will be produced by structures that include repeated games by a group, cooperative behavior when predetermining the winner of the bid, and cooperative behavior when setting the price because all players participate in the profit. The second depends on repeated games and cooperation predetermining the winner. The price is restrained by a competitive phase when agreeing on the profit that accrues only to the winner.

A change of the institutional arrangements of procurement processes is required if collusion is to be avoided; prosecution is not a sufficient deterrent. A word of warning is warranted at the end. Putting aside all arguments, collusion is an illegal practice. Prison sentences are not uncommon when collusion is uncovered. I know colleagues who went to prison, and their experiences were most unpleasant.

References

Akerlof, G. (1970). The market for "lemons": quality uncertainty and the market mechanism. *The Quarterly Journal of Economics* 84 (3): 488–500.

Brockmann, C. (2011). Collusion and corruption in the construction sector. In: *Modern Construction Economics* (ed. G. de Valence), 29–62. Abingdon: Spon Press.

Bülow, H., Zubeil, G., and Schröder, D. (1977). *Gutachten der Prognos AG zur Wettbewerbsordnung und Wettbewerbsrealität am Baumarkt, Forschungsreihe der Bauindustrie 39*. Basel: Prognos.

Coase, R. (1960). The problem of social cost. *Journal of Law and Economics* 3 (1): 1–44.

Dorée, A. (2004). Collusion in the Dutch construction industry: an industrial organization perspective. *Building Research & Information* 32 (2): 146–156.

European Commission (2021) EU Emissions Trading System (EU ETS). https://ec.europa.eu/clima/eu-action/eu-emissions-trading-system-eu-ets_en.

Kolmar, M. (2017). *Principles of Microeconomics: An Integrative Approach*. Cham: Springer.

Taylor, J. (1995). *Economics*. Boston: Houghton Mifflin.

Transparency International (2008). *Bribe Payers Index 2008*. Berlin: Transparency International.

Transparency International (2011). *Bribe Payers Index 2011*. Berlin: Transparency International.

Transparency International (2021a). What is corruption? https://www.transparency.org/en/what-is-corruption; accessed January 2022.

Transparency International (2021b). *Corruption Perception Index 2020*. Berlin: Transparency International.

Varian, H. (1999). *Intermediate Microeconomics: A Modern Approach*. New York, USA: Norton.

Zarkada-Fraser, A. and Skitmore, M. (2000). Decisions with moral content: collusion. *Construction Management and Economics* 18 (1): 101–111.

17

Government

The role of government differs around the world, from China to the USA and from Norway to South Africa. It is also hotly debated within several countries. There is in some aspects a considerable difference between the normative advice given by economists such as Sowell (2015) or Krugman and Wells (2017), to provide just one example. Few are the economists who do not prefer markets over government intervention. The differences arise mainly over the issues that economists believe require government intervention.

Each community must operate under the framework of regulations currently affecting the allocation of property rights, for example, as the functioning of markets depend on them. Then, governments have to pay attention to macroeconomic tasks such as stabilizing growth, keeping employment high, and keeping inflation low. They are players in microeconomic markets by government procurement. Very few are the conservative economists who deny the government a role in national defense and weapons procurement. Government's role in providing for infrastructure is a bit more contested. Hotly debated is income distribution through taxation and regulation of markets. Some see regulation as protection of a weaker market side against misuse of market power (e.g., Krugman and Wells); others see it as an unnecessary interference which lowers market efficiency (e.g., Sowell). The government's role as actor in markets is, of course, of more interest in a text on microeconomics than that of regulator.

The government determines not only the rules of the game in the economy but is by itself a very important economic actor; the latter is especially true for construction activities. In all countries, the government is a buyer as well as supplier of goods and services. With regard to construction activities, the statement needs further differentiation: In some countries, the government also acts as contractor with its own workforce. I will not consider this case further since this is not the norm. The government as contractor does not necessarily behave and follow the same goals as private contractors.

We will follow the government as actor in markets; they play an important role in construction (Section 17.1). Governments need an income before they can act on markets; they can raise it through taxation, borrowing (government bonds), and tariffs. From a microeconomic perspective it is interesting to analyze the influence of taxation and subsidies on market outcomes (Section 17.2). Construction is a heavily regulated sector, and this will increase in the future because of the large contribution of the built environment towards CO_2 emissions.

Construction Microeconomics, First Edition. Christian Brockmann.
© 2023 John Wiley & Sons Ltd. Published 2023 by John Wiley & Sons Ltd.

Construction was responsible for 39% of all energy and process-related CO_2 emissions in 2018 (Section 17.3) according to the International Energy Agency (2019). It is foreseeable that this high contribution to global warming will not be tolerated in many countries and that new regulations will be incorporated in building standards. The government also influences construction indirectly through the level of the interest rate (Section 17.4).

17.1 Government as Actor on Markets

The influence of the government in the economy is manifold besides setting the rules through legislation. The judicial system is also part of the government and it controls the keeping of the rules. This applies to property law, contract law, and tort law (Dorman 2014).

The government competes as a public enterprise with private firms. An example is the US Postal Service as public enterprise and UPS, FedEx, or DHL as private enterprises. In this way, the government plays a role in the overall economy as provider of goods and services.

In many other instances, the government acts as buyer on private markets through its procurement system. Large amounts are spent for national defense, education and research, construction, and real estate rent.

Government expenditures G are the sum of consumption (C) plus investments (I) plus interest payments (INT) plus transfers and subsidies (TS):

$$G = C + I + INT + TS \qquad (17.1)$$

Figure 17.1 provides data of government expenditures for OECD countries in 2020. What the figure shows is that the government is a serious actor in all countries. Government consumption for construction includes expenditures for maintenance of public infrastructure and buildings as well as payments for government employees acting as owner's representatives. Government investments cover inter alia expenses for new built structures.

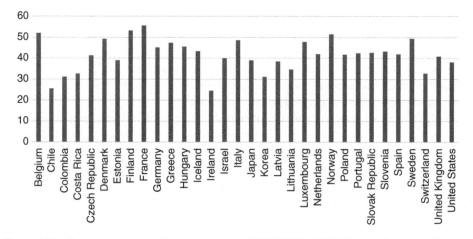

Figure 17.1 Government expenditures in percent of GDP, 2019 or 2020. *Source:* Data from General government spending 2021

To finance the expenditures, governments mostly rely on taxes. We also find construction subsidies in many countries. A typical example is support to families for financing their own home. This can be directly through cash payments or through tax deductions. So far, we have left taxes and subsidies out of our supply and demand model; therefore, it will be interesting to see their impact on the model.

Government regulations for construction are numerous, mostly regarding issues of public safety. By establishing zoning plans, municipalities as part of the government influence the urban environment. There are also many regulations related to protecting the environment. Regulations that significantly reduce CO_2 levels would substantially affect construction costs and building decisions. A final set of regulations aims at stabilizing the social fabric of cities by rent controls.

As Figure 17.1 shows, the governments in Chile and Ireland contribute around 25% to GDP while the figure in France is close to 55%. The respective figure for China is 37% (Organization for Economic Cooperation and Development 2022).

Governments also affect interest rates. In many countries, there exists a separation between the executive branch and the central bank responsible for managing the interest rates independently. These are highly important for construction goods. Many people can buy their daily bread by paying cash, some can finance a car in this way and very, very few their one-family home. The same applies to firms for investment in construction goods and to different degrees it also holds for governments.

We have already seen from Figure 17.1 that governments are important players in the economy in general, and this is true for the construction sector in particular. Construction spending in the US amounted to 1513 billion USD in 2020 and the government contributed 344 billion to this sum, i.e. 23%. Figure 17.2 shows a lesser involvement of the government in Germany in 2019. It contributed 4.1% spending for public buildings and 8.1% for public infrastructure, a total of 12.2% (Hauptverband der Deutschen Bauindustrie 2021).

Governments are bureaucracies, and their decision-making processes differs from households (one-family homes) or private firms (residential and industrial buildings, private infrastructure). Governments must answer to the public for their decisions. These are

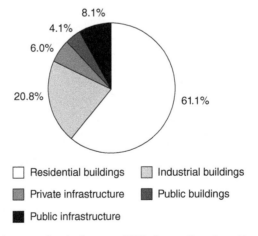

Figure 17.2 Construction spending in Germany 2019. *Source:* Data from Hauptverband der Deutschen Bauindustrie 2021

guided by laws, rules, and documentation of each step. One of the major problems is corruption, since government employees are susceptible to bribes in exchange for favorable zoning or contract awards. Any additional costs due to these decisions are passed on to the taxpayers. The procurement laws are thus intended to provide transparency and protect the citizens from unfair or unethical practices.

We have seen in Section 16.3.2 that public works and construction rank highest in a corruption survey of Transparency International in 2008 and 2011. One prerequisite for corruption is big financial purchase volume, and public works and construction are big-volume sectors.

Other problems in public projects are political influence and planning optimism. Private entrepreneurs must follow profit-maximizing, private goals, and their own ethical guidelines. The government consists of the elected politicians and public servants. Politicians face a vote at regular intervals, and they must garner enough votes at the next election. Such democratic institutions slow down construction while furthering public control and acceptance. However, this is not always beneficial for the project. An example is the Big Dig in Boston with its many time delays (Greiman 2013).

17.2 Taxes and Subsidies

It is rather easy to add taxes to the basic demand and supply model. Imagine an excise tax of 10% on all new-build one-family homes. This will increase the price at any given quantity. If we look at a subdivision with 100 identical homes at a price of 500,000 USD each, we can ask what happens to the quantity if all homes could be sold at 500.000 USD?

Figure 17.3 demonstrates that an excise tax increases the price from 500,000 USD for example to 525,000 USD and reduces the quantity of homes sold below 100. The constructed example also shows that buyer and developer exactly share the tax burden. The buyer has to pay 25,000 USD more and the developer receives 25,000 USD less (525,000 –

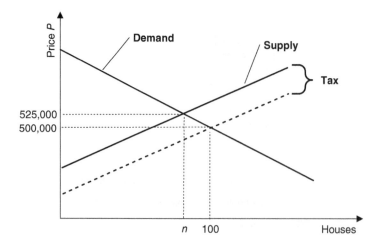

Figure 17.3 Influence of an excise tax on price and quantity.

50,000 = 475,000 USD). In general, it depends on the elasticity of the supply and the demand curve, who has to carry the larger part of the tax. Figure 17.4 shows a rather inelastic demand curve, i.e. the houses are in high demand. The supply curve is, on the other hand, highly elastic. The effect is that the buyer pays almost the whole excise tax. Should supply be inelastic and demand elastic, the opposite would happen: The developer would have to pay the tax.

Whatever the distribution, the buyer and developer are worse off with a tax. This is also a general note: A tax decreases consumer and producer surplus (Figure 17.5). In case of inelastic demand and elastic supply, we will find again that the overall surplus decreases and the consumer surplus suffers more than the producer surplus.

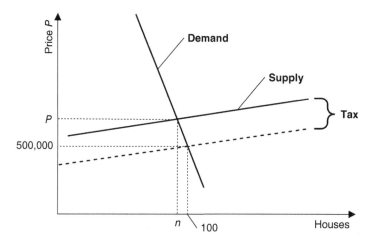

Figure 17.4 Tax distribution in case of inelastic demand and elastic supply curves.

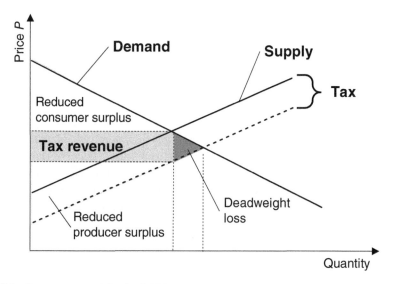

Figure 17.5 Tax revenue and deadweight loss.

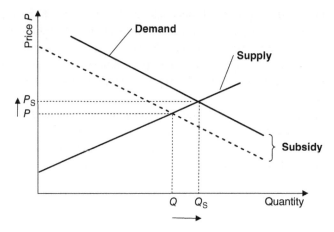

Figure 17.6 Effect of a subsidy to demand.

There is more to the story than a reduction in consumer and producer surplus. There is also a deadweight loss, a net reduction of welfare, and, of course, the tax revenue. The deadweight loss is an argument against taxes. It depends on the use of the tax revenue whether the deadweight loss can be justified. In many western countries, conservative voters might be happy with the tax if the revenue is spent on more law and order. Liberal voters would agree to more support for underprivileged families. As this example shows, the reception of a tax depends on values.

Governments can also grant subsidies to the contractor or the buyer. In case of a direct subsidy to families buying a home, we can see the effect in Figure 17.6. The subsidy increases the demand and shifts the demand curve to the right. Consequently, more homes are sold at a higher price.

Should the government decide to give a subsidy to the producer, then the supply curve would shift down. The effect would be an increase in quantity sold but a decrease of the price.

17.3 Regulations

Regulations are manifold in construction. They apply to materials, equipment, skills, and processes. Of course, details of regulations vary from country to country. Different regulations between countries are barriers to international trade. Many disputes about trade can be traced back to regulations. It is an important task for the European Union and every other integrated market to harmonize regulations. The application of regulations through bureaucracies can cause further problems.

Materials often need approval. Post-tensioning systems are an example. The system must be designed and tested by the producer. The test results and the system must be submitted for approval to a government agency. After approval, the post-tensioning system can be used in one area of regulation, typically within a country. An approval in Japan does not entail an approval in South Korea.

Equipment also has to pass government inspection to ensure safety. In this case, the product must follow more general regulations such as those pertaining to electricity, emissions, or handling.

Professional skills are also often regulated by governments. In the USA, only professional engineers can approve and sign drawings, usually through the department of the county engineer; the county engineer is often an elected position. Governments can perform the licensing by themselves or outsource it to an organization acting on behalf of the government.

Planning approval is probably the most cumbersome regulation in construction of large infrastructure projects. Some megaprojects in a democracy need 20 years from the first idea until all approvals are acquired and construction can start. There is a large difference between democratic and autocratic countries in this regard. It is a trade-off between speed and public acceptance. It is also a question of observance of individual property rights. Process regulations impact not only design but also construction. In the latter case, noise, dust, and dirt emissions are regulated, leading to limitations of work hours or emission levels.

Regulations are norming materials, equipment, skills, and processes. As such, they can be a barrier to innovation (Blayse and Manley 2004). They instill a tendency to stay with the tested and trusted approaches and the burden of proof for innovations might become too large to justify the innovation effort.

Rent controls are a form of more direct economic regulation. They provide a prime example of negative economic effects through regulation while (hopefully) pursuing social goals. From an economic point a view, rent controls lead to misallocations. A point in case is New York, where in one example three students share an open-market apartment for 1200 USD per month and a private investor lives with his wife at the same time in a more spacious and better-quality rent-controlled apartment for 350 USD per month. However justified access to rent-controlled apartments might be at the start, over time the allocation can beat the purpose. Rent-control leads to hoarding of apartments, to black markets, and to quality deterioration (Sowell 2015).

The economic argument is similar to that in case of taxes. Figures 17.7 and 17.8 show the situation before and after the introduction of rent controls. It is easy to see that with rent control (a price ceiling), prices for and quantity of apartments on the market are lower. The consumer surplus increases and the producer surplus decreases. In addition, there is a deadweight loss that is taken away from both consumers and producers. The deadweight loss in Figure 17.7 is smaller than the increase in consumer surplus. The actual situation depends on the elasticity of supply and demand.

The reduced producer surplus might lead developers and investors to refrain from adding to the building stock, aggravating the overall situation and to stop maintaining the apartment buildings, thus lowering the quality and the life span.

As we can see, rent controls are a two-sided sword. In the short-term, they might benefit underprivileged apartment seekers; in the long-run, they most likely will be detrimental to them.

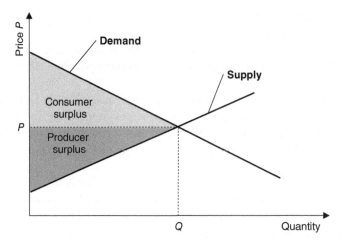

Figure 17.7 Market results before the introduction of rent controls.

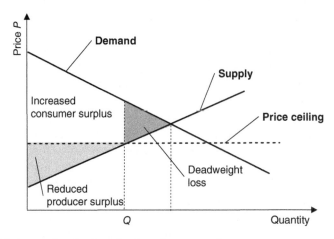

Figure 17.8 Market results after introduction of rent controls.

17.4 Interest Rates

Taxes and subsidies belong to the macroeconomic arsenal of fiscal policy; monetary policies influence the amount of money available in an economy and affect interest rates and inflation. Central Banks such as the US Federal Reserve (Fed), the Bank of England, or the European Central Bank determine a target for the federal funds rate several times per year. To achieve the target, the Fed (or another central bank) will issue treasury bills. By issuing treasury bills the Fed influences the money supply. Selling such bills will increase the amount of money available and buying back the bills will decrease the amount (Krugman and Wells 2018).

Figure 17.9 shows an equilibrium of the money market for a given interest rate i_{today}. The money demand curve is built on the premise that the low interest rates (the price of money), increase the demand for money. In the short run, money supply is fixed.

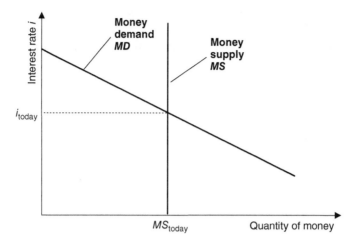

Figure 17.9 Equilibrium of the money market.

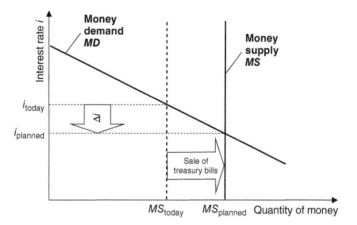

Figure 17.10 Effect of an increase in money supply.

Figure 17.10 illustrates what happens if the Fed wants to decrease the interest rate: It has to expand the money supply by selling a certain amount of treasury bills.

Interest rates impact investments in fixed assets considerably. An easy way to show this is the net present value criterion (NPV) with a series of cash flows A_t ($t = 0, 1, 2, \ldots, n$) and a discount (interest) rate i:

$$\text{NPV} = \sum_{t=0}^{n} A_t / (1 + i) \tag{17.2}$$

The advice to go ahead with an investment requires the NPV to be positive. In that case, the investment earns the assumed interest rate and this depends on opportunity costs. If some other investment offers a rate $i = 7\%$, then this rate should be considered for the NPV. If we compare the influence of different interest rates for discounting future cash flows, we can understand the importance of interest rates in construction. Assume an investment of 1 million USD and 10 years with an annual cash flow of 120,000 USD.

For an interest rate of 3%, we get:

$$NPV = -1000000 + 120000/(1.03) + 120000/1.03^2 + \ldots + 120000/1.03^{10} = 23624 \, USD$$

For an interest rate of 7% we get:

$$NPV = -1000000 + 120000/(1.07) + 120000/1.07^2 + \ldots + 120000/1.07^{10} = -151171 \, USD$$

While the first investment brings a higher return (23,624 USD) than the opportunity costs of 3% and the investment is advisable, this is no longer true for the rate of 7% (−151,171 USD). Should the Central Bank raise the interest rate, not only this investment would be cancelled but many others as well.

Table 17.1 calculates the mortgage payments for two cases, first with an interest rate of 4% and then with 7%. The loan is in both cases 400,000 USD and the annuity 30000. The annuity

Table 17.1 Repayment of a mortgage rate at two different interest rates.

	Case 1: interest = 4%				Case 2: interest = 7%			
Year	Loan	Annuity	Interest (4%)	Principal	Loan	Annuity	Interest (7%)	Principal
0	$400.000 00	$30.000 00	$16.000 00	$14.000 00	$400.000 00	$30.000 00	$28.000 00	$2.000 00
1	$386.000 00	$30.000 00	$15.440 00	$14.560 00	$398.000 00	$30.000 00	$27.860 00	$2.140 00
2	$371.440 00	$30.000 00	$14.857 60	$15.142 40	$395.860 00	$30.001 00	$27.710 20	$2.290 80
3	$356.297 60	$30.000 00	$14.251 90	$15.748 10	$393.569 20	$30.002 00	$27.549 84	$2.452 16
4	$340.549 50	$30.000 00	$13.621 98	$16.378 02	$391.117 04	$30.003 00	$27.378 19	$2.624 81
5	$324.171 48	$30.000 00	$12.966 86	$17.033 14	$388.492 24	$30.004 00	$27.194 46	$2.809 54
6	$307.138 34	$30.000 00	$12.285 53	$17.714 47	$385.682 69	$30.005 00	$26.997 79	$3.007 21
7	$289.423 88	$30.000 00	$11.576 96	$18.423 04	$382.675 48	$30.006 00	$26.787 28	$3.218 72
8	$271.000 83	$30.000 00	$10.840 03	$19.159 97	$379.456 77	$30.007 00	$26.561 97	$3.445 03
9	$251.840 87	$30.000 00	$10.073 63	$19.926 37	$376.011 74	$30.008 00	$26.320 82	$3.687 18
10	$231.914 50	$30.000 00	$9.276 58	$20.723 42	$372.324 56	$30.009 00	$26.062 72	$3.946 28
11	$211.191 08	$30.000 00	$8.447 64	$21.552 36	$368.378 28	$30.010 00	$25.786 48	$4.223 52
12	$189.638 72	$30.000 00	$7.585 55	$22.414 45	$364.154 76	$30.011 00	$25.490 83	$4.520 17
13	$167.224 27	$30.000 00	$6.688 97	$23.311 03	$359.634 59	$30.012 00	$25.174 42	$4.837 58
14	$143.913 24	$30.000 00	$5.756 53	$24.243 47	$354.797 02	$30.013 00	$24.835 79	$5.177 21
15	$119.669 77	$30.000 00	$4.786 79	$25.213 21	$349.619 81	$30.014 00	$24.473 39	$5.540 61
16	$94.456 56	$30.000 00	$3.778 26	$26.221 74	$344.079 19	$30.015 00	$24.085 54	$5.929 46
17	$68.234 83	$30.000 00	$2.729 39	$27.270 61	$338.149 74	$30.016 00	$23.670 48	$6.345 52
18	$40.964 22	$30.000 00	$1.638 57	$28.361 43	$331.804 22	$30.017 00	$23.226 30	$6.790 70
19	$12.602 79	$13.106 90	$504 11		$12.602 79	$325.013 51	$30.018 00	$22.750 95	$7.267 05
20	$-0 00	$-	$-0 00	$-	$317.746 46	$30.019 00	$22.242 25	$7.776 75

is the sum of interest and principal payments. The principal payments reduce the outstanding loan and accordingly the yearly interest payments become smaller. As the annuity remains constant, the principal payments increase. In the case of $i = 4\%$, the family has repaid the loan after 19 years and owns the house clear of debt. With a rate of $i = 7\%$, the same family would own the bank still 317 746.46 USD. This is quite a difference for a family!

The example is a bit simplified as it does not consider the effect of inflation or the possibility to deduct home mortgage interest from taxes.

17.5 Inflation

Most central banks use inflation targeting to help defining monetary policy and many set the target at 2% inflation. The influence is not as direct as that of monetary policy on interest rates. The macroeconomic argument is summarized in Figure 17.11. An increase in money supply lowers the interest rate and increases investments. This leads to higher incomes of the households and accordingly to a higher aggregate demand. If the supply remains constant in the short run, then the prices will rise and we are facing inflation. This is called an expansionary monetary policy and the opposite, contractionary fiscal policy, leads to a deflation (Krugman and Wells 2018).

Increases of the inflation rate favor those who hold a debt and punishes those who hold savings. Both are facing the real interest rate which equals the nominal interest rate minus inflation. In case of a mortgage of 4%, an increase in inflation from 1% to 3% lowers the real mortgage interest from 3% to 1% (4% nominal interest minus 3% inflation).

Saving rates of 3% are under the same circumstances reduced to zero. This is of interest when saving for an equity to buy a house.

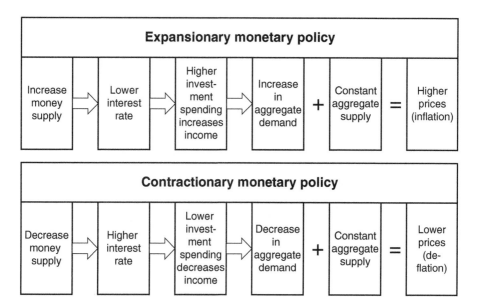

Figure 17.11 Monetary policy and inflation.

References

Blayse, A. and Manley, K. (2004). Key influences on construction innovation. *Construction Innovation* 4 (3): 143–154.

Dorman, P. (2014). *Microeconomics: A Fresh Start*. Heidelberg: Springer.

Greiman, V. (2013). *Megaproject Management: Lessons on Risk and Project Management from the Big Dig*. Hoboken: Wiley.

Hauptverband der Deutschen Bauindustrie (2021). *Bauwirtschaft im Zahlenbild*. Berlin: Hauptverband der Deutschen Bauindustrie.

International Energy Agency (ed.) (2019). *2019 Global status report for buildings and construction: Towards a zero-emissions, efficient and resilient buildings and construction sector*. Nairobi: United Nations Environment Programm.

Krugman, P. and Wells, R. (2017). *Economics*. New York: Worth.

Krugman, P. and Wells, R. (2018). *Macroeconomics*. New York: Worth.

Organization for Economic Cooperation and Development (2022). General Government Spending. https://data.oecd.org/gga/general-government-spending.htm#indicator-chart.

Sowell, T. (2015). *Basic economics: A common sense guide to the economy*. New York: Basic Books.

18

Public Goods

Samuelson (1954, p. 387) introduced a definition of a public good in the following terms: *"Therefore, I explicitly assume two categories of goods: ordinary private consumption goods $(X_1, \ldots, X_n)$ which can be parceled out among different individuals. . . and collective consumption goods $(X_n+1, \ldots, X_n+m)$ which all enjoy in common in the sense that each individual's consumption of such a good leads to no subtraction from any other individual's consumption of that good. . ."*

Krugman and Wells (2018, p. 483) provide an insightful example of a public good. Around 1850, London had a population of nearly 2.5 million and all the refuse of the households was discharged into the Thames. While the smell was bad, cholera and typhoid were deadly. No private company was willing or able to improve the situation and government did not see it as its task. Only when the situation became unbearable did the government act.

In 1858, after the Great Stink, parliament voted to commission the public works of creating a sewer system. Under the leadership of the civil engineer Bazalgette, 160 km of main sewers, 720 km of feeder sewers, and 21 000 km of local sewers were constructed. Estimates indicate that the construction of the sewer system extended the life span of the average Londoner by 20 years. Before government acted, there was a market failure from the private sector. London is not a lonely example; the senate in Hamburg, Germany, had decided 16 years earlier to implement a sewage system, and many cities worldwide followed. Still, sanitation is a problem in many parts of the world, especially where the government does not have the necessary financial means.

Governments provide public goods in the context of their allocation function because of shortcomings of private markets. Private firms supply goods to the market only if there is a chance of a profit. This requires buyers with an evident willingness to pay, and this is not given for public goods. These have two main characteristics: (i) Nobody can be excluded from using the public good. (ii) Use of the public good by someone does not hinder someone else to use it. In the former case we speak of excludability and in the latter of rivalry.

Construction Microeconomics, First Edition. Christian Brockmann.
© 2023 John Wiley & Sons Ltd. Published 2023 by John Wiley & Sons Ltd.

Section 18.1 discusses these two characteristics in more depth. These thoughts are followed by an economic analysis of demand for public goods, and we will find that it is very difficult to determine the demand (Section 18.2). Since demand is not clear, the possibility of making a profit becomes more risky and public firms might shy away from supplying a public good under such circumstances. If it is difficult or impossible to exclude someone from using a public good, it becomes possible to use it without paying. This problem is called free riding, and I will describe it in Section 18.3. Without knowledge of the demand function, it becomes impossible to use marginal analysis to determine the quantity produced. Instead, we have to use a cost–benefit analysis (Section 18.4). A cost–benefit analysis relies on assumptions, and they are prone to strategic misrepresentation and optimism bias (Section 18.5). I will summarize the ideas in Section 18.6.

18.1 Characteristics of Private Goods

Scene 1: A sailor is out on the sea; the night is frightfully dark. The boat is sailing against the wind, and the waves are high. This all happened before the invention of GPS and the sailor is a bit insecure about his position. He is close to the shore – but how close? Then, he sees the beam of the anticipated lighthouse in the right direction scanning over the horizon (Figure 18.1). Relief sets in; the harbor is not so far away – a very positive experience.

Scene 2: You need to renew your driver's license. You have known about it for weeks, but you kept putting it off. You are a good driver, and you've never even had a ticket, but you aren't looking forward to what you expect will be a long line. When you arrive at the DMV building, it looks shabby from the outside; there seems to be no money for maintenance. The lines inside are indeed terribly long. Frustration is turning to anger – a very negative experience.

Figure 18.1 Satisfaction by viewing a lighthouse.

Table 18.1 Categorization of goods based on rivalry and excludability.

	Rivalry	Nonrivalry
Excludability	Private goods	Artificially scarce goods
	• Food, clothes, cars	• Fire protection, cable TV
	• Toll roads with congestion	• Toll roads without congestion
Non-excludability	Public resources	Public goods
	• Fish, environment	• Alarm siren, national defense
	• Public roads with congestion	• Public roads without congestion

Source: Adapted from Mankiw and Taylor, 2014.

Both scenes describe experiences with public goods. The lighthouse is exemplary for a public good, as it displays the defining characteristics in a pure form: Nonexcludability and nonrivalry. Everybody on the sea can profit from the guiding beam of the lighthouse, the diligent taxpayer and phantoms like the Flying Dutchman. It does not matter if you add to the ships within the radius of a lighthouse a few or a hundred more, they do not become rivals.

The case of the DMV is a bit different. If there are more people than the employees can service in a period, a queue will form and rivalry will arise. At the DMV, we can also observe some form of excludability: You must have a certain age to become eligible for a driver's license. However, outside these boundaries, there exist nonexcludability and nonrivalry. Such goods are quasi-public goods and most public goods of the construction sector belong to this category: roads, bridges, and tunnels.

In many countries we can not only find roads, bridges, and tunnels as quasi-public goods but also as private goods; these are toll roads, toll bridges, and toll tunnels. Everyone who does not pay the toll is excluded from using the infrastructure.

These observations lead to a categorization of goods by using rivalry and excludability as categories (Table 18.1).

It is interesting to note that the categorization of roads depends on usage. Rivalry exists when a road is heavily used (congested). Much of our infrastructure is provided for by the government and paid by taxes as a public good or resource. There is, of course, a lively discussion in construction about the involvement of the private sector through public–private partnerships (PPPs). It should also be noted that to some people, a public good might be a public bad. Think of a diligent bicyclist who has to accept the construction of a new expressway passing close to his home. The situation becomes more obscure for the car driver who happily uses the expressway but would like to have in someone else's backyard. Looking back at externalities, it should also be clear that CO_2 emissions from cars on the expressway are public bads.

18.1.1 Rivalry

There is a continuous transition from a road with free-flowing traffic as a pure public good to a public resource in case of traffic jams on the same road. This might happen several times during the day (i.e. before and after rush hours). Evidently, there are different degrees of rivalry; there is no dualism between pure private and pure public goods.

Adding new users to a good with limited capacity leads to a diminished utility for all users. To keep the degree of utility for all users on the same level when users are added, additional investments are required. In the case of a road, this might be an extra lane.

If we take the case of a fixed investment, then the financial burden becomes smaller as more users are added. Each additional user, on the other hand, reduces the utility for all once a capacity is constant (i.e., there exists rivalry). These thoughts allow formulating the optimum size of a user collective: Users can be added until marginal benefits equal marginal costs (Erlei et al. 1999).

Now, we can define the degree of rivalry ρ. It is the fraction of relative changes in cost (C) over the relative change in users (n):

$$\rho = \frac{dC/C}{dn/n} \qquad\qquad 18.1$$

If a road has excess capacity, adding more users does not change the cost dC, thus $\rho = 0$.

The user elasticity γ of the quantity q of the good tells us how the quantity of the good must change with a change in users:

$$\gamma = \frac{dq/q}{dn/n} \qquad\qquad 18.2$$

For a pure public good, $\gamma = 0$ because no additional quantities are required $(dq = 0)$. If $\gamma < 1$, then it is advantageous to increase the number of users.

The quantity elasticity δ of the cost C is defined as the relative change in costs over the relative change in quantity:

$$\delta = \frac{dC/C}{dq/q} \qquad\qquad 18.3$$

If $\delta < 1$, then we have economies of scale; the additional costs per unit added are less than proportional.

It is easy to see that

$$\rho = \gamma \cdot \delta \qquad\qquad 18.4$$

We have already seen that $\rho = 0$ describes a pure public good with no rivalry. Adding a new user does not diminish the utility of the old users. The opposite is the case for $\rho = 1$, when every new user must replace an old user or the quantity must be increased by a 1 : 1 ratio. For values of $0 < \rho < 1$, it is possible to determine the optimum club size.

Goods can change their characteristics. A road can be a pure public good $(\rho = 0)$ and by an increase in traffic it can become a natural monopoly $(0 < \rho < 1)$. The task of the governments becomes to determine the optimum number of users for all public goods or natural monopolies.

18.1.2 Excludability

Private law and public law provide mechanisms for exclusion. Private law determines a clear assignment of property rights. An owner of a one-family home can exclude everyone from using his home unless he is willing to share or sell it. In the case of a sale, the next owner has exclusive user rights.

The options of public law are not so simple. A typical one is an enforced collective. Users of public TV in the UK have to pay a fee. This fee is the exclusion mechanism but it is only working if it can be enforced. For this purpose, detection vans are used. For private TV, payment for a decoding mechanism is required. TV reception is in both cases nonrival ($\rho = 0$).

18.2   Theory of Public Goods

Microeconomic theory provides evidence that markets allocate goods efficiently except in cases of market power, externalities, or other forms of market failures (Krugman and Wells 2018). As the example of the Great Stink shows, market failures exist. No private person or enterprise was willing to build a sewage system. Nonexcludability is the reason. It was not possible to exclude the Londoners unwilling to pay for the sewer system from the benefits of improved water quality in the Thames. If people can get improved sanitation for free, why pay for it? This is the free rider problem mentioned in the introduction, and it is a form of opportunism. Consequently, nonexcludable goods are produced at inefficiently low levels or not at all in private markets. Nonexcludability describes the fact that property rights cannot be clearly assigned and enforced. We saw already that this causes problems in case of externalities.

Nonrival but excludable goods such as toll roads without congestion also cause problems. Private firms are willing to provide the good because excludability allows them to make a profit. However, the marginal cost of letting one more motorist use the tollway are zero as there is no rivalry in using the toll road (no congestion). This means that the efficient price equals zero. This does not make sense to the toll road operator and he might charge a price of 10 USD. The difference between marginal cost and price keeps a number of users away from the toll road: Nonrival but excludable goods cause an inefficiently low consumption of the good.

Nonexcludable as well as nonrival goods cause inefficiencies in markets; only private goods are produced efficiently (Table 18.2). Fortunately, most goods are private goods, but the construction sector has to deal with a considerable number of public goods; they are typically commissioned by public works departments.

Table 18.2   Market outcomes for nonexcludable and nonrival goods.

	Rivalry	Nonrivalry
Excludability	Private goods • Efficient production by the private sector	Artificially scarce goods (produced by private sector) • Inefficiently low consumption
Non-excludability	Public resources (produced by private sector) • Inefficiently low production • No production	Public goods (produced by private sector) • Inefficiently low production • No production

18.2.1 Demand of a Public Good Based on Utility

An efficient production provides a good to consumers until marginal costs equal marginal utility ($\partial C/\partial y = \partial U/\partial y$). It should be clear that we are thinking about the marginal utility of a single consumer as the last car produced can only go to one consumer. There is a strict rivalry in consumption. Contrary to this, the production of public goods is advantageous as long as marginal cost equals marginal utility of all users. Thus, we have to optimize the following function L with U_i being the utility of a single consumer, y the quantity of the public good under consideration, n the number of consumers and C the cost:

$$L(y) = U_1(y) + U_2(y) + \ldots + U_n(y) - C(y) \qquad\qquad 18.5$$

We get the optimum of the function $L(y)$ by differentiating it and setting the value to zero:

$$\frac{\partial L}{\partial y} = \frac{\partial U_1}{\partial y} + \frac{\partial U_2}{\partial y} + \ldots + \frac{\partial U_n}{\partial y} - \frac{\partial C}{\partial y} = 0 \qquad\qquad 18.6$$

From this follows:

$$\sum_{i=1}^{n} \frac{\partial U_i}{\partial y} = \frac{\partial C}{\partial y} \qquad\qquad 18.7$$

This result means in a graphical representation that the marginal utility is not added horizontally but vertically (Figure 18.2).

This, unfortunately, is a rather theoretical analysis because it will be difficult to find out the preferences U of all the users of a highway which are required for the analysis. There are no markets for public goods and accordingly, demand is not clear. The possibility of free riding is for some even an incentive to hide demand. Accordingly, it can and does happen that governments provide public goods in an inefficient way.

PPPs face the same problems in defining demand as governments but the incentive structure is quite different. A private firm uses its own financial resources while the government has access to the taxpayers' money. Unfortunately, incentives work mostly on motivation and not on wisdom. The motivation of private firms can improve the efficiency of the investment. It will never improve the equity of the complete infrastructure. A rural

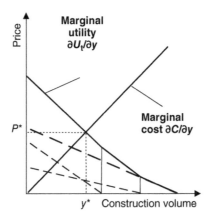

Figure 18.2 Optimal production of public goods.

road will very seldom generate a profit but this cannot be a reason not to invest in rural infrastructure. The fundamental economic problem remains the determination of demand; the fundamental political problem remains the equity of chances for participation in public life.

18.2.2 Demand for a Public Good Based on Willingness to Pay

It is often not a decision between the supply of a public good or not (either/or) but between more or less of a public good. This could be adding lanes to a highway to lower congestion, it could be street cleaning or garbage collection at shorter intervals. I will now introduce an example of street lighting. In a neighborhood street of 200 m, it is discussed among the neighbors whether there should be one or more streetlamps. The willingness to pay is taken as an indication of the marginal benefit. Figure 18.3 shows the

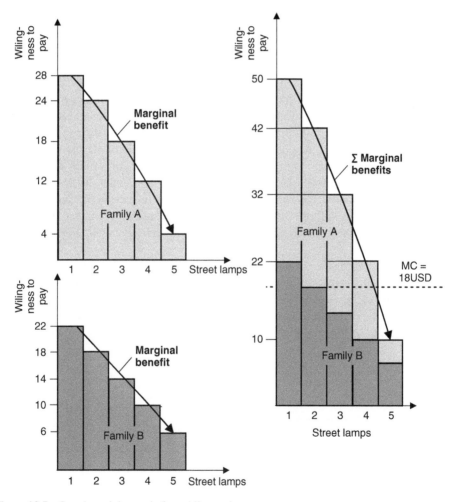

Figure 18.3 Supply and demand of a public good.

willingness to pay per month for street lighting of family A with four children and family B with two dogs. The left side of Figure 18.3 shows the willingness to pay (marginal benefit) per family and the right side the combined marginal benefits. If the monthly marginal costs are 18 USD, then the two families will decide to install four streetlamps at a distance of 50 m.

In this case, there are only two families and they are close neighbors. Concealing the true willingness to pay is not likely because of social interaction. The story would be very different if 100 families would live along the road stretched over 2 km.

18.3 Free Riding

Public goods are defined by nonexcludability and this opens the door for free riding. Users who do not contribute to covering the cost of a public good cannot be excluded from its benefits. Varian (2014) offers a game theoretical example for this problem. There are two user, A and B; the object is a public good, an access road. The payoff matrix is shown in Table 18.3. Although the game is limited to two players, the result will be the same for a larger number.

If neither player A nor B contributes, there will be no benefits even though the public access road would be very helpful to both. If A decides to contribute, then B would have a benefit of 100 by not contributing. The inverse holds true for player A, and the dominant strategy is for both not to contribute. In case of two players with different valuations of the access road, there is the possibility of trading to make both players better off. A contributes to the construction of the access road and has a loss of −50, in which case B has a gain of 100. If B were willing to give 75 to A, both would end up with a gain of 25, which is better than zero. However, this trading becomes impossible if there are many players. Thus, a private solution for the provision of the access road is not feasible and public investment is the last resource. There is a strong tendency to choose the option of free ridership.

If a private solution leads to hiding the true demand, then a possible solution is the indirect determination of demand by using analogical problems with market prices. Maybe there is a private access road in the vicinity? Then the public can find out the investment cost per user for the private road and determine the willingness to pay from those data. The result is based on a stringent use of a ceteris paribus clause. Intervening variables are not

Table 18.3 Payoff matrix for the free riding problem.

		Player B	
		Contribute	Do not contribute
Player A	Contribute	−50, −50	−50, 100
	Do not contribute	100, −50	0, 0

analyzed. It is also possible to ask users directly for their willingness to pay. Unfortunately, these have again a tendency not to disclose their true willingness.

18.4   Cost–Benefit Analysis

Given the problems of finding out the demand for a public good, most public agencies revert to cost–benefit analyses. The approach first list all the costs of a public invest-ment. This is not an easy task in construction because of the complexity and singularity of the projects. As in estimating, the public agency has to forecast future costs with all the contingencies that the future provides. Many public projects take a long time to implement, and inflation plays an important role. Political decisions such as the tariffs imposed in the United States in 2018 on steel imports are unforeseeable and increase costs considerably. Incomplete contracts lead to changes and claims: The list of prob-lems is long.

However, the determination of the benefits is even more difficult. Many projects have monetary and nonmonetary benefits; these can be social or ecological benefits. The bene-fits accrue over a long period, and they need to be discounted for comparison with invest-ment costs (maintenance cost must also be discounted). Thus, the assumption of a discount rate becomes crucial. Given the definition of costs as opportunity costs, the discount rates should compare to actual rates in the future. The biggest problem in transportation pro-jects, however, remains forecasting the number of users per period. This can only be esti-mated by analogy, but what is analog to a new rail line connecting points A and B?

18.5   Construction Goods as Public Goods

Once a government invests in a public good, it still has to decide how much involvement of the private sector is beneficial. This is a make-or-buy decision, and it typically involves a lot of private sector. Most often, the public works agency acts as owner and contractors carry out the works.

We can conceive of four phases, such as planning, construction, operation, and demoli-tion. The public works agency must come up with a concept and a cost–benefit analysis to determine the feasibility of the project. Then it needs to advance the design to a state that the required approvals can be sought. The design can be in-house or outsourced (private sector). Next, the tender documents must be prepared. Private contractors will submit a bid and both parties sign the contract.

The construction phase has the normal distribution of rights and responsibilities with the public works agency taking the role as owner. During the operation phase, the agency makes sure that use is safe and that necessary problems are detected timely (inspection). In the case of required repair works, it will again prepare a tender, sign a contract, and act as owner. A private contractor will carry out the works. Demolition follows a similar path (Figure 18.4).

While the decision to invest is not a market decision, most other steps involve markets, and we can therefore expect an efficient outcome.

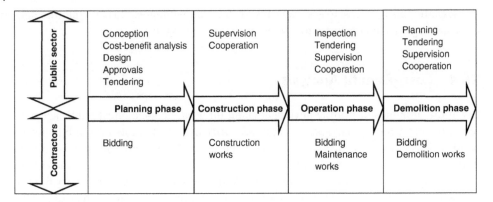

Figure 18.4 Involvement of the private sector in public works.

Table 18.4 Cost overruns for selected infrastructure projects.

Project	Cost overrun (%)
Boston artery, US	196
Humber bridge, UK	175
Boston–Washington–New York rail, USA	130
Great Belt tunnel, Denmark	110
Shinkansen Joetsu rail, Japan	100
Channel tunnel, UK and France	80
Mexico City metro, Mexico	60

18.6   Strategic Misrepresentation and Optimism Bias

The list of public projects with cost overruns is endless. Flyvbjerg et al. (2003) call this a calamitous history and provide some examples (Table 18.4). This list does not include the infamous example of the Sydney Opera House, which was projected to cost 7 million Australian dollars and ended up with a bill of 102 Australian dollars. It should be clear that published cost overruns are always the responsibility of the owner. If it were by the contractors' mistakes, the owner would refuse to pay and the losses would stay hidden in the accounts of the contractor.

Flyvbjerg et al. (2003) found two major reasons for cost overruns: Strategic misrepresentation (lying) and optimism bias (appraisal optimism). Public projects are often large-scale projects, and they draw the attention of a wider public. Their size implies high investment costs and thus high opportunity costs. A diverse public will face many investment projects where special interest groups find better use for the funds. In addition, many projects impact the quality of life of some people. This is the group of NIMBYs (not in my backyard). While this attitude is understandable, as few people want a highway next to their house or an airport in the vicinity, it decreases social welfare.

Proponents of large-scale projects sometimes use strategic misinterpretation to further the project chances. They might lie about the project costs or the project impacts. The special interest groups answer with a bias to overestimate costs and impacts. Optimism bias becomes most often evident by overestimating the benefits of the project, especially the number of users and accordingly the positive cashflow for amortization:

 Observation = (obs.) A new philharmonic concert hall opened in 2016 in Hamburg, Germany. It was way behind schedule and way over costs. The initial estimate was 75 million euros; this number was used in a presentation to get public approval by the Senate of the State of Hamburg. It seems to have been a strategic misrepresentation. Approval was given, construction started in in 2007. The start was followed by many problems, spiraling costs, and delays. The press was negative from day one. The final cost to the public ended around 800 million euros.

Almost immediately after the opening, the tone of the press changed, the acceptance of the building became overwhelmingly positive. The number of visitors exceeded all forecasts, thus there was no optimism bias in this regard, while the bias was documented for the time and cost overruns. The building generates a surprisingly strong positive cashflow.

The happy end did not include the contractor; he lost a very sizable amount of money.

References

Erlei, M., Leschke, M., and Sauerland, D. (1999). *Neue Institutionenökonomik*. Stuttgart: Schäffer-Poeschel.

Flyvbjerg, B., Bruzelius, N., and Rothengatter, W. (2003). *Megaprojects and Risk: An Anatomy of Ambition*. Cambridge: Cambridge University Press.

Krugman, P. and Wells, R. (2018). *Microeconomics*. New York: Worth.

Mankiw, N. and Taylor, M. (2014). *Economics*. Mason: Cengage Learning.

Samuelson, P. (1954). The pure theory of public expenditure. *The Review of Economics and Statistics* 36 (4): 387–389.

Varian, H. (2014). *Intermediate Microeconomics with Calculus: A Modern Approach*. New York: Norton.

19

Conclusion

I feel that some general remarks are helpful to place the previous chapters into a methodical context (Section 19.1). This will include observations about the necessity and possibility in construction microeconomics to generalize in order to build models. It will also highlight how this monography differs from others in its approach. The next chapters summarize the main findings on owners (Section 19.2), contractors (Section 19.3), construction goods (Section 19.4), construction markets (Section 19.5), and contracting (Section 19.6).

There is a quote from Philip Roth in his novel American Pastoral (1998, p. 63) that I love too much to omit in this place: *"Writing turns you into somebody who's always wrong. The illusion that you may get it right someday is the perversity that draws you on. What else could? As pathological phenomena go, it doesn't completely wreck your life."* I insist on dreaming *"to get it right someday"* and I am sure that we will be able to achieve that as a group of construction economists. What I submit here is a work in progress or just another construction site.

19.1 Methodical Context

A major problem is the heterogeneity of construction products. A look outside the window at the built environment will convince any sceptic. If it is too dark, you might consider Table 13.1. As there is a highly heterogenous output, so this also must apply to inputs and the connection between in- and output, the technology. This statement refers to product and process. Working at the level of full heterogeneity would not allow us to formulate abstract and general conclusions. As Newton had to abstract from the characteristics of an individual apple to arrive at his law of gravitation ($F = mg$), so must we as construction economists. Unfortunately, we do not have simple causes as gravitation and mass; the task of generalizing for description and modelling becomes much more complex. March put this into a catchy phrase (Dong et al. 2017): *"Unfortunately god gave all the easy problems to the physicists."* Falsification in the Popperian sense (Popper 1935) will not work in the context of construction economics, there is no rule without exception.

If we take the example of the owner and look at the characterization that I present, we find someone who has considerably more market power than a consumer. It might be a

Construction Microeconomics, First Edition. Christian Brockmann.
© 2023 John Wiley & Sons Ltd. Published 2023 by John Wiley & Sons Ltd.

Table 19.1 Economic first principles and construction.

	Content	Construction
Principle 1	Scarcity makes choices necessary	Applies fully
Principle 2	Opportunity costs	Applies fully
Principle 3	Marginal decisions	Does not apply to production
Principle 4	Incentives	Applies fully
Principle 5	Gains from trade	Applies fully
Principle 6	Markets move to equilibrium	Does not apply to contracting
Principle 7	Goal of efficiency	Applies fully
Principle 8	Markets lead to efficiency	Does not apply to contracting
Principle 9	Governments can improve outcome if markets are inefficient	Applies fully
Principle 10	One person's spending is another person's income	Applies fully
Principle 11	Spending and productive capacity are not always aligned	Applies fully
Principle 12	Government policies can change spending	Applies fully

cumbersome task but it is also possible to find in every country at least 1000 owners at the mercy of contractors. I think this is not a convincing counterargument. The owner with market power is a predominant type but not the general rule. While I admit that there are exceptions to each and every single statement that I forward in this monograph, I surely hope that I have captured the essential properties and left only the accidental ones. Ishii et al. (2020) provide a definition for both: *"P is an essential property of an object o just in case it is necessary that o has P, whereas P is an accidental property of an object o just in case o has P but it is possible that o lacks P."*

By concentrating on construction microeconomics, I have gained the possibility to present the general theory cohesively and in depth in Part A and to analyze the specific context of the construction sector and the way in which this context forces us to rethink construction microeconomics.

Going back to the first 12 principles of microeconomics as presented by Krugman and Wells (2018) and comparing those with the presented analysis shows three problematic areas (Table 19.1).

The first conflict concerns marginal decision-making in construction. The heterogeneity of the products and the predetermined scope by the owner's design does not allow contractors to use marginal analysis in production planning. Construction has to do without this powerful tool.

Construction markets on the level where prices are determined (i.e., in a submission) are never in an equilibrium; instead, there is a range of prices, often with a spread of 30 percent. The low bidder will have to get along with the minimum price. The monopsonistic market structure of a submission also leads to market inefficiencies; we cannot expect the results to be equivalent to those obtained in perfectly competitive markets. All three deviations are problematic and have a strong impact on construction.

This monography also differs from others because I allowed careful consideration of the effects of time. In many economic models, time plays no role. The supply and demand model is thought to result in an instantaneous equilibrium. I believe and bring forth arguments that time is an essential property in construction models; it is not an accidental one that we can leave away. Time has the following consequences in construction microeconomics when modeling:

- *Owner.* The power of the owner is different at different stages of a construction project. It depends on his ability to make independent decisions and on his information.
- *Contractor.* The statement about the owner is also true for the contractor. As owner and contractor sit on different sides of the table, this means that when one is relatively strong, the other is relatively weak.
- *Construction goods.* Exchange goods are traded instantaneously, not so contract goods. The idea of contract goods is central to understanding construction. It allows defining the good as a service as well as a product although at different times. The character is not fixed but transitional. The definition as contract good makes it also possible to analyze quality problems (product) and variants of principal–agent relationships (process). Contract goods are defined by the contract and as such they are heterogenous by necessity. As homogeneity is one of the essential conditions of the supply and demand model in perfect markets, this causes considerable problems of market analysis.
- *Construction markets.* Since contract goods are individually defined, the markets must also be individual markets. Prices are determined by sealed-bid auctions in individual transactions.
- *Contracting.* During a transaction of a sealed-bid auction one owner faces a number of contractors (bidders). This brings the owner at a most important moment of the process in a powerful position. The market characteristics are that of a monopsony at the time when the price is agreed. However, due to constant (horizontal) marginal costs for the contractor, the owner has limited monopsony power. A favorable price for the owner is more the result of sealed-bid auctions together with asymmetric information during contract negotiations. On the side of contractors, we typically find very strong competition.

19.2 Owners

An owner of a construction project is in command from the beginning to the end. During the development phase, he must make a considerable number of decisions setting the boundary conditions for the project (institutional framework); this includes the basic project idea, the location and financing. During the design phase he approves all the plans submitted by architects and engineers. For the tendering phase the owner provides the contract and determines the market design; he carries out the process according to his

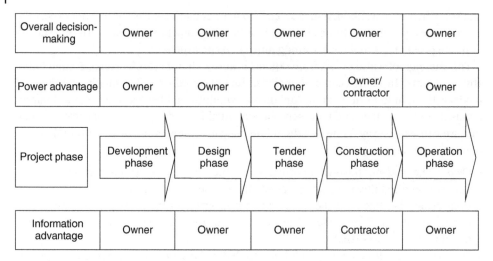

Overall decision-making	Owner	Owner	Owner	Owner	Owner

Power advantage	Owner	Owner	Owner	Owner/contractor	Owner

Project phase	Development phase	Design phase	Tender phase	Construction phase	Operation phase

Information advantage	Owner	Owner	Owner	Contractor	Owner

Figure 19.1 Decision-making, power, and information during a construction project.

design, contractors react. This changes during the construction phase when the contractors act and the owner takes the role of supervising and correcting the action. After handover, the owner uses the project to his own purposes or he sells it on the real estate market (Figure 19.1).

19.3 Contractors

Contractors find themselves most of the time in a highly competitive environment with strong pressures from better organized owners and suppliers. Competition centers on price leadership in an uncertain environment. Accordingly, profits as percentage of turnover are low. This perspective makes it almost incomprehensible why anybody would want to own a construction firm. Such a decision makes only sense if we look at opportunity costs and return on investment (ROI). Profits are small, but so is equity. The result is a reasonable ROI. Often, this beats alternatives such as savings accounts or other business ventures.

Contractors are the owner of the construction process; they own all the knowledge about managing the construction process including technology, labor, and capital. They use this comparative strength to their advantage whenever there are deviations from contract such as change orders or claims. These are not negotiated in a submission and accordingly, there is no submission price. The setup is a two-sided negotiation process with a knowledge advantage on the contractor's side. However, the owner is not clueless. The outcome is most likely a price slightly above market price.

A general contractor holds market power over subcontractors and often uses tendering to negotiate a subcontract. This reverses the position as the general contractor takes the place of the owner. The more subcontracts a general contractor signs, the more dependent he becomes on the subcontractors. This increases his risks as he is responsible to deliver works without defect on time to the owner. Contractors must carefully balance the price-cutting opportunities in subcontracting with the increasing risks toward their responsibilities.

19.4 Construction Goods

Construction goods are transitional contract goods where signing of the contract, production, and change of property rights at handover happen at different times. This opens the door for manifold problems during the transaction among which opportunism on both sides is most prominent. The owner makes the important decisions and keeps control during the whole life cycle of a structure from inception to demolition. He needs help to keep control by architects and engineers during the design, by contractors during construction, and by users during operation. This makes opportunistic behavior possible by different groups during the phases. Architects sometimes show little interest in the owner and want to increase their reputation at the owners' costs; structural and building systems engineers might design on the very safe side (over-dimensioning) to keep professional risks low; owners might use market power while tendering; contractors can take advantage of their information surplus during construction; finally, users can abuse the structure because they do not own it (don't be gentle; it's a rental!) (Figure 19.2)

19.5 Construction Markets

Regional, national, international, and global construction markets are virtual markets in the sense that no real construction goods are traded in them. Whatever data we have from those markets are average data for a specific type of structure. This could be a high-rise

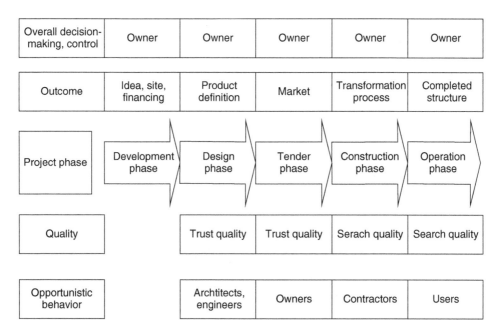

Overall decision-making, control	Owner	Owner	Owner	Owner	Owner
Outcome	Idea, site, financing	Product definition	Market	Transformation process	Completed structure
Project phase	Development phase	Design phase	Tender phase	Construction phase	Operation phase
Quality		Trust quality	Trust quality	Serach quality	Search quality
Opportunistic behavior		Archtitects, engineers	Owners	Contractors	Users

Figure 19.2 Outcomes of construction goods, qualities, and opportunistic behavior.

office building with 40 floors and above average quality. Even for such a project description, the price per square meter of usable space will be different between Melbourne and Belgrade, let alone Kampala. The same statement holds true for similar buildings in London.

These virtual markets have most of the time a structure of perfect competition. There are many owners and many contractors. The outcome should be textbook microeconomics and accordingly highly efficient without a chance for improvement. I did analyze several aspects why reality does not come close to the ideal. The problem is product heterogeneity and singularity; aggregated market information can only provide a guideline for action.

The only real market, where owners and contractors trade, is the project market. The market structure in project markets is far from ideal: One owner represents the demand side with a fixed quantity and a number of contractors the supply side. Thus, the market configuration is a monopsony at the time of signing a contract and determining a price with market power in the hands of owners; however, owners cannot use their monopsony power to the full extent because of the contractors' constant marginal costs (Figure 19.3).

Since contract goods are traded and since it is impossible to completely define construction contracts (theory of incomplete contracts), there arises most of the time the necessity to expand the contracts through change orders and claims. At this moment, two sides

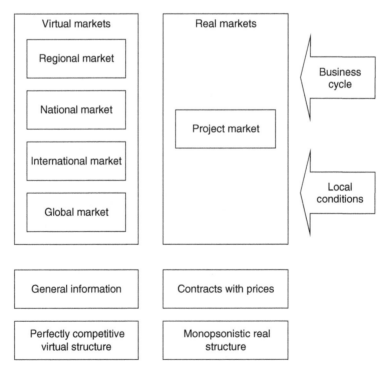

Figure 19.3 Construction market outcomes.

(owner and contractor) negotiate the contract changes in a setting of a two-sided monopoly. The use of the term monopoly instead of monopsony shall indicate that an information advantage and accordingly some negotiation leverage lies in the hands of the contractor. The final scope (product) and the final price are only clear at handover of the project from contractor to owner. Only at this time the full property rights change.

The business cycle and local conditions influence the general statements above. It happens that an owner is happy to receive a single offer during a boom so that a two-sided monopoly characterizes the whole transaction from beginning to the end. Typically, the owner will have to pay a higher price and might have to wait longer for the handover of the completed structure.

19.6 Contracting

Contracting describes the processes and the outcomes that evolve when all market influences act together. The nature of transitional contract goods signifies that construction markets are demand driven and the contracts are often of a large size for the contractor. As most contractors are small firms, a large contract must be defined in relation to the firm size. Both the dependence on demand and the large changes in employment at the beginning or end of a contract imply that flexibility is of utmost importance for the survival of contractors. The most common strategy to achieve this flexibility is outsourcing. This reduces the risks of changes in employment but increases the risks of efficient production. Typical ways of outsourcing are subcontracting of whole work packages, rental of equipment, and sometimes subcontracting of labor only. Outsourcing, in turn, leads to fragmentation of the supply side, and we have seen how this strategy leads to downsizing and fragmentation in recessions and to consolidation and insourcing during booms.

On the demand side, we have seen that owners possess considerable market power by designing the product, the organization, and the market structures, as well as by leading the tendering. The typical outcome is a contract price below equilibrium price. The lowest-bid award in sealed-bid auctions is not desirable for contractors; it is the starting point for an adversarial relationship during the contract period. Such an adversarial relationship finds its roots in the microeconomic goals that the parties to the contract pursue. If both are profit-maximizers, there is no way around fighting for the larger piece of the pie. Even if it is possible to increase the size of the pie (e.g., by reducing transaction costs through partnering), economic rationale remains to get the largest piece possible. There is no saturation point in competitive markets.

Profit maximization, market power with asymmetric information, and chances for opportunism lead to an adversarial relationship. At the same time, construction is based on team production with important contributions from both the owner and the contractor. The result of a contract award to the lowest bidder is the definition of negative standards. Since quality costs money, a rational contractor will aim to produce a quality just an undetectably bit below the minimum required by the contract. Low standards and adversarial relationships are the trademark of low performance teams. The result is an inefficient market outcome.

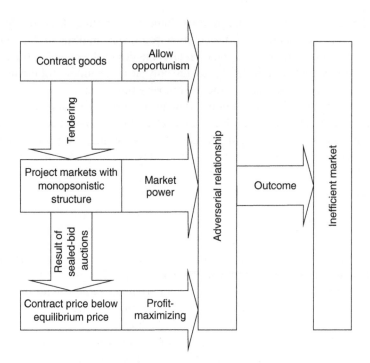

Figure 19.4 Cause and effect in contracting – inefficient markets.

Inefficiency means that there is space for improvement. However, all ideas for improvement of construction delivery will fail unless they take the microeconomic conditions into account: We need a solid understanding of construction microeconomics as the foundation on which we can build a better construction sector (Figure 19.4).

References

Dong, J., March, J., and Workiewicz, M. (2017). On organizing: an interview with James G. March. *Journal of Organization Design* 6 (1): Article number 14.

Ishii, R., Atkins, T., and Atkins, P. (2020). Essential vs. accidental properties. In: *The Stanford Encyclopedia of Philosophy* (ed. E. Zalta). https://plato.stanford.edu/archives/win2020/entries/essential-accidental/.

Krugman, P. and Wells, R. (2018). *Microeconomics*. New York: Worth.

Popper, K. (1935, 2002). *The Logic of Scientific Discovery*. Abingdon: Routledge.

Roth, P. (1998). *American Pastoral*. New York: Vintage International.

Index